Methods in Molecular Biology

Series Editor
John M. Walker
School of Life and Medical Sciences
University of Hertfordshire
Hatfield, Hertfordshire, UK

For further volumes:
http://www.springer.com/series/7651

For over 35 years, biological scientists have come to rely on the research protocols and methodologies in the critically acclaimed *Methods in Molecular Biology* series. The series was the first to introduce the step-by-step protocols approach that has become the standard in all biomedical protocol publishing. Each protocol is provided in readily-reproducible step-by-step fashion, opening with an introductory overview, a list of the materials and reagents needed to complete the experiment, and followed by a detailed procedure that is supported with a helpful notes section offering tips and tricks of the trade as well as troubleshooting advice. These hallmark features were introduced by series editor Dr. John Walker and constitute the key ingredient in each and every volume of the *Methods in Molecular Biology* series. Tested and trusted, comprehensive and reliable, all protocols from the series are indexed in PubMed.

Vascular Tissue Engineering

Methods and Protocols

Edited by

Feng Zhao

Department of Biomedical Engineering, Texas A&M University, College Station, TX, USA

Kam W. Leong

Department of Biomedical Engineering, Columbia University, New York, NY, USA

Humana Press

Editors
Feng Zhao
Department of Biomedical Engineering
Texas A&M University
College Station, TX, USA

Kam W. Leong
Department of Biomedical Engineering
Columbia University
New York, NY, USA

ISSN 1064-3745 ISSN 1940-6029 (electronic)
Methods in Molecular Biology
ISBN 978-1-0716-1707-6 ISBN 978-1-0716-1708-3 (eBook)
https://doi.org/10.1007/978-1-0716-1708-3

This Humana imprint is published by the registered company Springer Science+Business Media, LLC, part of Springer Nature.
The registered company address is: 1 New York Plaza, New York, NY 10004, U.S.A.

Preface

Cardiovascular disease (CVD) is the leading cause of death and is commonly associated with coronary artery disease, cerebrovascular disease, deep vein thrombosis, atherosclerosis, and other serious complications. Therapeutic interventions for CVD include lifestyle modification, pharmaceuticals, or surgery. Tissue-engineered vascular grafts (TEVGs) hold great promise as a source of grafting material when donor grafts are unsuitable or unavailable. Significant technological advances in vascular bioengineering have generated blood vessels at different scales as well as angiogenic biomaterials that can promote functional remodeling of vasculatures. This book covers detailed protocols for (1) deriving vascular cells from monocytes and induced pluripotent stem cells, on-site differentiation of mesenchymal stem cells into vascular cells; (2) engineering vascular grafts using various biomaterials and stem cells, stem cell-derived or primary vascular cells; (3) biomaterial modification by anticoagulation molecules; (4) vascular bioengineering technologies such as 3D bioprinting, electrospinning, and *in vivo* vascular graft maturation; and (5) fabrication of TEVGs with different geometry and multiphase structures. This book also includes protocols for grafting and evaluation of vascular grafts in animal models, vascular imaging in animals, the quantification of blood vessel permeability, force quantification of endothelial cells, and computational assessment of hemodynamic forces within diseased vasculature. This book is designed for biomedical engineers, tissue engineers, biomaterials scientists, biochemical engineers, biomechanical engineers, cell biologists, vascular surgeons, doctors, and nurses.

The editors are grateful to all the authors for their significant contributions to this book. The editors also thank the reviewers for their efforts in ensuring the high quality of all the chapters. The PhD candidate Dhavan Sharma and the postdoctoral researcher Dr. Weilue He of the Texas A&M University, Biomedical Engineering Department, are highly appreciated for their help with document management.

College Station, TX, USA — *Feng Zhao*
New York, NY, USA — *Kam W. Leong*

Contents

Contributors

NADIA O. ABUTALEB • *Department of Biomedical Engineering, Duke University, Durham, NC, USA*

YASMEEN AJAJ • *Biological and Biomedical Sciences Program, Yale University, New Haven, CT, USA*

JOSEPHINE B. ALLEN • *Department of Materials Science & Engineering, University of Florida, Gainesville, FL, USA*

DEIRDRE E. J. ANDERSON • *Department of Biomedical Engineering, Oregon Health & Science University, Portland, OR, USA*

STELIOS T. ANDREADIS • *Department of Biomedical Engineering, University at Buffalo, State University of New York, Amherst, NY, USA*

CHRIS A. BASHUR • *Department of Biomedical, Chemical Engineering, and Science, Florida Institute of Technology, Melbourne, FL, USA*

TAYLOR BERTUCCI • *The Neural Stem Cell Institute, Rensselaer, NY, USA*

MOHANISH K. CHANDURKAR • *Department of Biomedical Engineering, Michigan Technological University, Houghton, MI, USA*

ADRIAN J. Y. CHEE • *Schlegel Research Institute for Aging, Department of Electrical & Computer Engineering, University of Waterloo, Waterloo, ON, Canada*

GUOHAO DAI • *Department of Bioengineering, Northeastern University, Boston, MA, USA*

MATTHEW W. ELLIS • *Yale Cardiovascular Research Center, Section of Cardiovascular Medicine, Department of Internal Medicine, Yale School of Medicine, New Haven, CT, USA; Yale Stem Cell Center, New Haven, CT, USA; Department of Cellular and Molecular Physiology, Yale University, New Haven, CT, USA*

JUN FANG • *Department of Bioengineering, Department of Medicine, University of California, Los Angeles, Los Angeles, CA, USA*

BINGMEI M. FU • *Department of Biomedical Engineering, The City College of the City University of New York, New York, NY, USA*

PRERAK GUPTA • *Department of Biosciences and Bioengineering, Indian Institute of Technology Guwahati, Guwahati, Assam, India*

SANGYOON J. HAN • *Department of Biomedical Engineering, Michigan Technological University, Houghton, MI, USA*

JACQUELINE JIALU HE • *Division of Engineering in Medicine, Department of Medicine, Brigham and Women's Hospital, Harvard Medical School, Cambridge, MA, USA*

WEILUE HE • *Department of Biomedical Engineering, Michigan Technological University, Houghton, MI, USA*

MONICA T. HINDS • *Department of Biomedical Engineering, Oregon Health & Science University, Portland, OR, USA*

WILLIAM HO • *Department of Chemical and Materials Engineering, New Jersey Institute of Technology, Newark, NJ, USA*

ZHONGKUI HONG • *Department of Biomedical Engineering, University of South Dakota, Sioux Falls, SD, USA*

BRYAN D. JAMES • *Department of Materials Science & Engineering, University of Florida, Gainesville, FL, USA*

YEJIN JEONG • *Regenerative Nanomedicine Lab, Department of Chemical Engineering, University of Waterloo, Waterloo, ON, Canada*
WENKAI JIA • *Department of Biomedical Engineering, Texas A&M University, College Station, TX, USA*
JINGFENG JIANG • *Department of Biomedical Engineering, Michigan Technological University, Houghton, MI, USA*
SHRAVANI KAKARLA • *Department of Bioengineering, Northeastern University, Boston, MA, USA*
DIANA KIM • *Department of Bioengineering, Northeastern University, Boston, MA, USA*
DELING KONG • *Key Laboratory of Bioactive Materials, Ministry of Education, College of Life Sciences, Nankai University, Tianjin, People's Republic of China*
MAREK KUKUMBERG • *Department of Obstetrics & Gynaecology, Yong Loo Lin School of Medicine, National University of Singapore, Singapore, Singapore; Tissue Regeneration and Repair Laboratory, Department of Surgery, Yong Loo Lin School of Medicine, National University of Singapore, Singapore, Singapore*
FENGQIAO LI • *Department of Chemical and Materials Engineering, New Jersey Institute of Technology, Newark, NJ, USA*
MEI-LAN LI • *Biological and Biomedical Sciences Program, Yale University, New Haven, CT, USA*
NA LI • *Department of Biomedical Engineering, University of South Dakota, Sioux Falls, SD, USA; School of Biomedical Engineering, Dalian University of Technology, Liaoning IC Technology Key Lab, Dalian, China*
SONG LI • *Department of Bioengineering, Department of Medicine, University of California, Los Angeles, Los Angeles, CA, USA*
JIESI LUO • *Yale Cardiovascular Research Center, Section of Cardiovascular Medicine, Department of Internal Medicine, Yale School of Medicine, New Haven, CT, USA; Yale Stem Cell Center, New Haven, CT, USA*
LI LV • *Division of Engineering in Medicine, Department of Medicine, Brigham and Women's Hospital, Harvard Medical School, Cambridge, MA, USA*
SUSHILA MAHARJAN • *Division of Engineering in Medicine, Department of Medicine, Brigham and Women's Hospital, Harvard Medical School, Cambridge, MA, USA*
BIMAN B. MANDAL • *Department of Biosciences and Bioengineering, Indian Institute of Technology Guwahati, Guwahati, Assam, India; Centre for Nanotechnology, Indian Institute of Technology Guwahati, Guwahati, Assam, India*
ADAM C. MIDGLEY • *Key Laboratory of Bioactive Materials, Ministry of Education, College of Life Sciences, Nankai University, Tianjin, People's Republic of China*
KHANH P. NGUYEN • *Department of Surgery, Portland VA Medical Center, Portland, OR, USA; Division of Vascular Surgery, Oregon Health & Science University, Portland, OR, USA*
YIBING QYANG • *Yale Cardiovascular Research Center, Section of Cardiovascular Medicine, Department of Internal Medicine, Yale School of Medicine, New Haven, CT, USA; Yale Stem Cell Center, New Haven, CT, USA; Vascular Biology and Therapeutics Program, Yale University School of Medicine, New Haven, CT, USA; Department of Pathology, Yale School of Medicine, New Haven, CT, USA*
MUHAMMAD RAFIQUE • *Key Laboratory of Bioactive Materials, Ministry of Education, College of Life Sciences, Nankai University, Tianjin, People's Republic of China*
MUHAMMAD RIAZ • *Yale Cardiovascular Research Center, Section of Cardiovascular Medicine, Department of Internal Medicine, Yale School of Medicine, New Haven, CT, USA*

ALEX P. RICKEL • *Department of Biomedical Engineering, University of South Dakota, Sioux Falls, SD, USA*
ELLEN SALMON • *Department of Biomedical Engineering, Duke University, Durham, NC, USA*
MAHYAR SAMETI • *Department of Biomedical, Chemical Engineering, and Science, Florida Institute of Technology, Melbourne, FL, USA*
RICHARD SHANGGUAN • *Schlegel Research Institute for Aging, Department of Electrical & Computer Engineering, University of Waterloo, Waterloo, ON, Canada*
DHAVAN SHARMA • *Department of Biomedical Engineering, Texas A&M University, College Station, TX, USA*
RANDALL J. SMITH JR • *Department of Biomedical Engineering, University at Buffalo, State University of New York, Amherst, NY, USA*
KEVIN SUNDERLAND • *Department of Biomedical Engineering, Michigan Technological University, Houghton, MI, USA*
GEORGE A. TRUSKEY • *Department of Biomedical Engineering, Duke University, Durham, NC, USA*
DI WANG • *Division of Engineering in Medicine, Department of Medicine, Brigham and Women's Hospital, Harvard Medical School, Cambridge, MA, USA*
HE WANG • *State key Laboratory of Medicinal Chemical Biology & Key Laboratory of Bioactive Materials, Ministry of Education, College of Life Sciences, Nankai University, Tianjin, People's Republic of China*
KAI WANG • *Key Laboratory of Bioactive Materials, Ministry of Education, College of Life Sciences, Nankai University, Tianjin, People's Republic of China*
LIANYONG WANG • *Key Laboratory of Bioactive Materials, Ministry of Education, College of Life Sciences, Nankai University, Tianjin, People's Republic of China*
TINGTING WEI • *Key Laboratory of Bioactive Materials, Ministry of Education, College of Life Sciences, Nankai University, Tianjin, People's Republic of China*
XIAOYANG XU • *Department of Chemical and Materials Engineering, New Jersey Institute of Technology, Newark, NJ, USA; Department of Biomedical Engineering, New Jersey Institute of Technology, Newark, NJ, USA*
YUAN YAO • *Regenerative Nanomedicine Lab, Department of Chemical Engineering, University of Waterloo, Waterloo, ON, Canada*
EVELYN K. F. YIM • *Regenerative Nanomedicine Lab, Department of Chemical Engineering, University of Waterloo, Waterloo, ON, Canada*
BILLY Y. S. YIU • *Schlegel Research Institute for Aging, Department of Electrical & Computer Engineering, University of Waterloo, Waterloo, ON, Canada*
ALFRED C. H. YU • *Schlegel Research Institute for Aging, Department of Electrical & Computer Engineering, University of Waterloo, Waterloo, ON, Canada*
AUNG MOE ZAW • *Regenerative Nanomedicine Lab, Department of Chemical Engineering, University of Waterloo, Waterloo, ON, Canada*
XU ZHANG • *Department of Biomedical Engineering, Duke University, Durham, NC, USA*
YU SHRIKE ZHANG • *Division of Engineering in Medicine, Department of Medicine, Brigham and Women's Hospital, Harvard Medical School, Cambridge, MA, USA*
FENG ZHAO • *Department of Biomedical Engineering, Texas A&M University, College Station, TX, USA*
QIANG ZHAO • *State key Laboratory of Medicinal Chemical Biology & Key Laboratory of Bioactive Materials, Ministry of Education, College of Life Sciences, Nankai University, Tianjin, People's Republic of China*

Chapter 1

Differentiating Human Pluripotent Stem Cells to Vascular Endothelial Cells for Regenerative Medicine, Tissue Engineering, and Disease Modeling

Taylor Bertucci, Shravani Kakarla, Diana Kim, and Guohao Dai

Abstract

Vasculature plays a vital role in human biology as blood vessels transport nutrients and oxygen throughout the body. Endothelial cells (ECs), specifically, are key as they maintain barrier functions between the circulating blood and the surrounding tissues. ECs derived from human pluripotent stem cells (hPSCs) are utilized to study vascular development and disease mechanisms within in vitro models. Additionally, ECs derived from induced pluripotent stem cells (iPSCs) hold great promise for advancing personalized medicine, cell therapies, and tissue-engineered constructs by creating patient-specific cell populations. Here, we describe a xeno-free, serum-free differentiation protocol for deriving ECs from hPSCs. In brief, mesoderm progenitor cells are derived via WNT pathway activation. Following this, EC maturation is achieved with exogenous vascular endothelial growth factor A (VEGFA) and basic fibroblast growth factor 2 (bFGF2). We have characterized these cells as expressing mature EC markers and have illustrated their functionality in vitro.

Key words Human pluripotent stem cells, Endothelial cells, Endothelial differentiation, Wnt signalling

1 Introduction

Human vasculature is comprised of complex blood vessel networks responsible for transporting nutrients and oxygen to tissues throughout the body. Endothelial cells (ECs) line vessels, forming an anticoagulant barrier between tissue and blood. Endothelium dysfunction contributes to the pathogenesis of a large cohort of diseases across multiple tissues including but not limited to atherosclerosis [1–3], stroke [4–6], diabetes [7–9], and tumor growth and metastasis [10, 11]. There is a growing body of evidence

Taylor Bertucci and Shravani Kakarla contributed equally to this work.

Feng Zhao and Kam W. Leong (eds.), *Vascular Tissue Engineering: Methods and Protocols*, Methods in Molecular Biology, vol. 2375, https://doi.org/10.1007/978-1-0716-1708-3_1,

describing pathological changes associated with endothelial cells within various disease states, but complete disease mechanisms are widely unknown.

Human pluripotent stem cells (hPSCs) derived from ECs are an ideal cell source to further our understanding of the role the vascular system plays in disease states. They provide a tool to explore how genetic mutations contribute to disease states within in vitro modeling. Current vascular models often rely on primary cells and are therefore heavily handicapped due to the invasive nature of harvesting and limited cell quantity [12–14]. Other methods involve the use of small animal models; these experiments cannot accurately recapitulate all aspects of human biology [15]. Other important applications for hPSC-ECs include patient-specific cell therapies for personalized medicine such as tissue-engineered small-diameter grafts and revascularization strategies.

In development, stem cells commit to one of the three germ layers: ectoderm, mesoderm, or endoderm. After germ layers are formed, a subset of mesoderm progenitors further differentiates into vascular cells to form the vascular plexus, guided by temporal and spatial cell signalling. ECs express specific proteins on the cell surface that distinguish themselves from other cell fates or progenitor cells. Specifically, mature ECs distinctly express PECAM (CD31) and VE-cadherin.

In agreement with other labs' published works [16–22], this protocol targets key developmental pathways to drive stem cells into mesoderm-vascular lineage. Mesoderm progenitor cells are derived via WNT pathway activation using the small molecule CHIR99021. Then, exogenous vascular endothelial growth factor A (VEGFA) and basic fibroblast growth factor 2 (bFGF2) are added in concert to advance mesoderm progenitors into an EC fate. hPSC-ECs are purified using magnetic-activated cell sorting (MACS) technology by isolating the VE-cadherin + cell population. The resulting population has fully committed to the EC fate and retains its EC protein signature after multiple passages post-purification.

2 Materials

For all protocols using hPSCs, a centrifuge capable of holding 15-mL and 50-mL tubes, an inverted microscope for viewing cells, and a humidified, sterile incubator at 37 °C, 5% CO_2 are necessary.

Both human embryonic stem cells (hESCs) and human iPSCs can be used with this protocol. hESCs used were generously donated by the Daylon James laboratory (Weill Cornell Medical College); they contain an inserted VE-cadherin promoter

(VPR)-reporter cassette to track the temporal and spatial emergence of nascent ECs via the expression of the mOrange fluorescent protein [23]. IMR90-4 hiPSCs were also used (WiCell Institute, Madison, Wisconsin). Passage numbers under 40 are acceptable to use for this protocol, provided karyotype testing is carried out every 20–30 passages.

2.1 Matrigel Coating

1. Matrigel membrane matrix (Corning).
2. Dulbecco's Modified Eagle's Medium/nutrient mixture F-12 (DMEM/F12).
3. 35-mm culture dishes or 6-well plates.
4. Multi-well culture dishes.

2.2 Thawing hPSCs

1. mTESR or mTESR Plus medium (Stem Cell Technologies) (*see* **Note 1**).
2. Dulbecco's modified Eagle medium/nutrient mixture F-12 (DMEM/F12).

2.3 Maintenance of hPSCs

1. mTESR or mTESR Plus medium (Stem Cell Technologies).
2. Versene (Thermo Fisher Scientific).
3. Cell Lifters.
4. 35-mm culture dishes or 6-well plates.

2.4 Single-Cell Passaging of hPSCs

1. mTESR or mTESR Plus medium (Stem Cell Technologies).
2. ROCK inhibitor stock solution: PBS, 10 mM Y27632 dihydrochloride (Tocris Scientific).
3. StemPro™ Accutase™ cell dissociation reagent (Thermo Fisher Scientific).
4. Hemocytometer.

2.5 Differentiation Media

1. Basal differentiation media (BDM): advanced DMEM/F12, 60 μg/mL L-ascorbic acid 2-phosphate sesquimagnesium salt hydrate (Sigma) in deionized water, and 1% GlutaMAX (Gibco). Sterilize solution through a 0.22 μm filter before use. Store at 4 °C.
2. CHIR99021 stock solution: dimethyl sulfoxide, stock concentration of 6 mM CHIR99021 (Tocris Scientific).
3. VEGFA stock solution: PBS, 100 μg/mL recombinant human $VEGF_{165}$ (PeproTech).
4. bFGF2 stock solution: PBS, 100 μg/mL recombinant human FGF-basic (PeproTech).

2.6 Purification of hPSC-ECs

1. Phosphate-buffered saline (PBS).
2. MiniMACS™ Separator and Starting Kit (Miltenyi Biotec).
3. MS sorting columns (Miltenyi Biotec).

4. CD144 human microbeads kit (Miltenyi Biotec).
5. Gelatin-coating solution: sterile deionized water, 0.2% gelatin from porcine skin type A (Sigma). Sterilize solution through a 0.22 μm filter before use. Store at 4 °C.
6. Endothelial growth medium 2 (EGM2) (PromoCell).
7. Multi-well culture dishes.
8. Hemocytometer.

2.7 Maintenance and Passaging of hPSC-ECs

1. Endothelial growth medium 2 (EGM2) (PromoCell).
2. 0.25% Trypsin/EDTA (Gibco).
3. 1× Trypsin Neutralizer Solution (Gibco).
4. Multi-well culture dishes.
5. Hemocytometer.

3 Methods

The overall schematic of this method is outlined in Fig. 1. The resulting population displays hallmark EC functionality including acetylated low-density lipoprotein uptake and 3D vascular network capabilities (Fig. 2).

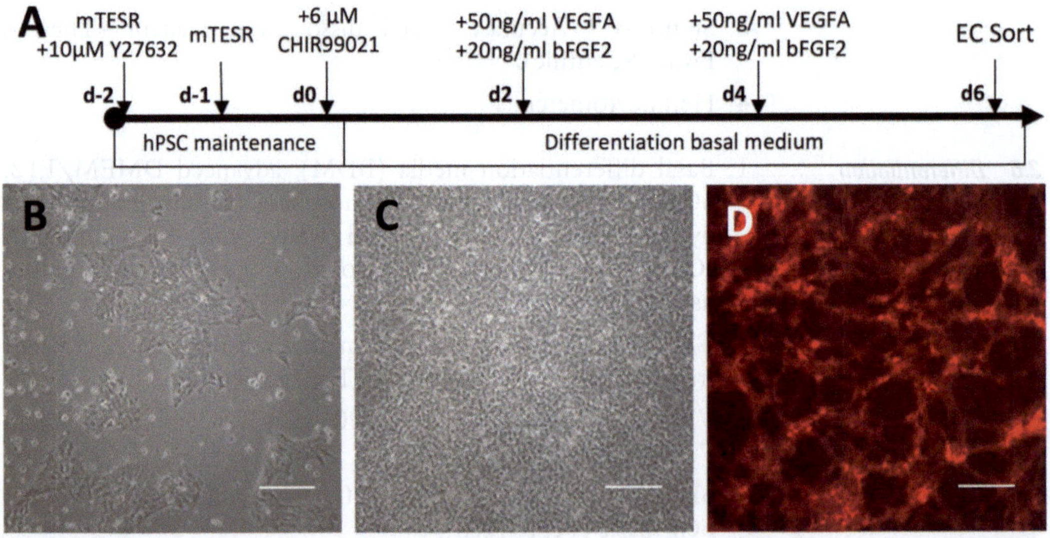

Fig. 1 (**a**) Schematic of hPSC derived EC differentiation protocol. (**b**) hPSC density at day 0 of differentiation. (**c**) Cell confluency at day 3 of differentiation. (**d**) VE-Cadherin+ cells (red) at day 6 proir to EC sort. Scale bar 200μm

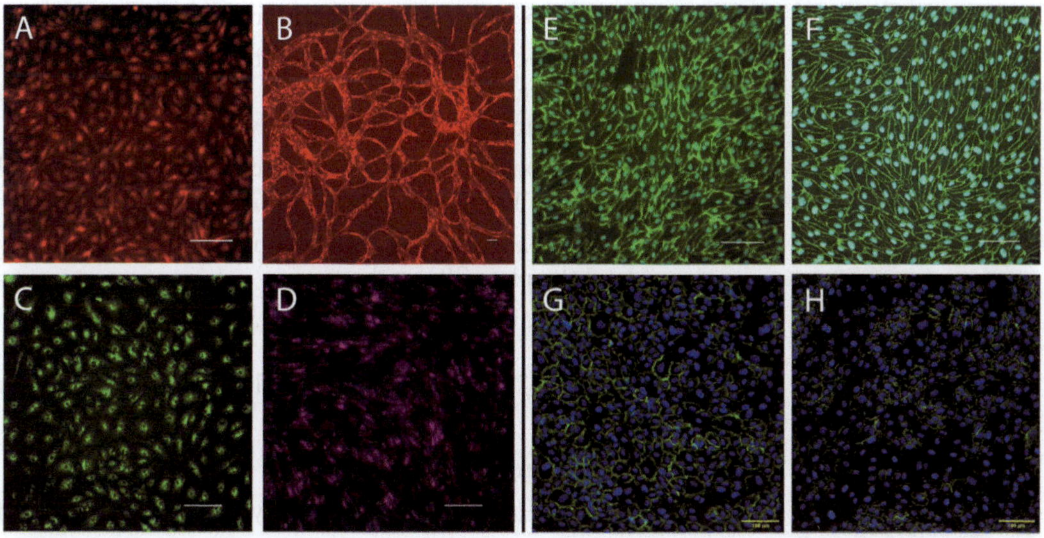

Fig. 2 (**a**) VE-CAD+ sorted monolayer derived from hESC-mOrange-VE-Cadherin reporter call line (red). (**b**) Tube formation assay in fibrin gel. (**c**) VECAD+ monolayer uptake of acetylated LDL FITC labeled (green). (**d**) Immunostaining of monolayer for von Willebrand Factor (purple). Comparison of endothelial maturation markers between hESCs (**top**) and HUVECs (**bottom**): (**e**, **g**) PECAM-1 (green), (**f**, **h**) VE-Cadherin (green). Scale bar 100μm. Please refer to Nakamura et al. Fig. 2A [24] to compare with hPSC PECAM-1 and VE-Cadherin staining

Table 1
Feed volumes for well plates

	35 mm dish	6-well plate	12-well plate	24-well plate	48-well plate
Added volume/dish or well (mL)	2	2	1	0.5	0.3

3.1 Preparation of Matrigel-Coated Tissue Culture Plastic Dishes (See Note 2)

1. Using pre-chilled pipette tips, add 150 μL Matrigel aliquot into 12 mL of cold DMEM/F12 on ice.
2. Pipette solution into pre-chilled culture dishes. Refer to the volumes shown in Table 1 for Matrigel coating volume.
3. Place Matrigel-coated dishes in a 4 °C refrigerator at least overnight prior to use. Matrigel plates are stable for about 1 month at 4 °C.

3.2 Thawing of hPSCs

1. Incubate Matrigel-coated dish at 37 °C for 20–30 min.
2. Remove a frozen vial of hPSCs from liquid nitrogen, and immediately swirl the bottom of the tube in a 37 °C water bath until only a small ice crystal remains.
3. Transfer the cell suspension into a 15 mL centrifuge tube, and slowly add 10 mL DMEM/F12.
4. Wash the frozen vial with 1 mL DMEM/F12, and add to the centrifuge tube.

5. Centrifuge for 5 min at 300 × *g* at room temperature.
6. Aspirate the supernatant without disturbing the cell pellet, and resuspend the cell pellet very gently in mTESR with minimal pipetting to maintain cell aggregates (*see* **Note 3**).
7. Remove solution from coated Matrigel dishes, and add cell suspension with the appropriate feed volume (Table 1).
8. In 37 °C incubator, move culture dish back and forth a few times, turn the plate 90°, and repeat.
9. Change the media everyday with pre-warmed mTESR.

3.3 Routine Clonal Passaging of hPSCs

When hPSCs are 75% confluent, clonally passage cells using Versene.

1. Incubate Matrigel-coated dishes at 37 °C for 20–30 min.
2. Briefly wash culture dish with a small volume of Versene and remove.
3. Add fresh Versene and incubate for no more than 3 min at 37 °C. Follow volumes from Table 1 for the appropriate amount of Versene (*see* **Note 4**).
4. Aspirate Versene from dish and replace with DMEM/F12.
5. Press the scraping edge of cell lifter onto surface of culture dish, and press down while rotating the culture dish to release hPSCs from the dish.
6. Transfer cell suspension into 15 mL centrifuge tube with 4 mL DMEM/F12. Wash the culture dish with 1 mL DMEM/F12 and add to the centrifuge tube.
7. Centrifuge for 5 min at 300 × *g*, room temperature.
8. Aspirate the supernatant without disturbing the cell pellet, and resuspend hPSCs in mTESR with minimal pipetting.
9. Remove solution from coated Matrigel dish.
10. For maintenance of hPSCs, split at a ratio 1:5–1:10 based on confluency into appropriate feed volume of mTESR (Table 1).
11. hPSCs must be cultured at least two passages after thaw and before performing differentiation protocol.

3.4 Single-Cell Passaging of hPSCs for EC Differentiation (Day 2 to Day 0)

1. When hPSCs are 75% confluent, add 10 μM of Y27632 to the medium. Incubate at 37 °C for 2 h.
2. Incubate Matrigel-coated well plates at 37 °C for 20–30 min.
3. Briefly wash culture dish with a small volume of Accutase™ and remove.
4. Add fresh Accutase™ and incubate for 10 min at 37 °C. Refer to Table 1 for volume of Accutase™ required.

Table 2
hPSC seeding density ranges for EC differentiation

	6-well plate	12-well plate	24-well plate	48-well plate
Surface area/well (cm^2)	9.6	3.5	1.9	1.1
Cells/cm^2	3.5×10^4	3.5×10^4	3.5×10^4	3.5×10^4
Total cells/well	3.36×10^5	1.225×10^5	6.65×10^4	3.85×10^4
Cell suspension conc. (cells/mL)	1.68×10^5	1.225×10^5	1.33×10^5	1.28×10^5
Added volume/well (mL)	2	1	0.5	0.3

5. Using a 1000 μL micropipette, pipet up and down to release single cells from the dish.
6. Transfer cell suspension into 15 mL centrifuge tube with 7 mL DMEM/F12. Wash the culture dish with 1 mL DMEM/F12 and add to the centrifuge tube.
7. Centrifuge for 5 min at 300 × *g*, room temperature.
8. Aspirate the supernatant without disturbing the cell pellet, and resuspend hPSCs in 5 mL DMEM/F12. Remove a small volume for cell counting.
9. Determine the total cell number using a hemocytometer.
10. Centrifuge for 5 min at 300 × *g*, room temperature.
11. Aspirate the supernatant without disturbing the cell pellet, and resuspend cells in mTESR with 10 μM Y27632 at a cell concentration based on well size (Table 2).
12. Remove solution from coated Matrigel wells, and add appropriate cell suspension volume to achieve 35,000 cells/cm^2 (Table 2) (*see* **Notes 5** and **6**).
13. The next day, replace medium with mTESR (without Y27632), and culture for 1 additional day.

3.5 Endothelial Cell Differentiation (Day 0 to Day 6)

1. To start the differentiation (day 0), replace mTESR medium with pre-warmed basal differentiation medium (BDM) with 6 μM of CHIR99021 (*see* **Note 7**).
2. On day 2—after 48 h of CHIR99021 treatment—remove 50% of cell culture medium, and replace with pre-warmed BDM containing 50 ng/mL VEGFA and 20 ng/mL bFGF2.
3. On day 4, remove 50% of cell culture medium, and replace with pre-warmed BDM containing 50 ng/mL VEGF and 20 ng/mL bFGF2.

3.6 Purification of hPSC-ECs Via Magnetic-Activated Cell Sorting (Day 6)

1. Briefly wash culture dish with a small volume of Accutase™ and remove.
2. Add fresh Accutase™ and incubate for 10 min at 37 °C. Refer to Table 1 for volume of Accutase™ required.
3. Using a 1000 μL micropipette, pipet up and down to release single cells from the dish.
4. Transfer cell suspension into 15 mL centrifuge tube with 7 mL DMEM/F12. Wash the culture dish with DMEM/F12 and add to the centrifuge tube.
5. Centrifuge for 5 min at 300 × *g*, room temperature.
6. Aspirate the supernatant without disturbing the cell pellet, and resuspend hPSCs in 5 mL DMEM/F12. Remove a small volume for cell counting.
7. Determine the total cell number using a hemocytometer.
8. Centrifuge for 5 min at 300 × *g*, room temperature.
9. Aspirate the supernatant without disturbing the cell pellet, and resuspend cell pellet in 80 μL of PBS per 10^7 total cells (*see* **Note 8**).
10. Add 20 μL of CD144 microbeads up to 10^7 total cells. Mix cell suspension well with microbeads by pipetting up and down. Incubate on ice for 15 min.
11. Wash cells with 1–2 mL of PBS up to 10^7 cells.
12. Centrifuge for 5 min at 300 × *g*, room temperature.
13. Aspirate the supernatant without disturbing the cell pellet, and resuspend up to 10^8 cells in 500 μL PBS.
14. Place a MS column in the magnetic field of a MiniMACS Separator. Refer to Fig. 3 for setup. Equilibrate column with 3 mL of PBS, and collect flow-through in a 15 mL centrifuge tube.
15. Apply 500 μL cell suspension onto the column. Flow-through will slowly drip out of the column. This flow-through will contain VE-cadherin-negative cells.
16. When the column reservoir is empty, add 2 mL of PBS. Repeat for a total of three washes.
17. Remove column from the magnetic field of the MiniMACS Separator, and place it on a new 15 mL centrifuge tube.
18. Add 1 mL of PBS onto the column. Wash column with 2 washes of 1 mL PBS (*see* **Note 9**).
19. Centrifuge for 5 min at 300 × *g*, room temperature.
20. Aspirate the supernatant without disturbing the cell pellet, and resuspend in EGM2.
21. Determine the cell number using a hemocytometer.

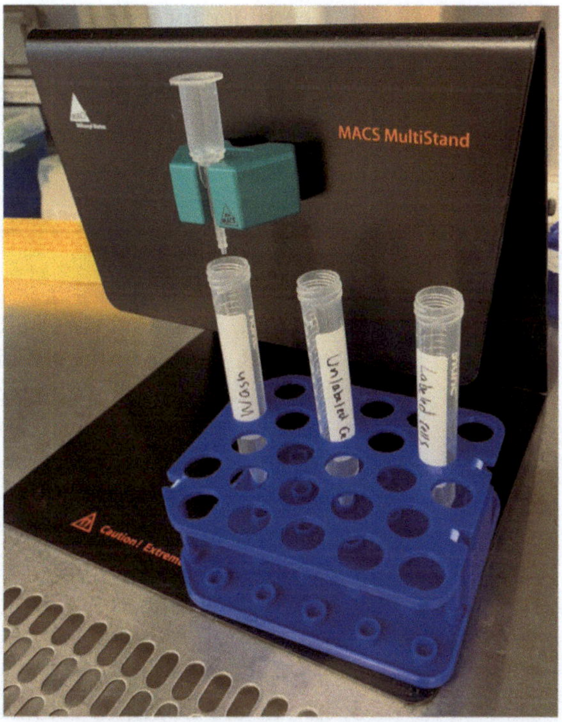

Fig. 3 Set-up for magnetic-activated cell sorting (MACS) in a sterile biosafety cabinet. As shown, the MS column is placed into the MiniMACS Separator, which is attached to the MACS MultiStand. When sorting VE-cadherin positive cells, the MS column is removed from the magnetic field of the MiniMACS Separator and MutiStand

22. Add gelatin-coating solution to well plate. Use Table 1 volumes for gelatin coating volumes.
23. Incubate for at least 15 min (up to 1 h) at room temperature.
24. Aspirate solution from gelatin-coated wells, and add 50,000 cells/cm^2 per well with appropriate medium volume (Table 1) (*see* **Note 10**).

3.7 Culture of hPSC-Derived VECAD+ Endothelial Cells

1. Change the media every 2 days with pre-warmed EGM2 until hPSC-ECs have reached 80–85% confluency (*see* **Note 11**).
2. To passage, briefly wash culture dish with a small volume of 0.25% Trypsin/EDTA and remove.
3. Add fresh 0.25% Trypsin/EDTA. Refer to Table 1 for volume of Trypsin required. Incubate well plate for 3–4 min at 37 °C.
4. Using a 1000 μL micropipette, pipet up and down to release single cells from the dish.
5. Transfer cell suspension into a 15 mL centrifuge tube with equal volume of Trypsin Neutralizer. Wash the culture dish with 1 mL Trypsin Neutralizer, and add to the centrifuge tube.

6. Centrifuge for 5 min at 300 × *g*, room temperature.
7. Aspirate the supernatant without disturbing the cell pellet, and resuspend hPSC-ECs in EGM2.

   Transfer hPSC-ECs into fresh gelatin-coated well plates at no more than a 1:3 split.

4 Notes

1. mTESR Plus and mTESR can be used interchangeably; mTESR Plus contains stabilized FGF2 to preserve cells for a longer time. Using with media, we recommend changing culture dish medium daily to avoid unwanted spontaneous differentiation of hPSCs.
2. Matrigel polymerizes at temperatures above 10 °C. Therefore, Matrigel must always be thawed slowly at 4 °C or on ice. We recommend aliquoting Matrigel and storing aliquots at −80 °C. All tips, tubes, and culture dishes used with Matrigel preparation must be pre-chilled. Do this by putting cultureware in freezer prior to use and work quickly.
3. hPSCs are frozen as colonies. Minimal pipetting throughout the thawing process is critical. Low viability will be observed if colonies are broken up into single cells.
4. Prolonged Versene exposure will result in colonies being passaged as single cells and thus result in low viability.
5. Cell density can be cell-line dependent and may need to be optimized for every new hPSC line; we recommend testing between 25,000 and 50,000 cells/cm^2.
6. Cells will settle to the bottom of the tube over time. Therefore, it is critical to continue to mix the cell suspension throughout the cell seeding process. This is particularly important when seeding a large number of wells.
7. CHIR99021 concentration may need to be adjusted across different hPSC cell lines. We recommend testing a working concentration range between 4 and 12 μM.
8. It is vital to have a single-cell suspension prior to sorting. Cell clumps will clog the sorting column and reduce your purity.
9. At this step, we prefer to not use the column plunger. The majority of the cells will flow freely out of the column without the additional shear force. Additional shear force can be used to eject the last remaining cells, but these cells may have lower viability.

10. Seeding density is important for post-sort viability and proliferation of hPSC-ECs. We do not recommend seeding at low cell densities [24].
11. hPSC-ECs should be passaged once they have reached 80–85% confluency and split at no more than a 1:3 ratio to remain proliferative.

References

1. Celermajer D, Sorensen K, Gooch V et al (1992) Non-invasive detection of endothelial dysfunction in children and adults at risk of atherosclerosis. Lancet 340:1111–1115
2. Bonetti PO, Lerman LO, Lerman A (2003) Endothelial dysfunction: a marker of atherosclerotic risk. Arter Thromb Vasc Biol 23:168–175
3. Davignon J, Ganz P (2004) Role of endothelial dysfunction in atherosclerosis. Circulation 109:III27–III32
4. Chen P, Wang P, Sheu W et al (2006) Changes of brachial flow-mediated vasodilation in different-ischemic stroke subtypes. Neurology 67:1056
5. Gormley K, Bevan S, Hassan A et al (2005) Polymorphisms in genes of the endothelin system and cerebral small-vessel disease. Stroke 36:1656–1660
6. Hassan A, Hunt BJ, Sullivan MO et al (2003) Markers of endothelial dysfunction in lacunar infarction and ischaemic leukoaraiosis. Brain 126:424–432
7. Beckman JA, Goldfine AB, Gordon MB et al (2003) Oral antioxidant therapy improves endothelial function in type 1 but not type 2 diabetes mellitus. Am J Physiol Heart Circ Physiol 285:2392–2398
8. Rizzoni D, Porteri E, Guelfi D et al (2001) Structural alterations in subcutaneous small arteries of normotensive and hypertensive patients with. Circulation 103:1238–1244
9. Endemann DH, Pu Q, De Ciuceis C et al (2004) Persistent remodeling of resistance arteries in type 2 diabetic patients on antihypertensive treatment. Diabetes Res Clin Pract 70:13–19
10. Suraj J, Kurpin A, Zakrzewska A et al (2019) Early and late endothelial response in breast cancer metastasis in mice: simultaneous quantification of endothelial biomarkers using a mass spectrometry-based method. Dis Model Mech 12:dmm036269
11. Franses JW, Drosu NC, Gibson WJ et al (2013) Dysfunctional endothelial cells directly stimulate cancer inflammation and metastasis. Int J Cancer 1345:1334–1344
12. Augustin-Voss H, Voss A, Pauli B (1993) Senescence of aortic endothelial cells in culture: effects of basic fibroblast growth factor expression on cell phenotype, migration, and proliferation. J Cell Physiol 157:279–288
13. Van Beijnum JR, Rousch M, Castermans K et al (2008) Isolation of endothelial cells from fresh tissues. Nat Protoc 3:1085–1091
14. Gui L, Ph D, Muto A et al (2009) Development of decellularized human umbilical arteries as small-diameter vascular grafts. Tissue Eng Part A 15:2665
15. Zaragoza C, Gomez-guerrero C, Martin-ventura JL et al (2011) Animal models of cardiovascular diseases. J Biomed Biotechnol 2011:497841
16. Gu M (2018) Efficient differentiation of human pluripotent stem cells to endothelial cells. Curr Protoc Hum Genet 98:1–16
17. Tan JY, Sriram G, Rufaihah AJ et al (2013) Efficient derivation of lateral plate and paraxial. Stem Cells Dev 22:1893
18. Bai H, Gao Y, Arzigian M et al (2011) BMP4 regulates vascular progenitor development in human embryonic stem cells through a Smad-dependent pathway. J Cell Biochem 109:363–374
19. Song S, Jung W, Li K et al (2013) Distinct transcriptional profiles of angioblasts derived from human embryonic stem cells. Exp Cell Res 319:1136–1145
20. Costa M, Sourris K, Mei S et al (2013) Derivation of endothelial cells from human embryonic stem cells in fully defined medium enables identification of lysophosphatidic acid and

platelet activating factor as regulators of eNOS localization. Stem Cell Res 10:103–117

21. Blancas AA, Shih AJ, Lauer NE et al (2011) Endothelial cells from embryonic stem cells in a chemically defined medium. Stem Cells Dev 20:2153
22. Lian X, Bao X, Al-Ahmad A et al (2014) Efficient differentiation of human pluripotent stem cells to endothelial progenitors via small-molecule activation of WNT signaling. Stem Cell Reports 3:804–816
23. James D, Zhan Q, Kloss C et al (2011) Lentiviral transduction and clonal selection of hESCs with endothelial-specific transgenic reporters. Curr Protoc Stem Cell Biol Chapter 1:Unit1F.12
24. Nakamura Y, Shimizu Y, Horibata Y et al (2017) Changes of plasmalogen phospholipid levels during differentiation of induced pluripotent stem cells 409B2 to endothelial phenotype cells. Sci Rep 7:1–9

Chapter 2

Generating Monocyte-Derived Endothelial-like Cells for Vascular Regeneration

Randall J. Smith Jr and Stelios T. Andreadis

Abstract

A major limitation in engineering vascular grafts is the lack of proper endothelium to prevent thrombosis and subsequent graft failure. Obtaining endothelial cells from patients' vasculature is intrusive and requires extensive culture time. Here we present an alternative strategy wherein abundant and easily accessible monocytes from peripheral blood are cultured and differentiated towards an endothelial-like state capable of preventing thrombosis through production of nitric oxide and formation of endothelial adherens junctions. Considering the plethora of monocytes present within peripheral blood, this method provides a robust alternative to generating endothelial cells required for vascular graft production.

Key words Monocytes, Endothelial cells, VEGF, Platelet-rich plasma, Fibronectin, Shear

1 Introduction

Endothelial cells are an essential component of the vasculature, without which tissue-engineered vessels thrombose and fail. Thus, obtaining endothelial cells either prior to or after implantation is a major obstacle. In adult humans, endothelial cells do not proliferate from anastomotic ends of implanted vessels [1] and must come from circulating endothelial progenitors (EPCs). However, EPCs are exceedingly rare within human blood, representing less than 0.01% of mononuclear cells in the blood [2]. However, we have shown that vascular endothelial growth factor (VEGF)-immobilized vascular grafts can capture circulating cells expressing the VEGF receptors, namely, monocytes, and that these cells can then differentiate towards an endothelial phenotype capable of maintaining patency [3–5]. Monocytes greatly outnumber potential EPCs at nearly 20% of peripheral blood mononuclear cells.

Previous studies have shown that monocytes can express endothelial genes after in vitro culture [6–11] but lack VE-cadherin-positive adherens junctions and the ability to form

Feng Zhao and Kam W. Leong (eds.), *Vascular Tissue Engineering: Methods and Protocols*,
Methods in Molecular Biology, vol. 2375, https://doi.org/10.1007/978-1-0716-1708-3_2,

tubes, indicating lack of endothelial (EC) functionality. To address these drawbacks, we developed a strategy to coax monocytes (MC) to differentiate into functional EC-like cells in vitro [12], and within this chapter we describe the methodology to differentiate this plentiful cell source into the crucial endothelial component of the vasculature.

2 Materials

All cells obtained from blood is following an approved IRB protocol or by purchasing platelet apheresis byproduct from a medical institution and further cell isolation.

2.1 Cell Media

All media used in this protocol are based on complete endothelial growth media (EGM2, Lonza; Allendale, NJ) with all supplements except for fetal bovine serum (FBS). The media is then supplemented with additional VEGF at a concentration of 50 ng/mL (*see* **Note 1**) and macrophage colony-stimulating factor (MCSF) at 1 ng/mL (Thermo Fisher; Grand Island, NY). Serum is replaced with 20% autologous platelet-rich plasma (PRP) (*see* **Note 2**).

1. *Attachment media*: Endothelial growth media supplemented with 10 μM Y-27632 (Sigma-Millipore; St. Louis, MO).
2. *Activation media*: Endothelial growth media supplemented with 10 μM CHIR-99021 (Sigma-Millipore).
3. *Platelet-rich plasma:* Blood is collected with anti-thrombotic agents such as heparin at 10 U/mL blood. Freshly drawn human blood is separated into the cellular component and plasma components by simple centrifugation at 400 × *g* for 10 min with no brake applied. Carefully separate the top, yellow-appearing plasma from the bottom red cellular component. Save both components. PRP can be frozen and thawed once with a shelf life of 1–2 months at −20 °C.

2.2 Culture Surface Preparation

Fibronectin surface: Fibronectin is dissolved at 1 mg/mL in phosphate-buffered saline (PBS). Non-tissue culture plates (hydrophobic) are coated for 12 h overnight at 4 °C with gentle rocking.

2.3 Other Materials

Red blood cell (RBC) lysis buffer (155 mM NH_4Cl, 12 mM $NaHCO_3$, 0.1 mM EDTA), chloroform, shaking platform (capable of 40–150 rotations per minute).

2.4 Shear Plate Preparation

The following is adapted from a similar procedure in White et al. [13].

1. Using a $10 cm^2$ diameter dish as a culture surface and a 3–5 cm^2 diameter dish as an insert ring, attach the smaller dish to the larger dish.
2. Dip the smaller dish into a bath of chloroform so that the edge is submerged.
3. Quickly place the treated edge of the smaller dish into the middle of the larger dish so that the smaller dish is centered onto the larger dish. Apply pressure as the chloroform seals the two plates together.

3 Methods

3.1 Monocyte Isolation

1. Obtain freshly drawn blood with 10 U/mL heparin to prevent thrombosis, OR obtain the leftover platelet apheresis cone (*see* **Note 3**). 50 mL peripheral blood yields ~10×10^6 monocytes and 20–25 mL of PRP (*see* **Note 4**). Keep the blood at room temperature to aid in component separation. Cold temperatures inhibit the separation.
2. Centrifuge the blood at $400 \times g$ for 10 min at room temperature with no brake. This separates the PRP from the cellular component containing monocytes.
3. Carefully remove the yellow layer containing PRP, and use this to prepare endothelial generation media; the remaining PRP can be frozen for future use.
4. The remaining red component containing RBC and white blood cells (WBC) is approximately half the initial volume. Use 1× RBC lysis buffer (or *see* **Note 5**) to lyse red blood cells for 5–10 min with occasional mixing by inverting the tube. Proper lysis is observed when the red component appears to darken to a very dark red color (*see* **Note 6**).
5. Centrifuge the lysed product at $300 \times g$ for 10 min with no brake at room temperature.
6. Carefully remove the supernatant—the cell pellet will appear white. Repeat the lysis procedure one time for complete lysis.
7. Resuspend cell pellet in PBS to wash the cells. Centrifuge the washed cells at $300 \times g$ for 5 min.
8. Repeat the wash until the supernatant is clear.
9. The resulting cells are all WBC—to obtain pure (90%+) monocytes, it is recommended to use adherence-based isolation. Seed cells on fibronectin-coated dishes at $2–3 \times 10^6$ cells per cm^2 in endothelial growth media without serum or PRP. Serum inhibits initial adhesion. Incubate for 1 h at 37 °C in a 5% CO_2 incubator.

10. Carefully remove unbound cells, and wash the plate with pre-warmed PBS to further remove unbound cells.
11. Monocyte purity can be verified with flow cytometry. A typical panel to determine purity would include the following: CD45, CD14, CD16, and CCR2. Such a panel would distinguish total monocytes as well as the subtypes: classical (CD14++/CD16-/CCR2+), non-classical (CD14-/CD16+/CCR2-), and intermediate (CD14+/CD16+/CCR2+) [14]. Inclusion of VEGFR2/KDR, CD133, CD144, and CD34 can determine if there is contaminating endothelial cells.

3.2 Monocyte Culture and Differentiation Under Static Conditions: See *Fig. 1 for Schematic of Time Course*

1. Immediately after adherence isolation add endothelial attachment media consisting of EGM2 containing Y-27632 (10uM) and PRP (20%).
2. Leave cells in the humidified incubator for 3 days without disturbing. Attachment and spreading occur slowly during this time.
3. Replace media with endothelial activation media on Day 3 consisting of EGM2 with CHIR99021 (10uM) (without Y-27632). Leave media for 2 days (*see* **Note** 7).
4. Replace media with endothelial growth media (20% PRP, 1 ng/mL MCSF, 50 ng/mL VEGF), and continue to culture for 7–10 days until endothelial-like cobblestone morphology begins to appear. Replace media every other day.
5. Monocyte-derived endothelial cells (MC-EC) can be further cultured under shear conditions that induce maturation.

3.3 MC-EC Culture Under Shear Conditions

1. MC-EC are detached from the surface using 0.1% trypsin in PBS and quenched with FBS once all the cells are detached.
2. Collect detached cells and centrifuge for 5 min at 1350 × *g* 300 × *g*.
3. Resuspend cell pellet in endothelial attachment media to allow cells to reattach with ease to the new surface.
4. Seed MC-EC at <80% confluency in the shear plates as prepared above.
5. Allow cells to spread overnight before removing endothelial attachment media and replacing with endothelial basal media.
6. Place shear plate on shaker device at the lowest setting possible to allow for around 20–40 RPM (*see* **Note 8**).
   (a) Shear strength is then changed by changing the rotation speed. We adapted this procedure with an orbital shaker of 1.9 cm radius. Shear was determined by the equation:

$$\tau = r\sqrt{\eta\rho(2\pi f)^3}$$

Fig. 1 Schematic of monocyte differentiation

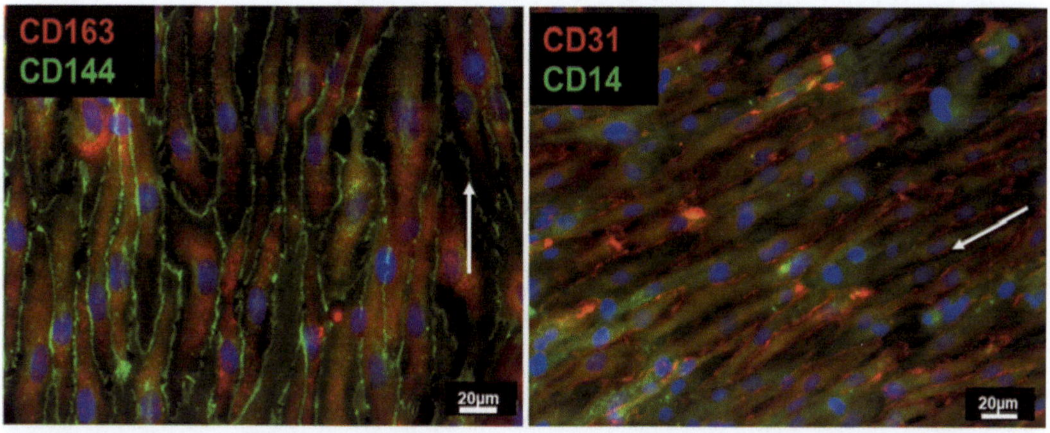

Fig. 2 Maturation of monocyte-derived endothelial cells under shear. Cells were fixed after 9 days of differentiation and 5 days of shear and assessed by immunocytochemistry for the endothelial markers CD144 and CD31 and monocyte markers CD14 and CD163. White arrows indicate direction of shear

where τ is the desired shear (dynes/cm^2), r is the radius of the orbital shaker, η is the viscosity (poise), ρ is the density (g/mL), and f is the rotation speed in rounds per second. Using this equation and converting to rotations per minute, approximately 29, 84, and 133 rotations per minute equated to 1, 5, and 10 dynes/cm^2 of shear, respectively.

7. Shear is slowly ramped up from 1 to 10 dynes/cm^2 over 2 days and maintained at that level for 3 days. Medium is replaced every other day during shear studies. Representative images of MC-EC after maturation under shear are depicted in Fig. 2.
8. Cells are now ready for use in vessel engineering.

4 Notes

1. VEGF can be produced as a recombinant protein as described [4]. Optimal concentration varies between 50 and 100 ng/mL.
2. PRP is best when autologous, but success has also been achieved with other human PRP and other non-human sources such as ovine. Low PRP concentrations increase attachment at the expense of survival. 10–20% PRP is best.
3. The cones are a component of platelet apheresis used in platelet donation. These cones accumulate WBC during the procedure

and are a rich source of cells, often containing up to 1×10^9 mononuclear cells.

4. 50 mL of peripheral blood generates enough PRP to make at least 100 mL of endothelial media if using 20% PRP. This volume of PRP can be adjusted down to 10% if PRP is a limiting reagent.
5. Alternatively, peripheral blood can be processed using Histopaque-1077 or other density-based methods to separate mononuclear cells out of the whole blood.
6. The concentration of the RBC lysis buffer components must remain near 1×—concentrations outside of this can inhibit lysis. Incomplete lysis is noticeable when the solution color remains lighter red.
7. During the CHIR activation step, the cells may round up and detach. Carefully collect unbound cells at the end of activation, and centrifuge them down at $300 \times g$ for 5 min. Re-seed these cells back onto the surface they came from in the presence of Y-27632. This will allow rounded up cells to re-spread if necessary. Y-27632 may be left in the media for up to 7 days if needed before cobblestone morphology is apparent.
8. During the shear process, cells may detach. Ideally, the MC-EC will proliferate into the empty spaces.

References

1. Ombrellaro MP, Stevens SL, Kerstetter K, Freeman MB, Goldman MH (1996) Healing characteristics of intraarterial stented grafts: effect of intraluminal position on prosthetic graft healing. Surgery 120(1):60–70. https://doi.org/10.1016/s0039-6060(96)80242-x
2. Chopra H, Hung MK, Kwong DL, Zhang CF, Pow EHN (2018) Insights into endothelial progenitor cells: origin, classification, potentials, and prospects. Stem Cells Int 2018:9847015. https://doi.org/10.1155/2018/9847015
3. Koobatian MT, Row S, Smith RJ Jr, Koenigsknecht C, Andreadis ST, Swartz DD (2016) Successful endothelialization and remodeling of a cell-free small-diameter arterial graft in a large animal model. Biomaterials 76:344–358. https://doi.org/10.1016/j.biomaterials.2015.10.020
4. Smith RJ Jr, Koobatian MT, Shahini A, Swartz DD, Andreadis ST (2015) Capture of endothelial cells under flow using immobilized vascular endothelial growth factor. Biomaterials 51:303–312. https://doi.org/10.1016/j.biomaterials.2015.02.025
5. Smith RJ Jr, Yi T, Nasiri B, Breuer CK, Andreadis ST (2019) Implantation of VEGF-functionalized cell-free vascular grafts: regenerative and immunological response. FASEB J 33(4):5089–5100. https://doi.org/10.1096/fj.201801856R
6. Seta N, Kuwana M (2007) Human circulating monocytes as multipotential progenitors. Keio J Med 56(2):41–47. https://doi.org/10.2302/kjm.56.41
7. Kuwana M, Okazaki Y, Kodama H, Satoh T, Kawakami Y, Ikeda Y (2006) Endothelial differentiation potential of human monocyte-derived multipotential cells. Stem Cells 24(12):2733–2743. https://doi.org/10.1634/stemcells.2006-0026
8. Zhang R, Yang H, Li M, Yao Q, Chen C (2005) Acceleration of endothelial-like cell differentiation from CD14+ monocytes in vitro. Exp Hematol 33(12):1554–1563. https://doi.org/10.1016/j.exphem.2005.08.011
9. Rehman J, Li J, Orschell CM, March KL (2003) Peripheral blood "endothelial progenitor cells" are derived from monocyte/macrophages and secrete angiogenic growth factors.

Circulation 107(8):1164–1169. https://doi.org/10.1161/01.cir.0000058702.69484.a0

10. Schmeisser A, Garlichs CD, Zhang H, Eskafi S, Graffy C, Ludwig J et al (2001) Monocytes coexpress endothelial and macrophagocytic lineage markers and form cord-like structures in Matrigel® under angiogenic conditions. Cardiovasc Res 49(3):671–680
11. Fernandez Pujol B, Lucibello FC, Gehling UM, Lindemann K, Weidner N, Zuzarte ML et al (2000) Endothelial-like cells derived from human CD14 positive monocytes. Differentiation 65(5):287–300. https://doi.org/10.1046/j.1432-0436.2000.6550287.x
12. Smith RJ Jr, Nasiri B, Kann J, Yergeau D, Bard JE, Swartz DD et al (2020) Endothelialization of arterial vascular grafts by circulating monocytes. Nat Commun 11(1):1622. https://doi.org/10.1038/s41467-020-15361-2
13. White LA, Stevenson EV, Yun JW, Eshaq R, Harris NR, Mills DK, et al (2016) The assembly and application of 'Shear Rings': a novel endothelial model for orbital, unidirectional and periodic fluid flow and shear stress. J Vis Exp (116):54632. https://doi.org/10.3791/54632
14. Franca CN, Izar MCO, Hortencio MNS, do Amaral JB, CES F, Tuleta ID et al (2017) Monocyte subtypes and the CCR2 chemokine receptor in cardiovascular disease. Clin Sci 131 (12):1215–1224. https://doi.org/10.1042/cs20170009

Chapter 3

Methods for Differentiating hiPSCs into Vascular Smooth Muscle Cells

Mei-Lan Li, Jiesi Luo, Matthew W. Ellis, Muhammad Riaz, Yasmeen Ajaj, and Yibing Qyang

Abstract

Despite numerous efforts to generate vascular tissues that recapitulate the physiological characteristics of native vessels, vascular cell source remains one of the principal challenges in the construction of tissue-engineered vascular grafts (TEVGs). Human pluripotent stem cells, therefore, represent an indispensable source to supply a large production of vascular smooth muscle cells (VSMCs) for cell-based therapy. In particular, human induced pluripotent stem cells (hiPSCs) generated from the same individual have opened up new avenues of achieving patient specificity through the derivation of autologous and immunocompatible VSMCs. This book chapter will detail three representative methods of differentiating hiPSCs into VSMCs that are structurally and functionally mature for TEVG engineering. Luo et al. reported an embryoid body (EB)-based approach to generate a robust, large-scale production of mature, functional hiPSC-derived VSMCs as a cell replacement for vascular tissue engineering. EB formation has an advantage of resembling early embryonic development and allowing cellular interactions in three dimensions. Cheung et al. established a system to produce embryological origin-specific hiPSC-derived VSMCs from the neuroectoderm, lateral plate mesoderm, and paraxial mesoderm lineages in a chemically defined manner. This allows site-specific vascular disease modeling. Moreover, Eoh et al. followed Wanjare et al.'s method to construct hiPSC-derived VSMCs using monolayer cultures of extracellular matrix proteins, with the addition of a pulsatile flow for the secretion of mature, organized elastic fibers. The generation of TEVGs, powered by the unlimited supply of hiPSC-derived VSMCs, has begun a new era in cellular therapy for vascular bypass and defective vessel segment replacement, aimed at addressing millions of cases of cardiovascular diseases across the globe.

Key words Vascular smooth muscle cells, Human induced pluripotent stem cells, Embryoid body differentiation method, Monolayer differentiation method, Tissue-engineered vascular grafts, Vascular bypassing

1 Introduction

Vascular bypass remains one of the major therapeutic treatments for cardiovascular disease [1]. While vascular grafting generally utilizes a patient's own vasculature for the most successful outcome, this

Feng Zhao and Kam W. Leong (eds.), *Vascular Tissue Engineering: Methods and Protocols*,
Methods in Molecular Biology, vol. 2375, https://doi.org/10.1007/978-1-0716-1708-3_3,

clinical procedure is limited by the availability and quality of native vessels. Synthetic vessels offer an alternative option for vascular grafting, but thrombosis and infection often lead to graft failure [2]. Such challenges gave rise to the generation of tissue-engineered vascular grafts (TEVGs), in which vascular cells are seeded on biodegradable, tubular scaffolds to form tissues that when transplanted mimic native vessels in morphology and function [3]. Recent advancements in TEVG engineering have demonstrated their remarkable mechanical strength, displaying no luminal dilation post-transplantation in an animal aortic model [4], thus representing a promising source of vascular grafts for clinical purposes.

Despite numerous efforts to advance TEVG engineering, cell source remains one of the biggest challenges [5]. Mature primary smooth muscle cells isolated from donors are hampered by limitations in quantity, proliferation, and reproducibility. Similarly, although adult stem cells are capable of replenishing vascular tissues, their proliferative and differentiative potentials decrease with increasing donor age [6]. Human induced pluripotent stem cells (hiPSCs), however, retain their characteristic self-renewal ability and differentiation potential into virtually all cell types and have adequately addressed the challenge of limited sources for vascular smooth muscle cells (VSMCs). hiPSCs reprogrammed from patient-specific somatic cells are capable of giving rise to cell types that are both autologous and immunocompatible, reducing the risk of graft rejection [7]. Recent progress in developmental biology has provided important insights into generating a large number of mature, functional VSMCs from hiPSCs over a short period of time. This book chapter will therefore discuss the major advancements in representative methods to derive robust, mature, and functional VSMCs from hiPSCs for future clinical applications of cell-based vascular therapy.

Past and recent efforts in generating hiPSC-derived VSMCs have been centered around two culture methodologies [8]: (1) the formation of embryoid bodies (EBs) [4, 9–12] and (2) the use of extracellular matrix (ECM)-coated monolayer cultures [13–18]. Here, we will present three approaches that represent major advancements in the derivation of mature, functional VSMCs from hiPSCs in the past two decades (*see* Fig. 1). We will further review the characteristics of the derived VSMCs from each approach based on their expression of VSMC markers and ECM proteins, their functional contractility in response to vasoconstrictors, and their application in robust, large-scale production for cell-based vascular engineering (***summarized in*** Table 1).

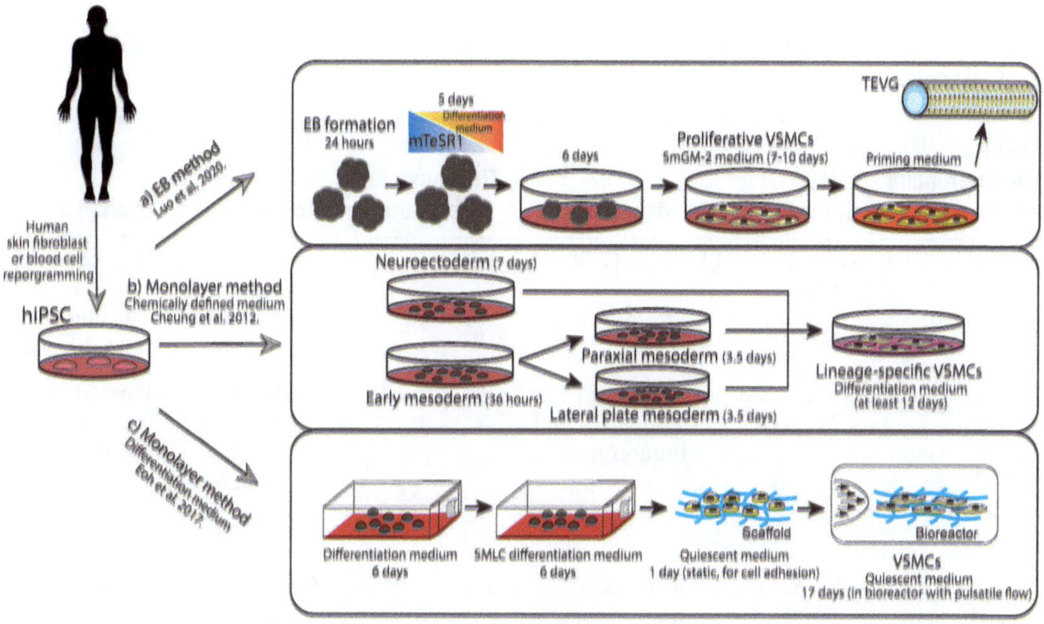

Fig. 1 Overview of three representative culture methodologies for VSMC differentiation. (**a**) Luo et al. describe an embryoid body (EB)-based approach to generate a large-scale production of VSMCs to supplement tissue-engineered scaffolds. (**b**) Cheung et al. employ a monolayer method with a chemically defined medium to derive lineage-specific VSMCs. (**c**) Eoh et al. utilize a monolayer method coupled with a pulsatile flow to generate VSMCs with enhanced contractility and elastin deposition. Drawing not to scale

2 Materials

2.1 In Vitro Differentiation of VSMCs Via EB Formation

The material preparations described below are derived from previous literature [4, 11, 12]. An overview of this method is illustrated in Fig. 1a. The characteristics of the derived VSMCs are summarized in Table 1. The advantages and limitations of this method are described in **Note 1**.

1. EB differentiation medium: 10% (v/v) fetal bovine serum (FBS; Gemini), 1% (v/v) non-essential amino acid (Thermo Fisher), 2 mM L-glutamine (Thermo Fisher), 0.012 mM 2-mercaptoethanol (Sigma-Aldrich), and 1% (v/v) penicillin-streptomycin (Thermo Fisher) in Dulbecco's Modified Eagle's Medium (DMEM), high glucose (Thermo Fisher) (*see* **Note 2**).
2. Priming medium: 10% (v/v) FBS, 1% (v/v) non-essential amino acid, 2 mM L-glutamine, 1 mM sodium pyruvate (Thermo Fisher), and 1% (v/v) Penicillin-Streptomycin in DMEM (*see* **Note 3**).

Table 1
A summary of the characteristics of hiPSC-derived VSMCs from three representative culture methodologies

VSMC differentiation method	VSMC markers	ECM deposition	Response to vasoconstrictor	Advantage	Limitation
Luo et al. 2020	α-SMA, CNN1, and MYH11	Elastin and collagen I	Yes	Optimal for large-scale production	Lack of mature elastin fiber production
Eoh et al. 2017	TAGLN and MYH11	Elastin, collagen I, collagen III, and fibronectin	Yes	Contains organized, mature elastin fibers and microfibrils	Limited proliferative capability
Cheung et al. 2012	α-SMA, CNN1, MYH11, TAGLN, and SMTN	Contains ECM gene expression, but no ECM deposition staining data collected	Yes	Generates lineage-specific VSMCs	Limited proliferative capability

3. Maturation medium: 1% (v/v) FBS, 1% (v/v) non-essential amino acid, 2 mM L-glutamine, 0.012 mM 2-mercaptoethanol, and 1 ng/ml transforming growth factor beta 1 (TGF-β1; PeproTech) in DMEM.

2.2 *In Vitro* Differentiation of Lineage-Specific VSMCs Via a Chemically Defined Method

The culture medium compositions described below are derived from previous literature [16, 17, 19, 20]. An overview of this approach is illustrated in Fig. 1b. The characteristics of the derived VSMCs are summarized in Table 1. The advantages and limitations of this approach are detailed in **Note 4**.

1. Chemically defined medium (CDM; 500 ml): 5 ml of chemically defined lipid concentrate (Life Technologies), 350μl of 10 mg/ml insulin (Roche), 250μl of 30 mg/ml transferrin (Roche), and 20μl monothioglycerol (Sigma-Aldrich) in 250 ml Iscove's Modified Dulbecco's Medium (IMDM; Life Technologies) and 250 ml F12 nutrient mix (Life Technologies) with an optional 5 ml of penicillin-streptomycin (Life Technologies). Dissolve 2.5 g of bovine serum albumin (BSA; Europa Bioproducts) in CDM (500 ml) to make the

CDM-BSA medium for hiPSC maintenance. Dissolve 0.5 g of polyvinyl alcohol (PVA; Sigma-Aldrich) in CDM to make the CDM-PVA medium for differentiation (*see* **Note 5**).

2. VSMC differentiation medium: 10 ng/ml platelet-derived growth factor BB (PDGF-BB; PeproTech) and 2 ng/ml TGF-β1 (PeproTech) in CDM-PVA.

2.3 Derivation of VSMCs on Monolayer Cultures of ECM Proteins with a Pulsatile Flow

The material preparations detailed below are derived from previous literature [13, 14]. An overview of this protocol is illustrated in Fig. 1c. The characteristics of the derived VSMCs are summarized in Table 1. The advantages and limitations of this protocol are described in **Note 6**.

1. 3D scaffold fabrication via electrospinning: Dissolve equal amounts of poly(ethylene glycol) dimethacrylate (PEGdma; Sigma-Aldrich) and poly(L-lactide) (PLA; Sigma-Aldrich) in 1,1,1,3,3,3-hexafluoro-2-propanol (Merck), with an addition of 1% photoinitiator (2-hydroxy-4′-(2-hydroxyethoxy)-2-methylpropiophenone; Sigma-Aldrich) for subsequent cross-linking. Electrospin the solution into a fiber mat in the dark using a 20-gauge nozzle, 18 kV, a flat collector, and a distance of 18 cm between the nozzle and collector. Cross-link the scaffolds with ultraviolet light at 365 nm [21] (*see* **Note 7**).
2. Differentiation medium: 10% FBS (HyClone) and 0.1 mM 2-mercaptoethanol (Invitrogen) in base medium alpha-MEM (Invitrogen).
3. Smooth muscle-like cell (SMLC) differentiation medium: 1 ng/ml TGF-β1 (R&D Systems) and 10 ng/ml PDGF-BB (R&D Systems) in the differentiation medium described above.
4. Quiescent medium: 1 ng/ml TGF-β1, 0.5% FBS, and 0.1 mM 2-mercaptoethanol in base medium alpha-MEM.
5. One-chamber perfusion bioreactor system [14, 22]: Connect the bioreactor to a closed tubing system with a 30 ml medium reservoir, a filter for gas exchange, and a peristaltic pump (Ismatec) for circulating the medium throughout the system.

3 Methods

3.1 In Vitro Differentiation of VSMCs Via EB Formation

This method is derived from previous literature [4, 11, 12]. An overview of this method is illustrated in Fig. 1a. The characteristics of the derived VSMCs are summarized in Table 1. The advantages and limitations of this method are described in **Note 1**. A brief overview of the regenerative applications of hiPSC-derived VSMCs is discussed in **Note 8**.

1. *hiPSC maintenance*:

   Culture hiPSC lines under feeder-free conditions in mTeSR™1 self-renewal medium (STEMCELL Technologies) on a 6-well tissue culture plate coated with Growth Factor Reduced (GFR)-Matrigel (Corning) (1:60) at 37 °C until 80% confluency.

2. *EB formation in suspension*:

   Day 0: Treat the culture with 0.5 mM ethylenediaminetetraacetic acid (EDTA; Thermo Fisher) for 3 min at 37 °C. Resuspend the dissociated cells in mTeSR™1 medium with 1:100 (v/v) GFR-Matrigel and 5μM ROCK inhibitor (Y-27632; Millipore). Plate the suspension on a 6-well low attachment plate (Corning). Seed in a ratio of 1 well of tissue culture plate to 1 well of low attachment plate. Uniform-sized EBs form in 24 h.

3. *EB differentiation*:

   Day 1: Remove the mTeSR™1 medium in the 6-well plate. Add a 2:1 mixture of mTeSR™1 medium and EB differentiation medium by volume to the EBs.

   Day 2: Replace the medium with 1:1 mixture of mTeSR™1 medium and EB differentiation medium by volume.

   Day 3: Replace the medium with 1:2 mixture of mTeSR™1 medium and EB differentiation medium by volume.

   Day 4–5: Replace the medium with EB differentiation medium.

   End of Day 5: Transfer the suspended EBs to a gelatin (Sigma-Aldrich)-coated 6-well plate. Culture for 6 days in EB differentiation medium with daily medium change.

4. *VSMC expansion*:

   Treat the EB-derived culture with 0.05% trypsin-EDTA (Thermo Fisher). Seed at 20,000 cells/cm^2 on a GFR-Matrigel-coated dish in Smooth Muscle Growth Medium (SmGM-2; Lonza). Culture until 80% confluency in approximately 7–10 days to generate proliferative VSMCs. Continue VSMC expansion in SmGM-2 medium with medium change every other day until ready to be primed for tissue engineering.

5. *Priming of proliferative VSMCs for vascular graft engineering*:

   Passage the proliferative VSMCs onto GFR-Matrigel-coated dishes in priming medium (10% FBS, 1% non-essential amino acid, 2 mM L-glutamine, 1 mM sodium pyruvate, and 1% penicillin-streptomycin in DMEM) for 1 or 2 passages with medium change every other day, followed by subsequent TEVG generation.

6. *VSMC maturation*:

Culture the proliferative VSMCs in maturation medium (1% FBS, 1% non-essential amino acid, 2 mM L-glutamine, 0.012 mM 2-mercaptoethanol, and 1 ng/ml TGF-β1 in DMEM) for 7 days to validate their properties and maturation.

3.2 In Vitro Differentiation of Lineage-Specific VSMCs Via a Chemically Defined Method

This approach is derived from previous literature [16, 17, 19, 20]. An overview of this approach is illustrated in Fig. 1b. The characteristics of the derived VSMCs are summarized in Table 1. The advantages and limitations of this method are detailed in **Note 4**. A brief overview of the regenerative applications of hiPSC-derived VSMCs is discussed in **Note 8**.

1. *hiPSC maintenance*:

   Culture hiPSC lines in CDM-BSA with 12 ng/ml fibroblast growth factor 2 (FGF2; R&D Systems) and 10 ng/ml activin A (R&D Systems). When the cell colonies reach about 70% confluency, digest them into small clumps. Plate the cells at 1:10 ratio onto gelatin (Sigma-Aldrich)-coated plates in CDM-PVA supplemented with 12 ng/ml FGF2 and 10 ng/ml activin A. Maintain the cells at 37 °C for 24 h.

2. *Induction of intermediate populations*:

   (a) *Neuroectoderm formation*:

   Remove old medium. Wash the cells with CDM-PVA once. Culture the hiPSCs in CDM-PVA supplemented with 12 ng/ml FGF2 and 10μM SB431542 (Sigma-Aldrich) for 7 days. Replace half of the medium with fresh medium every 2 days until day 7.

   (b) *Early mesoderm formation*:

   Remove old medium. Wash the cells with CDM-PVA once. Culture the hiPSCs in CDM-PVA supplemented with 20 ng/ml FGF2, 10μM LY294002 (Sigma-Aldrich), and 10 ng/ml bone morphogenetic protein 4 (BMP4; R&D Systems) for 36 h. Proceed with the following steps to differentiate into lateral plate mesoderm or paraxial mesoderm.

   (c) *Lateral plate mesoderm*:

   Remove old medium. Wash the cells with CDM-PVA once. Culture the early mesodermal cells in CDM-PVA supplemented with 20 ng/ml FGF2 and 50 ng/ml BMP4 for 3.5 days. Replace half of the medium with fresh medium after 2 days.

   (d) *Paraxial mesoderm*:

   Remove old medium. Wash the cells with CDM-PVA once. Culture the early mesodermal cells in CDM-PVA with 20 ng/ml FGF2 and 10μM LY294002 for 3.5 days. Replace half of the medium with fresh medium after 2 days.

3. *Induction of VSMC differentiation and maturation*:

   Remove old medium. Wash the cells with CDM-PVA once. Trypsinize to create single-cell suspension. Plate the neuroectoderm and lateral plate mesoderm intermediates at a seeding density of 20,000 cells per cm^2 onto gelatin-coated plates in VSMC differentiation medium at 37 °C. For paraxial mesoderm intermediates, plate at a seeding density of 30,000 cells per cm^2 instead. After 1 day, replace with fresh medium. Maintain the cells in VSMC differentiation medium for at least another 11 days. Replace half of the medium with fresh medium every 2–3 days (*see* **Note 9**). For subsequent expansion, adapt the differentiated VSMCs to DMEM (Life Technologies) with 10% (v/v) FBS (Biosera). Passage up to 10 times in culture with a splitting ratio of 1:3 to 1:6.

3.3 Derivation of VSMCs on Monolayer Cultures of ECM Proteins with a Pulsatile Flow

This protocol is derived from previous literature [13, 14]. An overview of this protocol is illustrated in Fig. 1c. The characteristics of the derived VSMCs are outlined in Table 1. The advantages and limitations of this method are detailed in **Note 6**. A brief overview of the regenerative applications of hiPSC-derived VSMCs is discussed in **Note 8**.

1. *hiPSC maintenance*:

   Culture hiPSC lines on irradiated mouse embryonic fibroblast feeder layer in 80% DMEM/F12 medium (GlobalStem) with 20% KnockOut Serum Replacement (Invitrogen) and 5 ng/ml FGF2 (Invitrogen). Replace medium daily. Passage cells every 3–5 days with 1 mg/ml Type IV collagenase (Invitrogen).

2. *Differentiation of smooth muscle-like cells (SMLCs) in monolayer cultures* [23, 24]:

   Coat T-150 cell culture flasks (Falcon) with collagen type IV (R&D Systems).

   Treat the culture with TrypLE (Invitrogen). Pass the digested culture through a 40μm mesh strainer (BD Biosciences). Seed the single-cell suspension onto the precoated T-150 cell culture flasks at a density of 5 × 10^4 cells/cm^2. Culture in differentiation medium for 6 days with daily medium changes.

   Coat T-75 cell culture flasks (Falcon) with collagen type IV.

   On day 6, treat the culture with TrypLE. Pass the digested culture through a 40μm mesh strainer to generate single-cell suspension. Seed onto the precoated T-75 cell culture flasks at a density of 1.25 × 10^4 cells/cm^2. Culture for additional 6 days in SMLC differentiation medium (*see* **Note 7**).

3. *Maturation of VSMCs in a pulsatile flow bioreactor*:

Prior to cell seeding, mount the PEGdma-PLA scaffolds onto the CellCrown 6-well inserts (Sigma-Aldrich) for better cell adhesion. Sterilize the scaffolds in 70% ethanol for 15 min, and rinse three times in Dulbecco's Phosphate-Buffered Saline (DPBS; Life Technologies). Coat with a 1:5 mixture of 1% hyaluronic acid (Advanced BioMatrix Inc.) and DPBS. Incubate the scaffolds at 37 °C for 3 h.

On day 12, treat the culture with TrypLE. Seed cells onto the prepared scaffolds at a density of 1.5×10^6 cells per scaffold in quiescent medium to induce maturation of SMLCs. Incubate for 1 day in static condition for cell adhesion.

On day 14, remove the CellCrown insert, and carefully transfer the cells seeded on scaffold to the bioreactor. Set the initial flow rate to 0.37 ml/min for the first day. Increase the flow rate to 0.74 ml/min the following day and keep it constant (*see* **Note 10**).

Incubate the sample in the bioreactor system with a peristaltic flow until day 30. Culture in quiescent medium with medium changes every 6 days.

4 Notes

1. As TEVGs demand a large starting input of cells, Luo et al. optimized an embryoid body (EB)-based approach to derive VSMCs from hiPSCs to supplement tissue-engineered scaffolds [4]. This VSMC differentiation method via EB formation has an advantage of resembling early embryonic development and allowing three-dimensional cellular interactions. Moreover, previous studies have shown that growth factors, such as fibroblast growth factor 2 (FGF2) and epidermal growth factor (EGF), in the commercially available Smooth Muscle Growth Medium (SmGM-2) can reduce the contractility and collagen production of the derived VSMCs [25, 26]. The priming medium employed in this method is designed to eliminate the effects of these exogenous growth factors on the characteristics of VSMCs while maintaining their proliferative capacity for large-scale production in vascular graft engineering. The resulting proliferative VSMCs with enhanced expression of VSMC markers are applicable for use in TEVG generation.

   Prior to the use of proliferative VSMCs after expansion for TEVG generation, Luo et al. characterized the maturation and function of the derived cells, relative to human primary VSMCs as a control. Compared to the native cells, these derived VSMCs demonstrated high expression of VSMC markers, such as α-smooth muscle actin (α-SMA), calponin (CNN1) and smooth muscle myosin heavy chain (MYH11), and ECM markers, such as elastin and collagen type I. Moreover, the decrease in cell surface area in response to vasoconstrictors observed in hiPSC-derived VSMCs served as an indicator of their contractile phenotype. Despite the lack of mature elastin fiber production in TEVGs, this methodology is unique in its ability to produce hiPSC-derived VSMCs on a large scale to meet the cell seeding requirement for TEVG development. The general yield post-expansion was reported to be 181.8 ± 10.3 million derived cells from approximately 4.5 million hiPSCs in about 4 weeks for the subsequent use in TEVG generation. Based on the expression of VSMC markers and their functional contractility, this EB-based method provides a robust, large-scale production of mature, functional VSMC replacement for future vascular tissue engineering.

2. Prepare this EB differentiation medium fresh after uniformly sized EBs form in suspension to induce differentiation.
3. Prepare this priming medium 72–96 h prior to cell seeding for tissue engineering.
4. Previous lineage mapping studies have shown diverse origins of smooth muscle cells in vascular development [27]. VSMCs with distinct embryological origins may contribute to the development of site-specific vascular diseases. Cheung et al. established an in vitro system to derive origin-specific VSMCs from hiPSCs [16, 17]. In this protocol, hiPSCs are first specified into the neuroectoderm, lateral plate mesoderm, or paraxial mesoderm lineage, followed by VSMC differentiation. This protocol employs a serum-free, chemically defined method that eliminates potential discrepancy caused by the use of serum with variable amounts of growth factors.

   After the 12-day VSMC differentiation induction period, Cheung et al. assessed the maturity and functionality of these hiPSC-derived VSMCs, using human aortic SMCs (ASMCs) as a positive control and human umbilical vein endothelial cells (HUVECs) as a negative control. Compared to ASMCs, these origin-specific VSMCs show similar gene expression levels of the early VSMC markers (α-SMA, TAGLN, CNN1) and later VSMC markers (SMTN, MYH11) by RT-qPCR analyses. Flow cytometric analysis further indicated cell populations with over 80% MYH11+/α-SMA+ in VSMCs derived from the three intermediates. Microarray analysis demonstrated upregulation of VSMC genes and downregulation of genes related to other cell types derived from the three intermediate lineages, such as neuronal, hematopoietic, and skeletal muscle cells. hiPSC-derived VSMCs and ASMCs displayed positive immunostaining for TAGLN and CNN1, while HUVECs were negative for these markers. Western blotting showed mature VSMC proteins (MYH11 and SMTN) were present in hiPSC-derived VSMCs and ASMCs, but absent in HUVECs. In addition, this protocol was reproducible in two hESC lines (H9 and H1) and a hiPSC line reprogrammed from wild-type human dermal fibroblasts.

The functionality of the hiPSC-derived VSMCs was assessed based on their contractility in response to cholinergic agonists, such as carbachol. Upon carbachol treatment, intracellular calcium flux into the derived VSMCs increased readily, followed by the reestablishment of physiological levels, the same trend observed in ASMCs. Furthermore, over 50% of the derived VSMCs contracted in response to the carbachol treatment. Notably, the use of a serum-free, chemically defined method results in derivation of cells with high contractile potential and limited proliferative capacity. As 10% FBS is generally required to magnify the hiPSC-derived VSMC cell number, these lineage-specific VSMCs derived without the use of serum remain quiescent with limited proliferative capacity. Expansion of these derived VSMCs requires subsequent use of a serum-containing medium to increase the scale of production. Nevertheless, this serum-free protocol generates mature, functional, origin-specific VSMCs that serve as a potential source of therapeutic cells for vascular regenerative medicine and a model for future studies of site-specific vascular diseases.

5. CDM is used as a base medium to make CDM-BSA and CDM-PVA. Add the chemicals or growth factors to CDM-BSA for hiPSC maintenance or to CDM-PVA when the cells are ready for the next step of lineage-specific differentiation.

6. As over-proliferation of cells can lead to restenosis of blood vessels, the ability of the hiPSC-derived VSMCs to maintain their mature, contractile phenotype with limited proliferative capacity in grafts poses a major challenge in TEVGs [28]. Previous studies have shown that the assembly and maturation of elastin into elastic fibers can maintain the contractile phenotype of VSMCs [29]. Eoh et al. followed Wanjare et al.'s protocol, engineering an effective approach to generate hiPSC-derived vascular smooth muscle tissue with mature, organized elastic fibers under a pulsatile flow [13, 14]. This method, detailed below, potentially addresses one of the key obstacles in TEVGs to maintain the specialization of contractile VSMCs in grafts.

   After exposing the derived cells in low-serum condition under a pulsatile flow, Eoh et al. examined the maturity and functionality of the hiPSC-derived VSMCs, relative to the static control. VSMCs matured in static or dynamic conditions not only exhibited TAGLN and MYH11 expression but also contained elastin, collagen types I and III, and fibronectin, indicating the contractile phenotype of VSMCs. Previous studies have indicated that native VSMCs align perpendicularly to an applied flow [30]. VSMCs exposed to flow demonstrated organized cytoskeletal and nuclear alignment transverse to flow, whereas static VSMCs showed random cellular alignment. Moreover, VSMCs under a pulsatile flow resulted in organized mature elastic fibers and microfibrils transverse to flow, resembling the characteristics of native VSMCs.

As contractility is a major feature of functional contractile VSMCs, Eoh et al. measured the contractility of the hiPSC-derived VSMCs under dynamic culture conditions in response to cholinergic agonists, compared to the static cultures. VSMCs exposed to flow had greater contractility, which was indicated by the greater change in area on both the cellular and tissue level and by an increase in intracellular calcium. Moreover, they exhibited a uniform contraction pattern and a higher mature, insoluble elastic fiber content, further supporting the highly organized cellular arrangement and maturity of VSMCs under dynamic conditions. Although this methodology enhances the elastin deposition of the derived cells, the use of low-serum culture medium restricts their proliferative capacity, limiting mass production of hiPSC-derived VSMCs for TEVG generation. Further research is needed to balance the proliferative potential and ECM deposition of the hiPSC-derived VSMCs in the hope of maximizing VSMC production with mature ECM deposition in the final engineered tissues.

7. Prepare the PEGdma-PLA scaffolds prior to the maturation of VSMCs in a pulsatile flow bioreactor.
8. hiPSC-derived VSMCs not only address the existing cell source problem of TEVGs for vascular bypass and defective vessel segment replacement but also contribute to various clinical applications, including vascular disease modeling and drug screening. Together with hiPSC-derived endothelial cells, TEVGs made with hiPSC-derived VSMCs offer a three-dimensional tissue platform that resembles the physiological characteristics of native vessels. This 3D culture system provides a more physiological representation of disease phenotype for the study of disease development and drug screening [31]. Moreover, recent efforts have successfully constructed "universal" hiPSCs that are potentially immunocompatible to any patient with the need of allogenic tissue transplantation [32–34]. Taken together, the use of hiPSC-derived VSMCs offers a new paradigm to advance vascular cellular therapy to address the global health burden of cardiovascular disease.
9. If cells become confluent, passage them onto gelatin-coated plates at a ratio of 1:2.
10. Setting the initial flow rate to a lower value helps to acclimate the cells to the flow and avoids shearing them off in the

bioreactor on the first day. The flow rate can then be increased the following day and kept constant for the remainder of the differentiation.

References

1. Laslett LJ, Alagona P, Clark BA et al (2012) The worldwide environment of cardiovascular disease: prevalence, diagnosis, therapy, and policy issues: a report from the American College of Cardiology. J Am Coll Cardiol 60:S1–S49. https://doi.org/10.1016/j.jacc.2012.11.002
2. Akoh JA, Patel N (2010) Infection of hemodialysis arteriovenous grafts. J Vasc Access 11:155–158. https://doi.org/10.1177/112972981001100213
3. Wu J, Hu C, Tang Z et al (2018) Tissue-engineered vascular grafts: balance of the four major requirements. Colloid Interface Sci Commun 23:34–44. https://doi.org/10.1016/j.colcom.2018.01.005
4. Luo J, Qin L, Zhao L et al (2020) Tissue-engineered vascular grafts with advanced mechanical strength from human iPSCs. Cell Stem Cell 26:251–261.e8. https://doi.org/10.1016/j.stem.2019.12.012
5. Bajpai VK, Andreadis ST (2012) Stem cell sources for vascular tissue engineering and regeneration. Tissue Eng Part B Rev 18:405–425. https://doi.org/10.1089/ten.teb.2011.0264
6. Kretlow JD, Jin Y-Q, Liu W et al (2008) Donor age and cell passage affects differentiation potential of murine bone marrow-derived stem cells. BMC Cell Biol 9:60. https://doi.org/10.1186/1471-2121-9-60
7. Ayoubi S, Sheikh SP, Eskildsen TV (2017) Human induced pluripotent stem cell-derived vascular smooth muscle cells: differentiation and therapeutic potential. Cardiovasc Res 113:1282–1293. https://doi.org/10.1093/cvr/cvx125
8. Dash BC, Jiang Z, Suh C, Qyang Y (2015) Induced pluripotent stem cell-derived vascular smooth muscle cells: methods and application. Biochem J 465:185–194. https://doi.org/10.1042/BJ20141078
9. Xie C-Q, Zhang J, Villacorta L et al (2007) A highly efficient method to differentiate smooth muscle cells from human embryonic stem cells. Arterioscler Thromb Vasc Biol 27:e311–e312. https://doi.org/10.1161/ATVBAHA.107.154260
10. Wang Y, Hu J, Jiao J et al (2014) Engineering vascular tissue with functional smooth muscle cells derived from human iPS cells and nanofibrous scaffolds. Biomaterials 35:8960–8969. https://doi.org/10.1016/j.biomaterials.2014.07.011
11. Dash BC, Levi K, Schwan J et al (2016) Tissue-engineered vascular rings from human iPSC-derived smooth muscle cells. Stem Cell Rep 7:19–28. https://doi.org/10.1016/j.stemcr.2016.05.004
12. Gui L, Dash BC, Luo J et al (2016) Implantable tissue-engineered blood vessels from human induced pluripotent stem cells. Biomaterials 102:120–129. https://doi.org/10.1016/j.biomaterials.2016.06.010
13. Wanjare M, Kuo F, Gerecht S (2013) Derivation and maturation of synthetic and contractile vascular smooth muscle cells from human pluripotent stem cells. Cardiovasc Res 97:321–330. https://doi.org/10.1093/cvr/cvs315
14. Eoh JH, Shen N, Burke JA et al (2017) Enhanced elastin synthesis and maturation in human vascular smooth muscle tissue derived from induced-pluripotent stem cells. Acta Biomater 52:49–59. https://doi.org/10.1016/j.actbio.2017.01.083
15. Bajpai VK, Mistriotis P, Loh Y-H et al (2012) Functional vascular smooth muscle cells derived from human induced pluripotent stem cells via mesenchymal stem cell intermediates. Cardiovasc Res 96:391–400. https://doi.org/10.1093/cvr/cvs253
16. Cheung C, Bernardo AS, Trotter MWB et al (2012) Generation of human vascular smooth muscle subtypes provides insight into embryological origin-dependent disease susceptibility. Nat Biotechnol 30:165–173. https://doi.org/10.1038/nbt.2107
17. Cheung C, Bernardo AS, Pedersen RA, Sinha S (2014) Directed differentiation of embryonic origin-specific vascular smooth muscle subtypes from human pluripotent stem cells. Nat Protoc 9:929–938. https://doi.org/10.1038/nprot.2014.059
18. Patsch C, Challet-Meylan L, Thoma EC et al (2015) Generation of vascular endothelial and smooth muscle cells from human pluripotent stem cells. Nat Cell Biol 17:994–1003. https://doi.org/10.1038/ncb3205
19. Vallier L, Touboul T, Chng Z et al (2009) Early cell fate decisions of human embryonic stem

cells and mouse epiblast stem cells are controlled by the same signalling pathways. PLoS One 4(e):6082. https://doi.org/10.1371/journal.pone.0006082
20. Bernardo AS, Faial T, Gardner L et al (2011) BRACHYURY and CDX2 mediate BMP-induced differentiation of human and mouse pluripotent stem cells into embryonic and extraembryonic lineages. Cell Stem Cell 9:144–155. https://doi.org/10.1016/j.stem.2011.06.015
21. Hinderer S, Seifert J, Votteler M et al (2014) Engineering of a bio-functionalized hybrid off-the-shelf heart valve. Biomaterials 35:2130–2139. https://doi.org/10.1016/j.biomaterials.2013.10.080
22. Hinderer S, Shena N, Ringuette L-J et al (2015) In vitro elastogenesis: instructing human vascular smooth muscle cells to generate an elastic fiber-containing extracellular matrix scaffold. Biomed Mater Bristol Engl 10:034102. https://doi.org/10.1088/1748-6041/10/3/034102
23. Vo E, Hanjaya-Putra D, Zha Y et al (2010) Smooth-muscle-like cells derived from human embryonic stem cells support and augment cord-like structures in vitro. Stem Cell Rev Rep 6:237–247. https://doi.org/10.1007/s12015-010-9144-3
24. Gerecht-Nir S, Ziskind A, Cohen S, Itskovitz-Eldor J (2003) Human embryonic stem cells as an in vitro model for human vascular development and the induction of vascular differentiation. Lab Investig 83:1811–1820. https://doi.org/10.1097/01.LAB.0000106502.41391.F0
25. Chen P-Y, Qin L, Li G et al (2016) Fibroblast growth factor (FGF) signaling regulates transforming growth factor beta (TGFβ)-dependent smooth muscle cell phenotype modulation. Sci Rep 6:33407. https://doi.org/10.1038/srep33407
26. Schlumberger W, Thie M, Rauterberg J, Robenek H (1991) Collagen synthesis in cultured aortic smooth muscle cells. Modulation by collagen lattice culture, transforming growth factor-beta 1, and epidermal growth factor. Arterioscler Thromb 11:1660–1666. https://doi.org/10.1161/01.atv.11.6.1660
27. Majesky MW (2007) Developmental basis of vascular smooth muscle diversity. Arterioscler Thromb Vasc Biol 27:1248–1258. https://doi.org/10.1161/ATVBAHA.107.141069
28. Beamish JA, He P, Kottke-Marchant K, Marchant RE (2010) Molecular regulation of contractile smooth muscle cell phenotype: implications for vascular tissue engineering. Tissue Eng Part B Rev 16:467–491. https://doi.org/10.1089/ten.teb.2009.0630
29. Ryan AJ, O'Brien FJ (2015) Insoluble elastin reduces collagen scaffold stiffness, improves viscoelastic properties, and induces a contractile phenotype in smooth muscle cells. Biomaterials 73:296–307. https://doi.org/10.1016/j.biomaterials.2015.09.003
30. Lee AA, Graham DA, Dela Cruz S et al (2002) Fluid shear stress-induced alignment of cultured vascular smooth muscle cells. J Biomech Eng 124:37–43. https://doi.org/10.1115/1.1427697
31. Atchison L, Zhang H, Cao K, Truskey GA (2017) A tissue engineered blood vessel model of Hutchinson-Gilford progeria syndrome using human iPSC-derived smooth muscle cells. Sci Rep 7:8168. https://doi.org/10.1038/s41598-017-08632-4
32. Deuse T, Hu X, Gravina A et al (2019) Hypoimmunogenic derivatives of induced pluripotent stem cells evade immune rejection in fully immunocompetent allogeneic recipients. Nat Biotechnol 37:252–258. https://doi.org/10.1038/s41587-019-0016-3
33. Xu H, Wang B, Ono M et al (2019) Targeted disruption of HLA genes via CRISPR-Cas 9 generates iPSCs with enhanced immune compatibility. Cell Stem Cell 24:566–578.e7. https://doi.org/10.1016/j.stem.2019.02.005
34. Gornalusse GG, Hirata RK, Funk S et al (2017) HLA-E-expressing pluripotent stem cells escape allogeneic responses and lysis by NK cells. Nat Biotechnol 35:765–772. https://doi.org/10.1038/nbt.3860

Chapter 4

On-Site Differentiation of Human Mesenchymal Stem Cells into Vascular Cells on Extracellular Matrix Scaffold Under Mechanical Stimulations for Vascular Tissue Engineering

Na Li, Alex P. Rickel, and Zhongkui Hong

Abstract

Small-diameter vascular grafts are considered to be a promising strategy to treat late-stage vascular diseases, one of the largest causes of morbidity and mortality worldwide. However, limited sources of functional vascular cells remain a major obstacle in vascular tissue engineering. Here we describe a novel approach whereby functional vascular cells were obtained by on-site differentiation of human mesenchymal stem cells on a vascular extracellular matrix scaffold under mechanical stimulations in a rotary bioreactor, which has the potential to work as an alternative source for robust implantable artificial vessel grafts.

Key words Mesenchymal stem cell, Differentiation, Vascular cells, Extracellular matrix scaffold, Mechanical stimulation

1 Introduction

Small-diameter vascular grafts are considered to be a promising strategy to treat vascular diseases, which represents one of the largest causes of morbidity and mortality worldwide [1–4]. Currently, autologous vessels are the gold standard for small-diameter vessel grafts but suffer from associated morbidity at the donor site and a general scarcity due to disease and previous harvest [5, 6]. Synthetic blood vessels have shown promise for medium-/large-diameter vessel grafts, but current synthetic small-diameter vessel substitutes suffer from low patency due to acute thrombogenicity and rapid occlusion [5, 7–9]. These events are attributed to poor endothelialization and a mismatch of mechanical properties compared to native vessels [9]. Thus, fabricating functional biomimetic vascular conduits that are identical to their native counterparts remains a great challenge [10].

Feng Zhao and Kam W. Leong (eds.), *Vascular Tissue Engineering: Methods and Protocols*,
Methods in Molecular Biology, vol. 2375, https://doi.org/10.1007/978-1-0716-1708-3_4,

Native blood vessels are composed of three distinct layers: the intima, a monolayer of endothelial cells (ECs) on the luminal surface; the media, the middle layer made of several layers of vascular smooth muscle cells (VSMCs) embedded into extracellular matrix (ECM); and the adventitia [11, 12]. Therefore, regeneration of functional VSMCs and endothelium as well as a vessel scaffold that can recapitulate the mechanical properties and biocompatibility of the native blood vessel wall are three major elements in tissue-engineered blood vessels (TEBV). Compared to synthetic scaffolds, which can provide desirable mechanical support to accommodate vascular cells and maintain durability in vivo [13], native decellularized ECM possesses not only excellent mechanical strength and biocompatibility but also beneficial bioactivity due to its three-dimensional ultrastructure and the large amount bioactive components naturally present [13–16], leading to its wide acceptance in tissue engineering and regenerative medicine. Three approaches have been developed to deliver TEBVs based on decellularized ECM including acellular grafts for small vessels, direct implantation of cell-seeded scaffolds using the body as a natural bioreactor, and cell-seeded scaffolds grown in bioreactors in vitro [14]. Intriguingly, we found in our previous study that the native vascular ECM was able to facilitate the differentiation of stem cells into VSMCs [17].

The following protocol aims to describe a novel approach, where functional ECs and VSMCs are obtained by on-site differentiation of human bone marrow-derived mesenchymal stem cells (hBM-MSCs) on decellularized vascular ECM scaffold under mechanical stimulations in a rotary bioreactor (Fig. 1). Our results showed that soluble growth factor-directed differentiation of hBM-MSCs towards both VSMCs and ECs is enhanced by the stimulation of native arterial ECM and mechanical cues generated in bioreactor (Figs. 2 and 3). This on-site differentiated vascular phenotype integrated into native vascular ECM could not only be

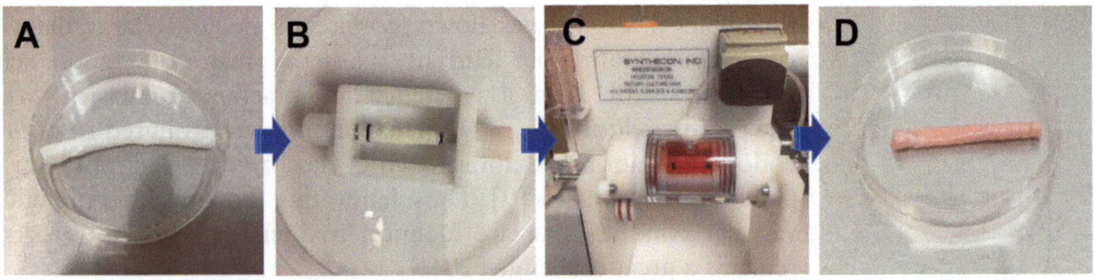

Fig. 1 Schematic of the ex vivo bioreactor incorporated with a decellularized carotid artery scaffold. (**a**) Decellularized ECM scaffold. (**b**) MSCs were seeded on either intimal or adventitial side of decellularized carotid artery scaffold and mounted on the bioreactor insert. (**c**) The ECM-MSCs were cultured in vascular bioreactor with VEGF medium inside of the vessel scaffold and TGFβ1 medium outside of vessel scaffold. (**d**) The tissue-engineered blood vessel (TEBV). (Adapted from ref. 4)

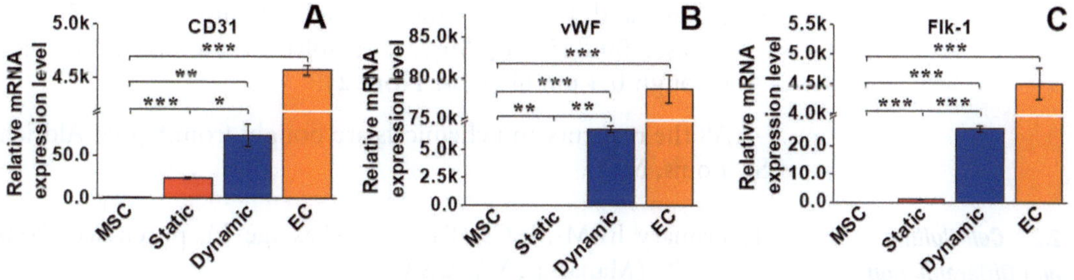

Fig. 2 Real-time RT-PCR analysis for the EC marker gene expressions. Each marker gene was significantly upregulated for both static and dynamic groups compared to negative control (MSCs), and the mRNA expression level of each marker gene in dynamic culture system was significantly higher than that in static culture system. Data are expressed as mean ± SEM ($n = 3$). *$P < 0.05$, **$P < 0.01$, ***$P < 0.001$. (Adapted from ref. 4)

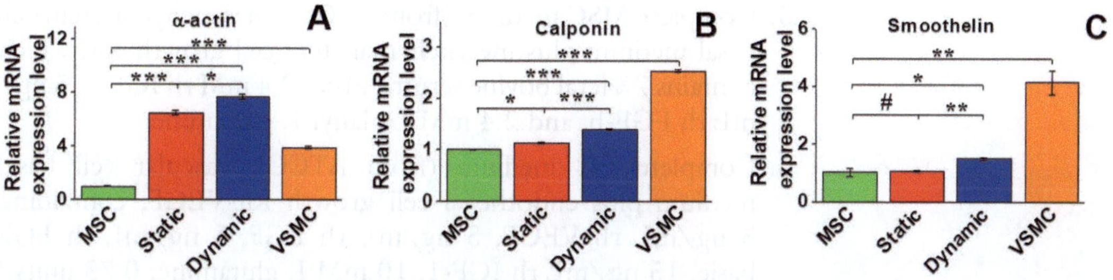

Fig. 3 Real-time RT-PCR analysis for the VSMC marker gene expressions. The mRNA expression level of each marker protein in the dynamic group was significantly higher than that in the static group, and α-actin expression levels in both the dynamic and static groups were even higher than that in positive control VSMCs (A). Data are expressed as mean ± SEM ($n = 3$). # $P > 0.05$, *$P < 0.05$, **$P < 0.01$, ***$P < 0.001$. (Adapted from ref. 4)

an advanced cell resource for tissue-engineered arteries but also provide an alternative approach to assembling a robust implantable artificial vessel graft.

2 Materials

2.1 Decellularization

1. Fresh porcine carotid arteries.
2. Phosphate-buffered saline (PBS).
3. Scissor and forceps.
4. 3-[(3-Cholamidopropyl)dimethylammonio]-1-propanesulfonate (CHAPS) buffer: weigh 983.808 mg CHAPS, 11.688 g NaCl, 1.8612 g EDTA, dissolve in 200 mL PBS, filter through 0.4 μm filter, store at 4 °C.
5. Sodium dodecyl sulfate (SDS) buffer: weigh 103.82 mg SDS, 11.688 g NaCl, 1.8612 g EDTA, dissolve in 200 mL PBS, filter through 0.4 μm filter, store at 4 °C (*see* **Note 1**).

6. DNase and RNase solution: weigh 10 mg DNase, 50 mg RNase, 50.825 mg $MgCl_2$, dissolve in 50 mL PBS, filter through 0.4 μm filter (*see* **Note 2**).

All the reagents and chemicals are bought from Sigma-Aldrich (St. Louis, MO).

2.2 Cell Culture and Differentiation

1. Primary hBM-MSCs (Passage 1-Passage 5), purchased from ATCC (Manassas, VA, USA).
2. Primary human VSMCs (Passage 1-Passage 3), purchased from ATCC.
3. Primary human aortic ECs (Passage 1-Passage 3), purchased from ATCC.
4. Vascular endothelial growth factor (VEGF) and transforming growth factor-beta 1 (TGFβ1) (*see* **Note 3**).
5. Complete MSC medium (from ATCC): mesenchymal stem cell basal medium plus mesenchymal stem cell growth Kit, which contains 7% fetal bovine serum (FBS), 2.4 mM rh IGF-1, 5 ng/mL rh FGF-b, and 2.4 mM L-alanyl-L-glutamine.
6. Complete EC medium (from ATCC): vascular cell basal medium plus endothelial cell growth Kit-VEGF, containing 5 ng/mL rh VEGF, 5 ng/mL rh EGF, 5 ng/mL rh FGF basic, 15 ng/mL rh IGF-1, 10 mM L-glutamine, 0.75 units/mL heparin sulfate, 1 μg/mL hydrocortisone hemisuccinate, 2% FBS, and 50 μg/mL ascorbic acid.
7. Complete VSMC medium (from ATCC): vascular cell basal medium plus vascular smooth muscle cell growth kit, which contains 5 ng/mL rh FGF-basic, 5 μg/mL rh insulin, 50 μg/mL ascorbic acid, 10 mM L-glutamine, 5 ng/mL rh EGF, and 5% FBS.
8. 0.25% trypsin-EDTA solution.
9. 24-well plate.
10. *Ex vivo* blood vessel bioreactor system (*see* **Note 4**).
11. Suture.
12. Syringe.

All the reagents are purchased from Invitrogen (Carlsbad, CA).

2.3 Real-Time RT-PCR Analysis

1. TRIzol reagent.
2. High-Capacity cDNA Reverse Transcription Kit containing RT Buffer, 25 × dNTP Mix, 10 × RT Random Primers, and MultiScribe™ Reverse Transcriptase.
3. Chloroform.
4. 70% ethanol.

5. Isopropanol.
6. RNase-free water.
7. TaqMan Gene Expression Assays (all from Applied Biosystems) for CD31 (Hs01065279), von Willebrand factor (vWF, Hs01109446), fetal liver kinase-1 (flk-1, Hs00911700), α-actin (Hs00426835), calponin (Hs00206044), and smoothelin (Hs00199489).
8. TaqMan Universal PCR Master Mix.
9. Ultraviolet spectrophotometer.
10. PCR thermal cycler.
11. Applied Biosystems 7500 Fast Real-Time PCR System.

All the reagents and chemicals are purchased from Invitrogen (Carlsbad, CA) or Applied Biosystems (Forster City, CA).

3 Methods

3.1 Preparation of Decellularized ECM Scaffolds (See Note 5)

1. Trim excess connective tissues and fat from fresh porcine carotid arteries (around 5 cm length), and wash the arteries completely with PBS.
2. Treat the cleaned carotid arteries with CHAPS buffer for 22 h followed by brief washes with PBS.
3. Further treat the carotid vessels for 24 h with SDS buffer followed by complete washes with PBS (*see* **Note 6**).
4. Incubate the carotid vessels with DNase and RNase solution for 16 h at 37 °C, and wash thoroughly with PBS. The decellularized vessel ECM scaffolds can be stored at −80 °C (*see* **Note 7**).

3.2 Cell Culture and On-Site Differentiation of MSCs on Decellularized Vascular ECM Scaffold in a Rotary Bioreactor

3.2.1 Cell Maintenance Culture

MSCs are cultured in complete MSC medium. Human aortic SMCs are grown in complete VSMC medium. Human aortic ECs are cultured in complete EC medium. All the three types of cells are subcultured when they reach approximately 90% confluence, and the medium is refreshed every other day. All the cells are incubated at 37 °C in a 5% CO_2 and 95% humidity environment.

3.2.2 MSCs Differentiation into ECs

Three conditions are set as control: (1) MSCs cultured in complete MSC medium are defined as MSC group; (2) MSCs seeded onto the intimal surface of decellularized vessel ECM scaffolds (1×10^6 cells/scaffold) and cultured in complete EC medium supplemented with 50 ng/mL VEGF are defined as static group; (3) ECs cultured in complete EC medium are set as positive control (*see* **Note 8**).

1. Thaw the decellularized pig carotid artery scaffold (around 5 cm in length, 2.5 mm in diameter) at room temperature and immerse it in 75% ethanol for 30 min, wash the scaffold with PBS completely, and allow it to equilibrate in complete MSC medium overnight (*see* **Note 9**).
2. Cut the decellularized pig carotid artery scaffold into 2.5 cm length and mount it onto the insert of bioreactor with suture under sterile conditions, and then place the insert back into the bioreactor (Fig. 1) (*see* **Note 10**).
3. Remove the fill port plug on the vessel wall of bioreactor and place it on an alcohol swab. Fill the chamber of the vessel with complete EC medium until it reaches the fill port.
4. Trypsinize and collect 1×10^6 MSCs in 2 mL EC medium supplemented with 50 ng/mL VEGF, fill a small syringe with the cells and attach it to one of the ports on the rotating coupling of bioreactor, and inject the cells into the lumen of vessel scaffold from the port.
5. Attach an empty syringe to the port on the other rotating coupling of bioreactor (*see* **Note 11**).
6. Pick up the vessel of bioreactor and hold it vertically while injecting cells into the lumen from the bottom. Continue to inject cells until media is observed entering the empty syringe at the top (*see* **Note 12**).
7. Close the valves on the rotating couplings so the media and cells in the lumen are confined, and remove the syringes from the rotating coupling.
8. Start to rotate the rotary cylinder 90° every 30 min for 4 h to promote cell attachment to the vessel scaffold (*see* **Note 13**).
9. Open the cover of the peristaltic pump so that the tubing is not pinched. Fill the media bottle with 50 mL complete EC medium supplemented with 50 ng/mL VEGF. Place a sterile syringe on one of the luer ports on the rotating coupling, and turn the valve to open the port connected to the syringe. Using the syringe, prime the tubing by drawing media from the bottle until it reaches the syringe. Rotate the valve so that the media flow is directed into the vessel through the perfusion core. Close the cover of the peristaltic pump (*see* **Note 14**).
10. Start the perfusion through the whole circuit at a lower flow rate of 0.5 mL/min for the first 24 h followed by 1.5 mL/min for the rest of the culture period (13 days) (*see* **Note 15**). The cells cultured in this method are defined as dynamic group.

3.2.3 MSCs Differentiation into VSMCs

The differentiation of MSCs into VSMCs in the bioreactor is also studied as similar strategy described above. (1) MSCs cultured in complete MSC medium are defined as MSC group; (2) MSCs seeded onto the adventitial surface of decellularized vessel ECM scaffolds (1×10^6 cells/scaffold) and maintained in complete MSCs medium supplemented with 10 ng/mL TGFβ1 are defined as static group; (3) VSMCs cultured in complete VSMC medium are set as positive control (*see* **Note 8**).

1. Thaw the decellularized pig carotid artery scaffold (around 5 cm in length, 2.5 mm in diameter) at room temperature and immerse it in 75% ethanol for 30 min, wash the scaffold with PBS completely, and allow it to equilibrate in MSCs medium overnight (*see* **Note 9**).
2. Cut the decellularized pig carotid artery scaffold into 2.5 cm length, and mount it onto the insert of bioreactor with suture under sterile conditions.
3. Trypsinize and collect 1×10^6 MSCs in 300 μL MSCs medium supplemented with 10 ng/mL TGFβ1, and seed them onto the adventitial side of the vessel scaffold manually.
4. Place the insert back into the bioreactor (*see* **Note 10**), and remove the fill port plug on the vessel wall and place it on an alcohol swab. Fill the chamber of the vessel with MSCs medium supplemented with 10 ng/mL TGFβ1 until it reaches the fill port.
5. Start to rotate the rotary cylinder 90° every 30 min for 4 h to promote cell attachment to the vessel scaffold (*see* **Note 13**).
6. Open the cover of the peristaltic pump so that the tubing is not pinched. Fill the media bottle with 50 mL complete MSC medium. Place a sterile syringe on one of the luer ports on the rotating coupling, and turn the valve to open the port attached to the syringe. Using the syringe, prime the tubing by drawing media from the bottle until it reaches the syringe. Rotate the valve so that the media flow is directed into the vessel through the perfusion core. Close the cover of the peristaltic pump (*see* **Note 14**).
7. Start the perfusion through the whole circuit at a lower flow rate of 0.5 mL/min for the first 24 h followed by 1.5 mL/min for the rest of the culture period (13 days). The cells cultured in the vascular bioreactor are defined as dynamic group.

3.3 Real-Time RT-PCR Analysis

1. Remove the vessel scaffold seeded with cells from the bioreactor, and rinse it twice with PBS to remove residual growth medium; cut 0.5 cm graft into small pieces with scissors, and add 1 mL Trizol reagent to collect cell lysate.
2. Remove the leftover pieces of scaffold, centrifuge for 5 min, and transfer the supernatant into another tube. After 5 min at room temperature, add 1/5 chloroform, mix fully with Trizol reagent, and wait for 5 min at room temperature again. Centrifuge at $12{,}000 \times g$ for 15 min at 4 °C.

3. Transfer the supernatant carefully into another tube, and then add same volume of isopropanol, mix fully, and keep the mixture for 10 min at room temperature. Centrifuge at 12,000 × *g* for 10 min at 4 °C.
4. Transfer the supernatant carefully, and add 70% ethanol (the same volume to Trizol reagent) to wash the tube wall. Centrifuge at 7500 × *g* for 5 min at 4 °C.
5. Discard the supernatant, and then keep the cap open for a couple of minutes at room temperature to dry the RNA.
6. Add 10 μL RNase-free water to dissolve the RNA, and then add 98 μL TE buffer into 2 μL RNA; measure OD260 value using UV instrument. The RNA concentration (μg/μL) = OD260 × 50 × 0.04.
7. Pipette 2.0 μL 10 × RT Buffer, 0.8 μL 25 × dNTP Mix, 2.0 μL 10 × RT Random Primers, 1.0 μL MultiScribe™ Reverse Transcriptase, 14.2 μL (RNase-free water volume + RNA volume) into individual tube; the total volume is 20 μL, and total RNA should be less than 2 μg (*see* **Note 16**).
8. Seal the tubes, and briefly centrifuge the tubes to spin down the contents and to eliminate any air bubbles. Load the tubes into the thermal cycler to start the reverse transcription.
9. After reverse transcription, dilute the cDNA 10 times with RNase-free water.
10. Pipette 10 μL TaqMan Universal PCR Master Mix, 1 μL TaqMan Gene Expression Assay, and 9 μL cDNA template into individual tube.
11. Cap the tubes, and centrifuge the tubes to spin down the contents and to eliminate any air bubbles. Load the tubes into 7500 Fast Real-Time PCR system, and start to run the PCR.

4 Results and Discussion

Our results showed that the transcript expression level of each EC marker gene in the dynamic culture system was significantly higher than that of the static culture system, suggesting mechanical forces played a substantial role in facilitating the VEGF-directed differentiation of MSCs towards EC-like phonotype (Fig. 2). As shown in Fig. 3, α-actin and calponin were significantly upregulated in each experimental group compared to the negative control (MSCs). The mRNA expression levels of all three VSMC marker proteins in the dynamic culture group were statistically higher than those in the static group, and α-actin expression levels in both dynamic and static groups were even higher than that of the positive control group (VSMCs).

Functional vascular cells are the necessary constituents to construct an implantable vessel graft in vitro. In this protocol, we described an approach to obtaining functional ECs or VSMCs for vascular tissue engineering by on-site differentiation of hBM-MSCs on the natural blood vessel ECM scaffold under biomechanical stimulation. Growth factors VEGF and TGFβ1 were used as chemical cues to induce the differentiation of MSCs towards ECs and VSMCs respectively in previous studies [4, 17]. We found both VEGF-directed differentiation towards ECs and TGFβ1-directed differentiation towards VSMCs of MSCs were enhanced by the native blood vessel ECM scaffold and mechanical cues provided by a perfusion rotary bioreactor. Similar findings have revealed that decellularized ovine artery matrix could serve as an efficient culture system for promoting differentiation of adipose-derived MSCs toward ECs and decellularized swine carotid vessel scaffold is able to expedite MSCs-VSMCs differentiation [17, 18]. The bioactive components present in native vascular ECM including collagen, fibronectin, and laminin could provide a reasonable explanation for the inductive effect exerted by ECM scaffold during the differentiation process. For example, laminin was reported to guide highly efficient endothelial commitment from human pluripotent stem cells, and collagen IV can efficaciously induce the differentiation of mouse embryonic stem cells into smooth muscle cells [19, 20]. Furthermore, a report that a combination of substrate topography and VEGF treatment could increase the endothelial differentiation in MSCs is also consistent with our result [21]. Additionally, the physiological mechanical stimulations from both decellularized ECM scaffold and rotary bioreactor might also work as a potential promoter to elevate the differentiation efficiency. This point coincides with the results reported by Colazzo et al. that shear stress and VEGF enhance endothelial differentiation of human adipose-derived stem cells [22]. Other researchers have also observed the upregulation of VSMC markers when MSCs are exposed to mechanical stretch [21].

In conclusion, we provide a novel approach to attaining functional vascular cells for constructing an autologous vessel graft with physiologically relevant mechanical properties and biological functions.

5 Notes

1. SDS buffer should be prepared under heat to dissolve SDS completely; SDS would be precipitated at room temperature, so SDS buffer should be warmed at 37 °C before use.
2. To maximize the effect of DNase and RNase, DNase and RNase solution should be prepared before use; it cannot be stored for a long time.

3. VEGF and TGFβ1 should be dissolved following the protocols provided by the manufacturer and stored at −20 °C or −80 °C.
4. *Ex vivo* blood vessel bioreactor system was purchased from Synthecon (Houston, TX), and its structure was described by Shi et al. [23]. Do not autoclave the rotator base/upright assembly. Storage of the rotator base/upright assembly in an incubator while not in use will corrode the motors eventually resulting in loss of function. Placing the power supply/control box inside an incubator will result in loss of function.
5. All decellularization steps were performed with agitation and under sterile conditions.
6. The decellularization step should be performed at 50 °C to avoid SDS precipitation.
7. The enzymatic process should be completed in a shaker at 37 °C.
8. The cells in all control groups should be seeded at the same time as the cells in the dynamic group were seeded, and they are all from the same batch of MSCs.
9. Some decellularized vessel scaffolds would be contaminated during the decellularization process, and any scaffolds that are suspected to be contaminated should be discarded after culture in MSC medium overnight.
10. The whole procedure should be performed under strict sterile conditions.
11. Make sure that both ports are in the open position for the syringes.
12. This procedure is necessary to prevent bubble from being trapped in the lumen, which will hinder cell attachment. If there is any doubt that all bubbles have been removed, work the cell-containing media back and forth between the syringes until no bubbles can be seen.
13. Make sure that the peristaltic pump is off; otherwise, pressure may build up and blow the tubing off the plastic fittings.
14. Place a syringe with a plastic cannula on the sampling port (wiping the septum with a sterile alcohol swab whenever it is to be punctured is recommended). Maneuver the bubble in the cell chamber beneath the sampling port, and draw the bubble out. Media will be pulled through the core to replace the volume of the bubble. During the culture, if bubbles appear in the vessel, they can be removed by this procedure.
15. Changing perfusate media is done by replacing the media bottle with a bottle containing fresh media. This should always be done in a laminar flow hood.
16. The extra RNA should be stored at −80 °C.

Acknowledgments

This work was supported in part by the American Heart Association (15SDG25420001) and the South Dakota Board of Regents (UP1600205).

References

1. Unal B, Critchley JA, Capewell S (2003) Missing, mediocre, or merely obsolete? An evaluation of UK data sources for coronary heart disease. J Epidemiol Community Health 57:530–535
2. Isenberg BC, Williams C, Tranquillo RT (2006) Small-diameter artificial arteries engineered in vitro. Circ Res 98:25–35
3. Laslett LJ, Alagona P Jr, Clark BA III et al (2012) The worldwide environment of cardiovascular disease: prevalence, diagnosis, therapy, and policy issues a report from the American College of Cardiology. J Am Coll Cardiol 60: S1–S49
4. Li N, Rickel AP, Sanyour HJ et al (2019) Vessel graft fabricated by the on-site differentiation of human mesenchymal stem cells towards vascular cells on vascular extracellular matrix scaffold under mechanical stimulation in a rotary bioreactor. J Mater Chem B 7:2703–2713
5. Masuda T, Ukiki M, Yamagishi Y et al (2018) Fabrication of engineered tubular tissue for small blood vessels via three-dimensional cellular assembly and organization ex vivo. J Biotechnol 276:46–53
6. Nemeno-Guanzon JG, Lee S, Berg JR et al (2012) Trends in tissue engineering for blood vessels. J Biomed Biotechnol 2012:956345
7. Kurobe H, Maxfield MW, Breuer CK et al (2012) Concise review: tissue-engineered vascular grafts for cardiac surgery: past, present, and future. Stem Cells Transl Med 1:566–571
8. Pevec WC, Darling RC, Litalien GJ et al (1992) Femoropopliteal reconstruction with knitted, nonvelour Dacron versus expanded polytetrafluoroethylene. J Vasc Surg 16:60–65
9. Venkatraman S, Boey F, Lao LL (2008) Implanted cardiovascular polymers: natural, synthetic and bio-inspired. Prog Polym Sci 33:853–874
10. Meiring M, Khemisi M, Laker L et al (2017) Tissue engineered small vessel conduits - the anti-thrombotic effect of re-endothelialization of decellularized baboon arteries: a preliminary experimental study. Med Sci Monit Basic Res 23:344–351
11. Xu JY, Shi GP (2014) Vascular wall extracellular matrix proteins and vascular diseases. Biochim Biophys Acta 1842:2106–2119
12. Zhang L, Zhou J, Lu Q et al (2008) A novel small-diameter vascular graft: in vivo behavior of biodegradable three-layered tubular scaffolds. Biotechnol Bioeng 99:1007–1015
13. Zhao Y, Zhang S, Zhou J et al (2010) The development of a tissue-engineered artery using decellularized scaffold and autologous ovine mesenchymal stem cells. Biomaterials 31:296–307
14. Catto V, Farè S, Freddi G et al (2014) Vascular tissue engineering: recent advances in small diameter blood vessel regeneration. ISRN Vasc Med 2014:27
15. Negishi J, Hashimoto Y, Yamashita A et al (2017) Evaluation of small-diameter vascular grafts reconstructed from decellularized aorta sheets. J Biomed Mater Res A 105:1293–1298
16. Syazwani N, Azhim A, Morimoto Y et al (2015) Decellularization of aorta tissue using sonication treatment as potential scaffold for vascular tissue engineering. J Med Biol Eng 35:258–269
17. Li N, Sanyour H, Remund T et al (2018) Vascular extracellular matrix and fibroblasts-coculture directed differentiation of human mesenchymal stem cells toward smooth muscle-like cells for vascular tissue engineering. Mater Sci Eng C Mater Biol Appl 93:61–69
18. Zhang W, Huo Y, Wang X et al (2016) Decellularized ovine arteries as biomatrix scaffold support endothelial of mesenchymal stem cells. Heart Vessel 31:1874–1881
19. Liu D, Shao X, Zhong Z et al (2018) All-trans retinoic acid in combination with collagen IV induces the differentiation of mouse embryonic stem cells into smooth muscle cells. Int J Clin Exp Med 11:166–172

20. Ohta R, Niwa A, Taniguchi Y et al (2016) Laminin-guided highly efficient endothelial commitment from human pluripotent stem cells. Sci Rep 6:35680
21. Henderson K, Sligar AD, Le VP et al (2017) Biomechanical regulation of mesenchymal stem cells for cardiovascular tissue engineering. Adv Healthc Mater 6:1700556
22. Colazzo F, Alrashed F, Saratchandra P et al (2014) Shear stress and VEGF enhance endothelial differentiation of human adipose-derived stem cells. Growth Factors 32:139–149
23. Shi Q, Hodara V, Simerly CR et al (2013) Ex vivo reconstitution of arterial endothelium by embryonic stem cell-derived endothelial progenitor cells in baboons. Stem Cells Dev 22:631–642

Chapter 5

End-Point Immobilization of Heparin on Electrospun Polycarbonate-Urethane Vascular Graft

Jun Fang and Song Li

Abstract

There is a tremendous clinical need for synthetic vascular grafts either for bypass procedure or vascular access during hemodialysis. However, currently, there is no small-diameter vascular graft commercially available to meet long-term patency requirement due to frequent thrombus formation and intimal hyperplasia. This chapter describes the fabrication of electrospun small-diameter polycarbonate-urethane (PCU) vascular graft with a biomimetic fibrous structure. Additionally, the surface of the vascular graft is aminated via plasma treatment for the subsequently end-point heparin immobilization to enhance antithrombosis property.

Key words Vascular grafts, Electrospinning, Polyurethane, Surface modification, Plasma treatment, Heparin

1 Introduction

Cardiovascular diseases (CVDs) are the leading cause of morbidity and mortality in the world, accounting for around 17.9 million deaths each year, according to the data from the World Health Organization (WHO) [1]. The majority of CVDs, including coronary heart diseases, cerebrovascular diseases, and peripheral vascular diseases, are mainly caused by the formation of plaques and blockages in arteries. Currently, vascular bypass reconstruction is performed to restore the blood flow of diseased vessels in CVD patients [2, 3]. In addition, vascular access grafts is widely used for hemodialysis in more than 2.62 million patients with end-stage kidney disease in 2010 globally [4]. Autologous blood vessels (e.g., saphenous vein) have been used extensively for bypass surgery with superior patency rates and are considered as the gold standard. However, such autografts are hampered by invasive harvesting, limited availability, and poor quality from diseased patients [5]. Alternatively, vascular grafts become a promising option for

Feng Zhao and Kam W. Leong (eds.), *Vascular Tissue Engineering: Methods and Protocols*,
Methods in Molecular Biology, vol. 2375, https://doi.org/10.1007/978-1-0716-1708-3_5,

vascular replacement. In the clinic, the replacement of large-diameter vessels (≥6 mm internal diameter), such as the aorta, has been performed successfully with synthetic vascular grafts, such as expanded polytetrafluoroethylene (ePTFE) or polyethylene terephthalate (PET). In contrast, clinical results appear disappointing in the replacement of small-diameter vessels with inner diameter (ID) less than 6 mm. The small-diameter vascular grafts usually occlude rapidly due to thrombosis in the early stage and intimal hyperplasia in the late stage [6, 7]. Thus, there is a great clinical demand for small-diameter vascular grafts with improved long-term patency and performances.

Over the past two decades, much effort has been devoted to developing various small-diameter vascular grafts, including synthetic grafts, decellularized-matrix grafts and tissue-engineered vascular grafts [8–11]. Among them, synthetic small-diameter vascular grafts without cell seeding are attractive for clinical applications as they are available off the shelf. Many manufacturing techniques have been developed to fabricate vascular grafts, including electrospinning [12, 13], thermally induced phase separation [14], weaving [15, 16], three-dimensional printing [17, 18], or their combination [19]. Among them, electrospinning provides a straightforward and versatile technique to fabricate biomimetic porous scaffolds that have been extensively explored in a variety of advanced biomedical applications, such as the pharmaceutical delivery system, tissue engineering scaffolds, and biosensors [20, 21]. The basic principle of electrospinning is producing micro- or nano-fibers by charging a viscoelastic polymer solution through a spinneret under a high-voltage electric field, which forms solid fiber on a collector after solvent evaporation [21]. Generally, the electrospinning system consists of a high-voltage power supply, a spinneret, and a collector. The tubular vascular graft can be fabricated by collecting fibers on a rotating steel rod with a small diameter (Fig. 1). By choosing polymer materials (e.g., degradable or non-degradable natural and synthetic polymers) and engineering electrospinning parameters (e.g., solution concentration, voltage, flow rate, and distance between tip and collector), the physical and biological properties of electrospun vascular grafts can be adjusted to match the structure and function of native blood vessels.

Since the 1960s, polyurethanes (PU) become one of the most commonly used materials in blood-contacting devices ranging from artificial arteries to artificial heart components because of their diverse chemical structures, outstanding physical properties (elasticity, compliance), and excellent biocompatibility [22, 23]. Recently, biodegradable polyurethanes have gained extensive investigations for fabricating non-cellular tissue-engineered vascular grafts, which allow cell infiltration and remodeling for regeneration [12, 13, 24–27]. On the other hand, non-biodegradable thermoplastic polycarbonate-urethane (PCU)

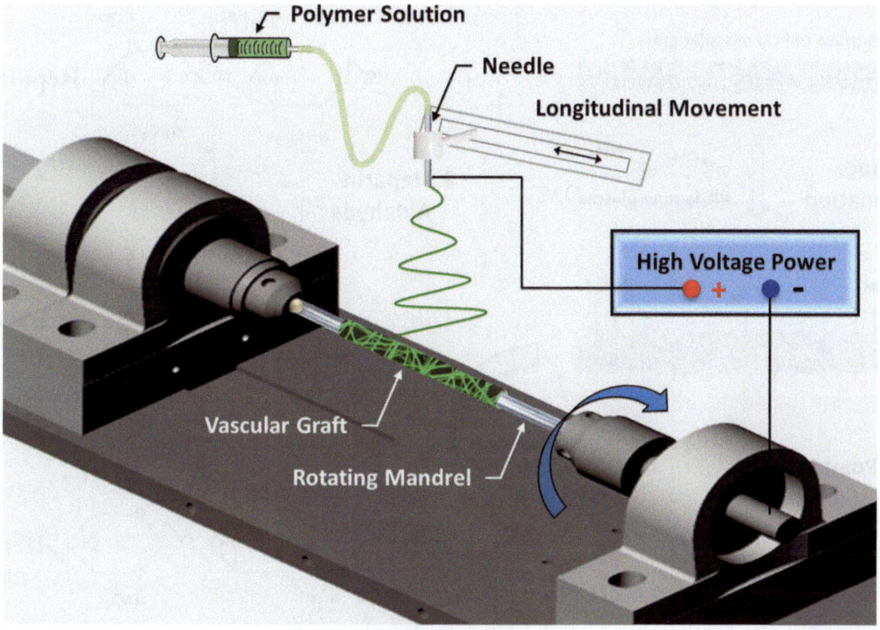

Fig. 1 Fabrication of fibrous vascular graft via electrospinning

has also been applied to make biostable vascular grafts, which is valuable for hemodialysis as a puncture-resistant arteriovenous vascular access and for bypass procedure [28, 29]. Non-degradable materials have the advantage of maintaining a stable structure and integrity, but the materials need to be carefully selected and optimized to minimize chronic foreign body responses. In addition, thrombosis and neointimal hyperplasia remain the major reasons leading to the failures of small-diameter polyurethane vascular scaffolds. Thus, it is vital to recapitulate a hemocompatible blood-contacting interface on such vascular grafts [30, 31].

Currently, there are many approaches for the antithrombogenic surface modification of blood-contacting medical devices. For more than three decades, the anticoagulant heparin has been employed as a covalently immobilized surface coating on cardiovascular devices [32]. Heparin is a natural polysaccharide, the most commonly used anticoagulant in the clinic, which produces an anticoagulant effect by binding to antithrombin III (AT) and is used to inactivate thrombin, factor Xa, and other proteases. In general, heparin can be covalently immobilized through a carboxyl-to-amine reaction, such as using carbodiimide to activate carboxyl groups on the repeating disaccharide unit and subsequently conjugated to amines on the target surface. However, such conjugation usually obstructs the binding ability of heparin to AT, leading to reduced or eliminated anticoagulant activity and insufficient thromboresistance [33]. Alternatively, an approach of end-point heparin immobilization has been developed to modify vascular graft, which

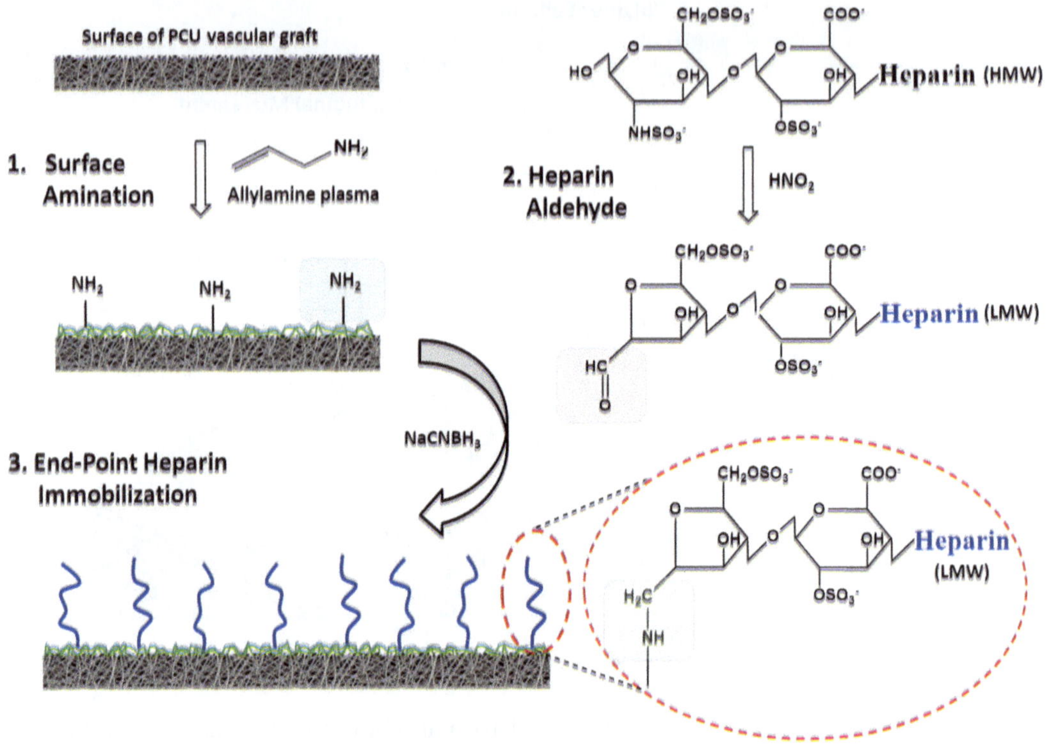

Fig. 2 End-point immobilization of heparin on PCU vascular graft. Step 1—surface amination of PCU vascular graft treated by allylamine plasma. Step 2—preparation of aldehyde-terminated heparin. Step 3—end-point heparin immobilization of aminated vascular graft. The aldehyde (-CH=O) group of heparin can react with the amino group (-NH_2) of the aminated surface to form imine group (-CH=N-) and can be reduced into second amine group (-CH_2-NH-) by sodium cyanoborohydride ($NaCNBH_3$). HMW (high molecular weight); LMW (low molecular weight)

can retain heparin affinity and activity to enhance the long-term graft patency [34, 35]. Briefly, heparin is partially depolymerized by nitrous acid deamination to generate low molecular weight heparin that contains single reactive aldehyde groups at the end-point. The depolymerized heparin is subsequently covalently "grafting-to" surface aminated substrate.

In this chapter, we described the detailed procedures for the end-point immobilization of heparin on PCU vascular graft (Fig. 2), including the fabrication of electrospun small-diameter vascular graft, surface amination by plasma treatment, preparation of aldehyde-terminated heparin for covalent end-point heparin conjugation, amine detection and quantification of plasma-treated grafts and characterizations of aminated and heparinized modification, and heparin quantification and antithrombogenic evaluation of heparin-modified vascular grafts. Similarly, the surface modification protocol also can be applied to other biodegradable small-diameter polyurethane vascular grafts for vascular regeneration.

2 Materials

2.1 Fabrication of Electrospun PCU Vascular Graft

1. Polycarbonate-urethane (PCU) (medical-grade).
2. Dimethylformamide (DMF).
3. 70% ethanol (EtOH).
4. Glass vials with screw threads (20 mL).
5. Parafilm.
6. Blunt-ended needles (16 G).
7. Luer-Lok syringes (10 mL).
8. Laboratory magnetic stirrer and magnetic stirring bar.
9. Silicon tubing extender (1.5 mm ID).
10. High-voltage generator (positive and negative).
11. Syringe pump.
12. Conventional aluminum foil.
13. Stainless steel mandrel with PTFE coating (1–6 mm in diameter, 30 cm in length).
14. Vacuum oven.
15. Vernier caliper.
16. Scanning electron microscopy (SEM).

2.2 Surface Amination of PCU Vascular Graft by Plasma Treatment

1. Low-Pressure Plasma System Aurora (Plasmatreat, USA).
2. Allylamine vapor.
3. Oxygen tank.
4. Argon tank.
5. Vacuum oven.
6. Self-sealing sterilization pouch.

2.3 Preparation of Aldehyde-Terminated Heparin

1. Heparin sodium salt (from porcine intestinal mucosa, $\geq$180 USP units/mg).
2. Ice bath.
3. Deionized water (DI water).
4. Sodium nitrite.
5. Acetic acid.
6. 1 M sodium hydroxide (NaOH) solution.
7. Round-bottom glass flask (one neck, 500 mL).
8. Laboratory magnetic stirrer and magnetic stirring bar.
9. Dialysis membrane (MWCO = 1 kDa).
10. Argon tank.

2.4 End-Point Heparin Immobilization of Aminated Vascular Grafts

1. Sodium cyanoborohydride.
2. Sodium phosphate.
3. Sodium chloride.
4. 10 M NaOH solution.
5. 4 M hydrochloric acid (HCl).
6. Cyanoborohydride coupling and reducing buffer (pH 5.5, 0.02 M sodium phosphate, 0.2 M sodium chloride, and 3 mg/mL sodium cyanoborohydride).
7. 30 mg/mL aldehyde-terminated heparin in cyanoborohydride coupling and reducing buffer.

2.5 Amine Detection and Quantification of Plasma-Treated Grafts

1. Orange II dye.
2. 1 M HCl solution.
3. 1 M NaOH solution.
4. NovaPEG amino resin.
5. Vacuum oven.
6. Black clear bottom 96-well plate.
7. Plate reader.
8. 5 mL glass bottles.

2.6 Heparin Quantification of Heparin-Modified Vascular Grafts

1. PBS buffer (pH = 7.4).
2. Toluidine blue.
3. Toluidine blue solution: 0.005 wt% in 0.2% NaCl aqueous solution.
4. Heparin sodium salt (from porcine intestinal mucosa).
5. Heparin standard solutions: 0–30 μg/mL heparin sodium salt in PBS buffer (pH = 7.4).
6. Hexane (reagent grade, > 99%).
7. Corning Falcon 15 mL conical centrifuge tubes.
8. Black clear bottom 96-well plate.
9. Plate reader.

2.7 Antithrombogenic Evaluation of Heparin-Modified Vascular Grafts

1. Thrombin from human plasma.
2. Human thrombin solution: 6 IU/mL in 0.15 M NaCl solution.
3. Antithrombin III from human plasma (4.38 IU).
4. Antithrombin III solution: 2 IU/mL in DI water.
5. Chromogenix S-2238™ (Diapharma).
6. S-2238 solution: 0.625 mg/mL in DI water.

7. Heparin standard solutions: 0, 0.25, 0.5, 0.75, 1, 1.5, 2 IU/mL in pH 8.4 Tris buffer.
8. Tris buffer pH 8.4: 50 mM Tris, 175 mM NaCl, 7.5 mM Na2EDTA, in distilled water.
9. 20% acetic acid in DI water.
10. Black clear bottom 96-well plate.
11. Plate reader.
12. Water bath heater.

3 Methods

3.1 Fabrication of Electrospun PCU Vascular Graft

1. Prepare 20 mL 16.5 wt/v % PCU solutions by dissolving the predetermined weight of PCU polymer in DMF in a glass vial. Seal via screwed cap with parafilm, and then magnetic stir the solution overnight at room temperature (*see* **Note 1**).
2. Wipe and clean the electrospinning devices and areas with 70% EtOH and Kim wipes (*see* **Note 2**).
3. Set the electrospinning temperature at 20–25 °C and humidity at 50–52%.
4. Draw the PCU solution into a syringe, and connect the syringe with a blunt needle by a silicon tubing extender; then push the solution to fulfill the silicon tube, and ensure that there are no air bubbles in the solution (*see* **Note 3**).
5. Fasten the syringe on the syringe pump, and set the electrospun parameters with flow rate at 1 mL/h, the distance between tip and collector at 16–18 cm, and spinneret sweep speed at 5 cm/s. Place the steel mandrel (1–6 mm in diameter, 30 cm in length) as a collector, and set the mandrel rotation speed around 100 rpm (*see* **Note 4**).
6. Attach the positive electrode clamp to the needle tip, and attach the negative electrode to the mandrel collector. Set positive voltage at +16 kV and negative voltage at −2 kV (*see* **Note 5**).
7. Run the device at room temperature, monitor the electrospinning process by collecting fibers with a long insulated fork made by two plastic pipettes, and roughly observe the fiber morphology with the optical microscope.
8. Stop electrospinning when the wall thickness of the vascular graft reaches the desired value (need around 30 min to make a 1 mm (ID) graft with a thickness of 130 μm and need around 10 h to make a 6 mm graft (ID) with a thickness of 500 μm); then, turn off the powers, and clean the device.
9. Dry the electrospun vascular grafts in a vacuum oven for more than 2 days to remove any residual DMF.

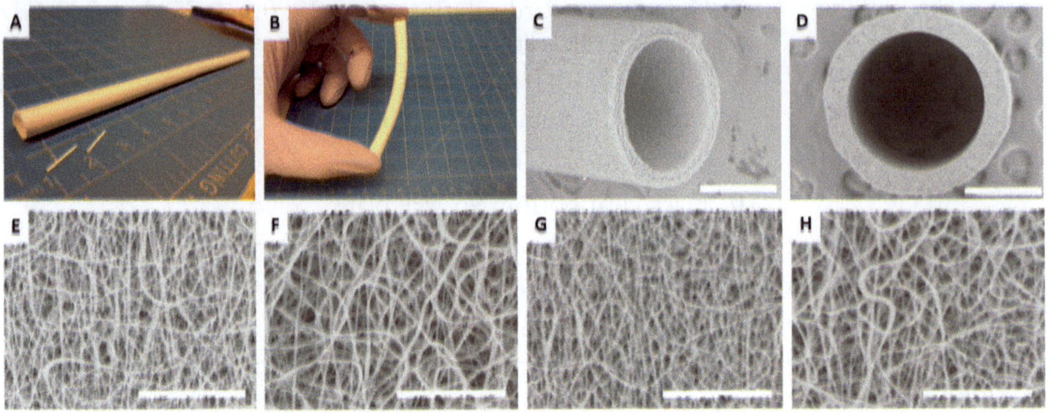

Fig. 3 Structural properties of nanofibrous PCU vascular graft. (**a**) PCU vascular grafts of various IDs (e.g., 1 mm and 6 mm) were fabricated. (**b**) PCU graft shows a flexible bendability. (**c–h**) Representative SEM images of vascular grafts with fibrous structures. (**c**) Side and (**d**) Top view of the vascular graft. (**e**) Inner and (**f**) The outer surface of the vascular graft before the plasma treatment. (**g**) Inner and (**h**) The outer surface of the vascular graft after the plasma treatment. Scale bar = 500 μm in **c**, **d**; scale bar = 30 μm in **e–h**

10. Measure the vascular thickness with a Vernier caliper, and characterize the fibrous structural properties (fiber orientation, fiber diameter, pore size, and porosity) of PCU graft using SEM (Fig. 3). Different electrospun PCU vascular grafts with varying IDs (e.g., 1 mm and 6 mm) and uniform wall thickness (e.g., 130 ± 5 μm for 1 mm (ID) graft) can be fabricated. Besides, the grafts have fibrous architectures with random fiber orientation, fiber diameter around 500 nm, inter-fibrous pore size ranging from 1 to 20 μm, and porosity around 35%. After plasma treatment, the fibrous structures of both the inner and outer surfaces of the graft are maintained.

3.2 Surface Amination of PCU Vascular Graft by Plasma Treatment

1. Cut the vascular grafts with a length of 1 cm.
2. Position PCU grafts atop Whatman™ glass microfiber filter suspending on screen trays spanning between the power electrodes in the reactor chamber, with the sides of the graft pointing toward the gas phase (*see* **Note 6**).
3. Evacuate the chamber with a rotary pump to a base pressure of 1×10^{-5} Pa.
4. Close the valve to the vacuum, and open the other valve, which is connected to argon and oxygen gas.
5. Decontaminate and activate the PCU grafts with oxygen/argon plasma (500 W, 13.56 MHz, 50 Pa, 100 sccm O_2, and 25 sccm Ar) for 1 min (*see* **Note 7**).
6. Connect the allylamine solution with the chamber by opening the corresponding valve, and pour the solution into the

chamber. The allylamine evaporation is assisted by a 70 °C hot tube and liquid injection system.

7. Amine functionalization of graft surface with allylamine plasma vapor (500 W, 13.56 MHz, 50 Pa, 50 sccm allylamines, and 50 sccm Ar) for 9 min (*see* **Notes 8** and **9**).
8. Vent the chamber with room air to the atmosphere.
9. Wash the membranes and grafts with methanol for 5 min, and then repeat the washing procedure with water three times.
10. Dry the grafts in vacuum at room temperature for at least 2 days, and package the grafts in a self-sealing sterilization pouch (*see* **Note 10**).

3.3 Preparation of Aldehyde-Terminated Heparin

1. Add 1 g heparin in 300 mL deionized (DI) water into a 500 mL glass flask, and cool the solution to 0 °C in an ice bath.
2. Add and dissolve 10 mg sodium nitrite into a heparin solution during magnetic stirring.
3. Drop 2 mL acetic acid in cold heparin solution to pH 4.0, and react for two more hours (*see* **Note 11**).
4. Neutralize heparin solution to pH 7 by 1 M NaOH.
5. Purify the product by dialysis (MWCO = 1 kDa) against 1 L DI water for 2 days at room temperature, with the water replaced every 6 h (*see* **Note 12**).
7. Dry the solution by lyophilization to obtain aldehyde-terminated heparin.
8. Store the aldehyde-terminated heparin in a tightly sealed container filled with inert gas (argon), and place it in a −20 °C refrigerator for further application (*see* **Note 13**).

3.4 End-Point Heparin Immobilization of Aminated Vascular Grafts

1. Prepare a fresh pH 5.5 cyanoborohydride coupling and reducing buffer.
2. Dissolve 30 mg/mL aldehyde-terminated heparin solution in pH 5.5 cyanoborohydride coupling and reducing buffer, freshly prepared.
3. Immerse aminated vascular grafts in aldehyde-terminated heparin solution for 24 h on a shaker at room temperature.
4. After one day reaction, adjust the pH to 10 by 4 M NaOH, and keep the solution at 40 °C for 1 h.
5. Then, adjust pH to 7 with 4 M HCl, followed by washing in DI water overnight under shaking.
6. Dry heparin-modified grafts in vacuum at room temperature for further applications.

3.5 Amine Detection and Quantification of Plasma-Treated Grafts

1. Immerse plasma-treated grafts (1 cm length) in 14 mg/mL orange II dyes dissolved in pH 3 HCl solution in glass bottles for 30 min at 40 °C (*see* **Note 14**).
2. Wash the grafts thoroughly, and carefully use the pH 3 HCl solution to remove any unbound dyes.
3. Dry the grafts in a vacuum oven at room temperature for 2 days.
4. Immerse the dried grafts in 1 mL pH 12 NaOH solution to desorb the physically absorbed orange II dyes for 10 min under shaking (*see* **Note 15**).
5. Adjust the pH of the solution containing the desorbed dyes to pH 3 using a 1 M HCl solution.
6. Transfer 100 μL of each sample into black clear bottom 96-well plate, and measure the absorbance at 490 nm with a plate reader.
7. The standard solutions containing 0–20 mg of NovaPEG amino resin are prepared and used to determine the amine density on the surface of plasma-treated grafts.

3.6 Heparin Quantification of Heparin-Modified Vascular Grafts

1. Immerse heparin-modified vascular grafts in 2.5 mL PBS and 2.5 mL toluidine blue solution in a 15 mL Corning centrifuge tube for 2 h at 37 °C under mild shaking.
2. Add 5 mL hexane to each centrifuge tube for another 15 min under shaking, which leads to phase separation of the mixture.
3. Remove the upper organic layer and the grafts.
4. Transfer 100 μL aqueous bottom layer solution into the clear black bottom 96-well plate, and measure the absorbance at 630 nm with a plate reader (*see* **Note 16**).
5. Prepare a series of heparin standard solutions to make a standard curve to calculate the heparin density on the vascular grafts or membranes.

3.7 Antithrombogenic Evaluation of Heparin-Modified Vascular Grafts

1. Add 100 μL antithrombin III (2 U/mL) to heparin-modified grafts (1 mm ID, 1 cm in length) immersed in 900 μL Tris buffer (pH 8.4) or to 900 μL heparin standard solutions, and incubate at 37 °C for 4 min.
2. Transfer 200 μL of up-prepared solutions into 1.5 ml centrifuge tubes (each group has five parallel samples), separately; then, add 100 μL thrombin (6 IU/mL) into each sample, and incubate at 37 °C for 30 s.
3. Add 200 μL S-2238 solution (0.625 mg/mL) into step 2 solution, and incubate at 37 °C for 8 min.
4. Lastly, stop all the reactions by adding 300 μL 20% acetic acid.

5. Transfer 100 μL of each sample solution immediately into the clear black bottom 96-well plate, and measure the absorbance at 405 nm with a plate reader.
6. Then, calculate the heparin concentrations of heparin-modified grafts by plotting their absorbance values to the heparin standard curve.

4 Notes

1. DMF is a toxic, flammable, and corrosive organic solvent. When handling DMF and polymer solutions, be sure to wear appropriate personal protective equipment, including long pants, close-toed shoes, lab coats, and gloves. Additionally, organic solvent and solution should be handled in the chemical fume hood and stored in the appropriate chemical cabinet.
2. A clean electrospinning environment is critical to avoid particle contamination and to generate a vascular graft with a smooth surface.
3. The electrospun polymer solution is very viscous. It takes a large force to push the solution to fill the silicone connecting tube. Thus, make sure to hold the weakly connected location to avoid risks due to the tube slipping off and the solution spraying out.
4. The sweep of the spinneret is required to make a vascular graft with uniform thickness.
5. High voltage is highly dangerous, although the current is low. Don't touch any conductive parts of the electrospinning device when the high voltage is on. Make sure to put a warning card or sign, such as "Risk, High Voltage On" to remind others when the electrospinning progress is ongoing.
6. Placing the graft sides toward the gas phase is required to modify the inner lumen surface.
7. Plasma interactions predominantly occur on the sub-micron or atomic scales. Physical debris and contaminants on the surface may compromise the performance of surface functionalization. Therefore, a short plasma cleaning step is required to ensure an atomically clean substrate surface.
8. To achieve plasma treatment in various vascular grafts with different materials and physical properties (ID, wall thickness, pore size, and length), the plasma parameters (plasma power, gas ratio, vapor flow, chamber pressure, and process time) need to be systematically optimized.

9. Monitor the plasma process by observing the plasma color. A color change may indicate gas flow issues, air leaks, excessive outgassing, or an unclean chamber.
10. Pay attention to avoid or limit contamination of the plasma-treated product during transfer, handling, and storing.
11. To maintain the anticoagulant activity, it is critical to carry out heparin depolymerization by nitrous acid at pH 4.0, to produce aldehyde-terminated heparin with a molecular weight around 8 kDa after dialysis.
12. Long-time dialysis in DI water with frequent change is necessary to remove small fragments to get pure products.
13. Aldehydes are very reactive groups and can easily be oxidized to form carboxylic acids by oxygen in the air. Thus, it is required to keep the aldehyde-terminated heparin in a tightly sealed container with inert gas protection and store at low temperature for long-term application.
14. The container for the amine detection should avoid plasma-treated tissue culture plate or tubes.
15. Shaking is required to desorb the physically absorbed dyes from the grafts as completely as possible.
16. Make sure to transfer the aqueous bottom layer solution without organic layer solution. The resident organic reagent can affect the absorbance data.

Acknowledgments

The authors acknowledge funding from the National Health Institute (HL121450).

References

1. Cardiovascular diseases. World Health Organisation. Available at: https://www.who.int/news-room/fact-sheets/detail/cardiovascular-diseases-(cvds)
2. Weintraub WS, Grau-Sepulveda MV, Weiss JM et al (2012) Comparative effectiveness of revascularization strategies. N Engl J Med 366 (16):1467–1476
3. Thuijs D, Kappetein AP, Serruys PW et al (2019) Percutaneous coronary intervention versus coronary artery bypass grafting in patients with three-vessel or left main coronary artery disease: 10-year follow-up of the multicentre randomised controlled SYNTAX trial. Lancet 394(10206):1325–1334
4. Luyckx VA, Tonelli M, Stanifer JW (2018) The global burden of kidney disease and the sustainable development goals. Bull World Health Organ 96(6):414–422D
5. Pashneh-Tala S, MacNeil S, Claeyssens F (2016) The tissue-engineered vascular graft-past, present, and future. Tissue Eng Part B Rev 22(1):68–100
6. Begovac PC, Thomson RC, Fisher JL et al (2003) Improvements in GORE-TEX (R) vascular graft performance by Carmeda (R) bioactive surface heparin immobilization. Eur J Vasc Endovasc 25(5):432–437
7. Chlupac J, Filova E, Bacakova L (2009) Blood vessel replacement: 50 years of development

and tissue engineering paradigms in vascular surgery. Physiol Res 58:S119–S139
8. Seifu DG, Purnama A, Mequanint K et al (2013) Small-diameter vascular tissue engineering. Nat Rev Cardiol 10(7):410–421
9. Dahl SL, Kypson AP, Lawson JH et al (2011) Readily available tissue-engineered vascular grafts. Sci Transl Med 3(68):68ra69
10. Song HHG, Rumma RT, Ozaki CK et al (2018) Vascular tissue engineering: progress, challenges, and clinical promise. Cell Stem Cell 22(3):340–354
11. Bos GW, Poot AA, Beugeling T et al (1998) Small-diameter vascular graft prostheses: current status. Arch Physiol Biochem 106 (2):100–115
12. Hasan A, Memic A, Annabi N et al (2014) Electrospun scaffolds for tissue engineering of vascular grafts. Acta Biomater 10(1):11–25
13. Rocco KA, Maxfield MW, Best CA et al (2014) In vivo applications of electrospun tissue-engineered vascular grafts: a review. Tissue Eng Part B Rev 20(6):628–640
14. Guan JJ, Fujimoto KL, Sacks MS et al (2005) Preparation and characterization of highly porous, biodegradable polyurethane scaffolds for soft tissue applications. Biomaterials 26 (18):3961–3971
15. Shin'oka T, Imai Y, Ikada Y (2001) Transplantation of a tissue-engineered pulmonary artery. N Engl J Med 344(7):532–533
16. Singh C, Wong CS, Wang XG (2015) Medical textiles as vascular implants and their success to mimic natural arteries. J Funct Biomater 6 (3):500–525
17. Giannopoulos AA, Mitsouras D, Yoo SJ et al (2016) Applications of 3D printing in cardiovascular diseases. Nat Rev Cardiol 13 (12):701–718
18. Melchiorri AJ, Hibino N, Best CA et al (2016) 3D-printed biodegradable polymeric vascular grafts. Adv Healthc Mater 5(3):319–325
19. Soletti L, Hong Y, Guan J et al (2010) A bilayered elastomeric scaffold for tissue engineering of small diameter vascular grafts. Acta Biomater 6(1):110–122
20. Ding JX, Zhang J, Li JN et al (2019) Electrospun polymer biomaterials. Prog Polym Sci 90:1–34
21. Xue J, Wu T, Dai Y et al (2019) Electrospinning and electrospun nanofibers: methods, materials, and applications. Chem Rev 119 (8):5298–5415
22. Boretos JW, Pierce WS (1967) Segmented polyurethane: a new elastomer for biomedical applications. Science 158(3807):1481–1482
23. Akindoyo JO, Beg MDH, Ghazali S et al (2016) Polyurethane types, synthesis and applications - a review. RSC Adv 6 (115):114453–114482
24. Fang J, Zhang J, Du J et al (2016) Orthogonally functionalizable polyurethane with subsequent modification with heparin and endothelium-inducing peptide aiming for vascular reconstruction. ACS Appl Mater Interfaces 8(23):14442–14452
25. Hong Y, Ye SH, Nieponice A et al (2009) A small diameter, fibrous vascular conduit generated from a poly(ester urethane)urea and phospholipid polymer blend. Biomaterials 30 (13):2457–2467
26. Fang J, Ye SH, Shankarraman V et al (2014) Biodegradable poly(ester urethane)urea elastomers with variable amino content for subsequent functionalization with phosphorylcholine. Acta Biomater 10(11):4639–4649
27. Fang J, Ye SH, Wang J et al (2015) Thiol click modification of cyclic disulfide containing biodegradable polyurethane urea elastomers. Biomacromolecules 16(5):1622–1633
28. Qiu X, Lee BL, Ning X et al (2017) End-point immobilization of heparin on plasma-treated surface of electrospun polycarbonate-urethane vascular graft. Acta Biomater 51:138–147
29. Ferraresso M, Bortolani EM, Amnon G (2016) A two-year experience with a rapid access, self-sealing, polycarbonate urethane nanofiber vascular access graft for hemodialysis. J Vasc Access 17(3):210–214
30. Ren XK, Feng YK, Guo JT et al (2015) Surface modification and endothelialization of biomaterials as potential scaffolds for vascular tissue engineering applications. Chem Soc Rev 44 (15):5745–5745
31. Li S, Henry JJ (2011) Nonthrombogenic approaches to cardiovascular bioengineering. Annu Rev Biomed Eng 13:451–475
32. Biran R, Pond D (2017) Heparin coatings for improving blood compatibility of medical devices. Adv Drug Deliv Rev 112:12–23
33. Gore S, Andersson J, Biran R et al (2014) Heparin surfaces: impact of immobilization chemistry on hemocompatibility and protein adsorption. J Biomed Mater Res B 102 (8):1817–1824
34. Olsson P, Sanchez J, Mollnes TE et al (2000) On the blood compatibility of end-point immobilized heparin. J Biomater Sci Polym Ed 11(11):1261–1273
35. Daenens K, Schepers S, Fourneau I et al (2009) Heparin-bonded ePTFE grafts compared with vein grafts in femoropopliteal and femorocrural bypasses: 1-and 2-year results. J Vasc Surg 49 (5):1210–1216

Chapter 6

Microfluidic Coaxial Bioprinting of Hollow, Standalone, and Perfusable Vascular Conduits

Sushila Maharjan, Jacqueline Jialu He, Li Lv, Di Wang, and Yu Shrike Zhang

Abstract

Three-dimensional bioprinting represents promising approach for fabricating standalone and perfusable vascular conduits using biocompatible materials. Here we describe a step-by-step method by using a multichannel coaxial extrusion system (MCCES) and a blend bioink constituting gelatin methacryloyl, sodium alginate, and eight-arm poly(ethylene glycol)-acrylate with a tripentaerythritol core for the fabrication of standalone circumferentially multilayered hollow tubes. This microfluidic bioprinting method allows the fabrication of perfusable vascular conduits with a core lumen, an inner endothelial layer resembling the *tunica intima*, and an outer smooth muscle cell layer resembling the *tunica media* of the blood vessel. Biocompatible and perfusable blood vessels with a widely tunable size range in terms of luminal diameters and wall thicknesses can be successfully fabricated using the MCCES.

Key words Multichannel coaxial extrusion system, Microfluidic bioprinting, Gelatin methacryloyl, Sodium alginate, Eight-arm poly(ethylene glycol)-acrylate with a tripentaerythritol core, Vascular conduits, Lumen, Endothelial cells, Smooth muscle cells

1 Introduction

Vascular diseases such as atherosclerosis and thrombosis represent one of the largest causes of death worldwide [1]. Both the autologous vascular grafts, harvested from the internal thoracic artery or saphenous vein [2, 3], and the synthetic vascular grafts, such as those made of polyethylene terephthalate (Dacron) [4, 5], are frequently used as bypass vascular grafts for the treatment of vascular diseases. However, a wide range of drawbacks limit the application of existing vascular grafts, for instance, invasive harvesting, deficiency of autologous vessels, and mismatched biomechanical properties of synthetic vessels [6–10]. Therefore, there is an increasing demand for bioengineered vascular conduits with the potential to restore patency to the occluded blood vessels [11].

Feng Zhao and Kam W. Leong (eds.), *Vascular Tissue Engineering: Methods and Protocols*,
Methods in Molecular Biology, vol. 2375, https://doi.org/10.1007/978-1-0716-1708-3_6,

In addition, bioengineered vascular conduits can be used as in vitro models for basic vascular research and drug screening studies [12].

To this end, the three-dimensional (3D) bioprinting techniques have been employed for the fabrication of vascular conduits using either cell-laden bioinks or acellular bioinks [13]. These include inkjet bioprinting [14–16], extrusion bioprinting [16–19], laser-assisted bioprinting [16, 20, 21], and vat polymerization-based bioprinting [22–25]. Using different bioprinting approaches, artificial blood vessels and/or vasculature are created in the form of standalone hollow tubes [26, 27], integrated channels within bulk hydrogel matrices [28, 29], or via self-assembly of multicellular spheroids or cylinders patterned into tubular structures [30]. Alginate and gelatin-based hydrogels have been widely used in bioink formulations for vascular bioprinting due to their biocompatibility, printability, and favorable gelation properties [31, 32]. To increase the stability and controllability of the bioprinted vessels, gelatin methacryloyl (GelMA) is usually introduced to the bioinks [33, 34]. Alginate undergoes rapid ionic crosslinking with calcium chloride ($CaCl_2$), whereas GelMA undergoes irreversible photocrosslinking under ultraviolet (UV) or visible-light illumination.

The standalone perfusable vascular tubes are usually fabricated by extrusion bioprinting, in which the bioink is extruded through a coaxial nozzle along with a physical crosslinking agent, thus forming the stable hollow tubes in a single step, in the absence or presence of a secondary chemical crosslinking method. Here, we describe a protocol for bioprinting standalone, multilayered vascular conduits, using a combination of a multichannel coaxial extrusion system (MCCES) consisting of three concentric needles of different diameters, and a blend bioink comprised of GelMA, sodium alginate, and eight-arm poly(ethylene glycol)-acrylate with a tripentaerythritol core (PEGOA) (Fig. 1a).

The blend bioink was shown to possess the suitable rheological properties (Fig. 1b). When the endothelial cell (EC)-encapsulated bioink was extruded through the middle layer, smooth muscle cell (SMC)-encapsulated bioink through the outer layer, and $CaCl_2$ solution through the inner layer of the coaxial nozzle, a blood vessel-mimetic perfusable conduit could be generated (Fig. 2). We have obtained multilayered vascular conduits comprised of a core lumen, an inner *tunica intima*, and an outer *tunica media*, with an average inner diameter of 663 ± 52μm and an outer diameter of 977 ± 63μm; the inner and the outer wall thicknesses were 62 ± 10μm and 94 ± 10μm, respectively (Fig. 3). The viability of ECs in the inner layer and SMCs in the outer layer remained high (Fig. 4), and immunocytochemical analyses showed the expressions of CD31 and VE-cadherin by the ECs and alpha-smooth muscle actin (α-SMA) by the SMCs (Fig. 5). Similarly, the structural integrity of the vascular tubes was retained during continuous perfusion (Fig. 6). Furthermore, the luminal diameters and wall thicknesses

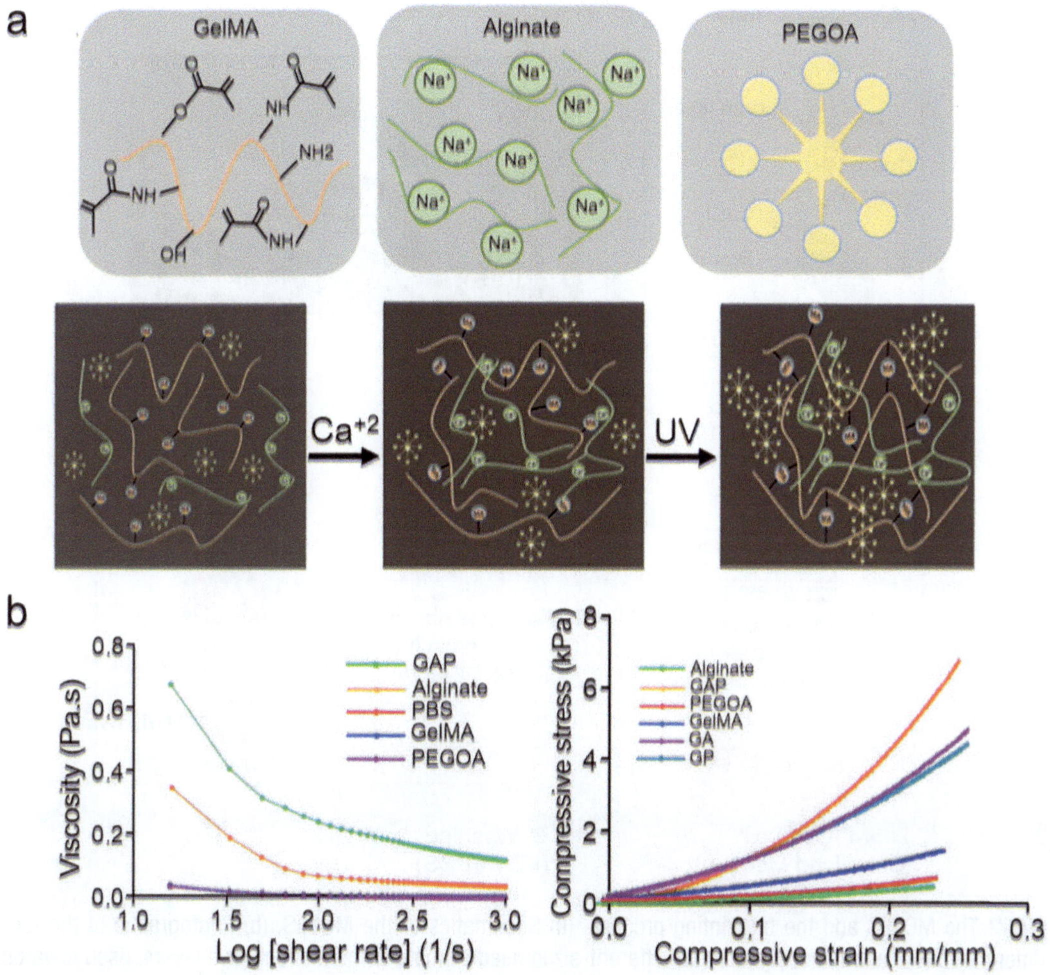

Fig. 1 The bioink formulation. (**a**) Schematic structures of GelMA, sodium alginate, and PEGOA (upper panel) and two-step crosslinking of GelMA-based blend bioink (GAP) (lower panel). (**b**) Rheological behavior of the bioink and its components as the function of shear rate (left) and compressive stress–strain curves for the crosslinked bioink and its components (right)

of the bioprinted vascular conduits can be readily tuned by modifying the parameters of the MCCES and the bioink delivery options.

2 Materials

Store all reagents at room temperature unless indicated otherwise. Follow hazardous waste identification regulations when disposing waste materials.

2.1 Bioink

1. PEGOA (M_w 20 kDa).
2. 3% (w/v) calcium chloride ($CaCl_2$) stock solution.

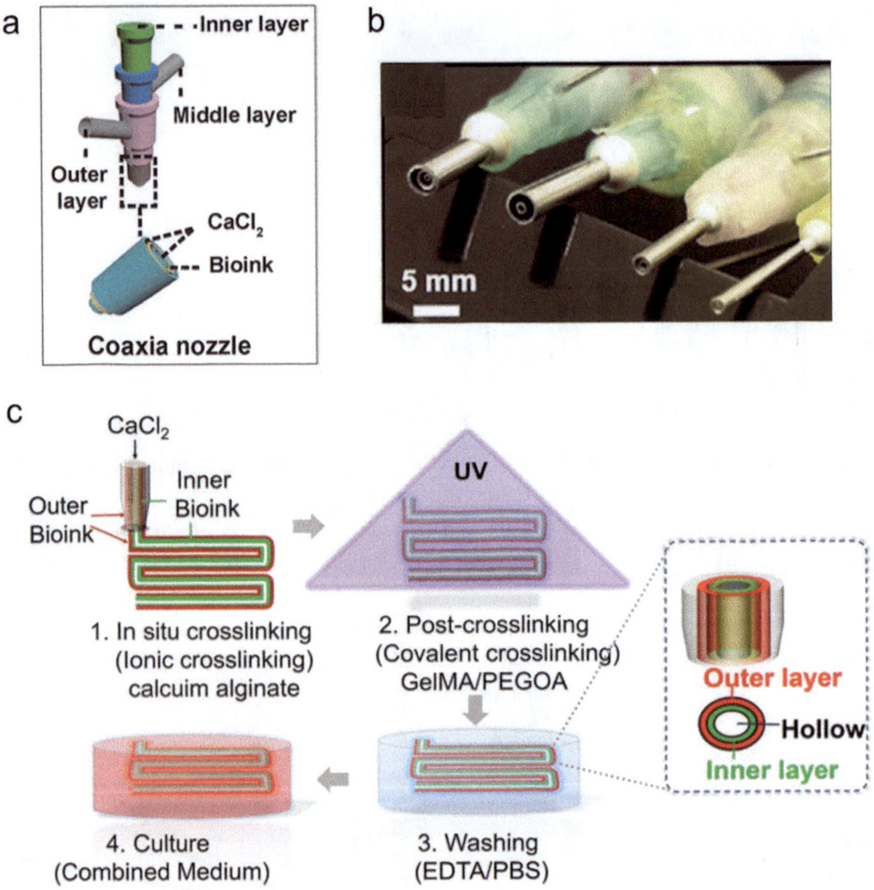

Fig. 2 The MCCES and the bioprinting process. (**a**) Schematics of the MCCES. (**b**) Photographs of the four different coaxial nozzles designed with different-sized needles (note some had only two layers, used towards bioprinting of single-layer tubular constructs). (**c**) Schematics of different steps involved in the bioprinting of double-layered vascular conduits using the MCCES

3. Dialysis membrane (MWCO 12,000–14,000 Da).
4. 20-mM ethylenediaminetetraacetic acid (EDTA) stock solution.
5. 10% (w/v) gelatin type A (from porcine skin, 300 bloom) stock solution.
6. 2% (w/v) 2-hydroxy-1-[4-(2-hydroxyethoxy) phenyl]-2-methyl-1-propanone (Irgacure 2959) stock solution.
7. 25-mM 2-[4-(2-hydroxyethyl) piperazin-1-yl]-ethanesulfonic acid stock solution (HEPES buffer, 25 mM, pH 7.4).
8. 8% (v/v) methacrylic anhydride stock solution.
9. Phosphate-buffered saline (PBS) tablets.
10. 10% (w/v) sodium alginate (low viscosity) stock solution.

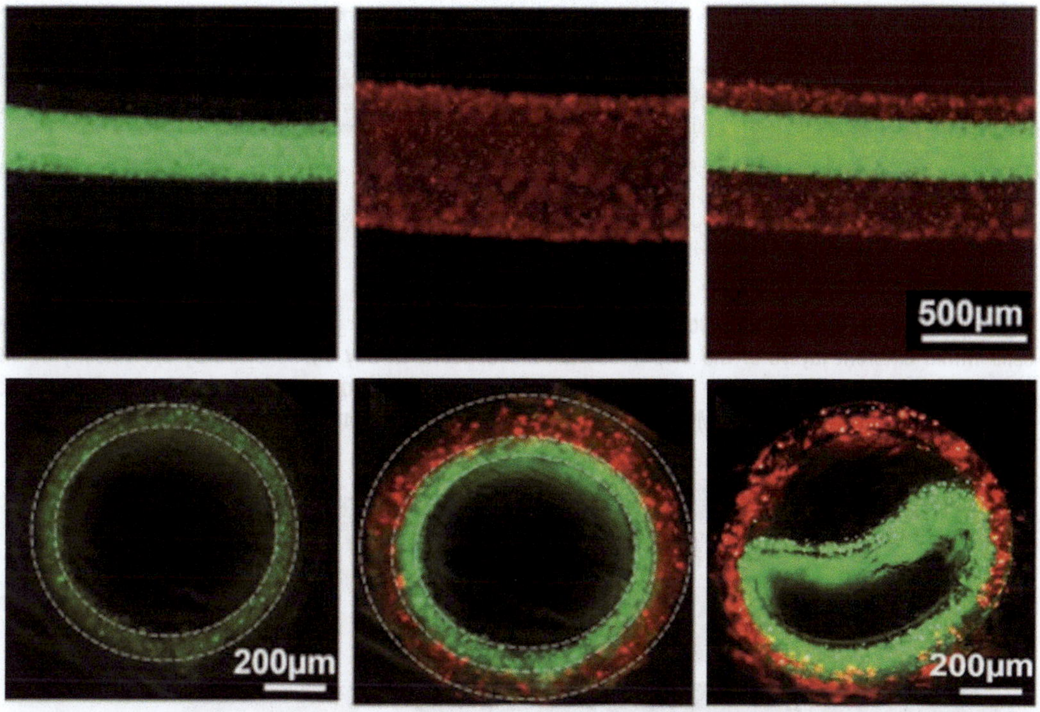

Fig. 3 The bioprinted tubular constructs. Fluorescence microscopic images of double-layered perfusable hollow tubes, bioprinted with green fluorescence beads in the inner layer and red fluorescence beads in the outer layer of the wall. Lower panel shows the cross-sectional views of the single-layered (left) and double-layered (central and right) hollow tubes. The fluorescence image on the right intentionally demonstrates the two separable layers of the bioprinted hollow tube

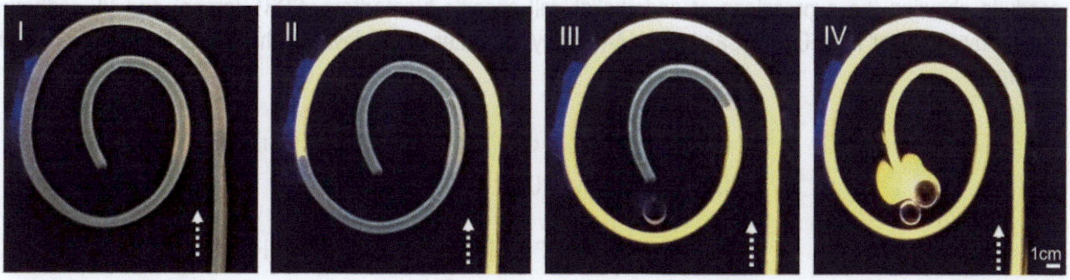

Fig. 4 Perfusability of the bioprinted tubular structure. Time-lapse fluorescence images showing the perfusion of a bioprinted tubular structure

2.2 Cells

1. Human umbilical vein ECs (HUVECs).
2. Human SMCs (hSMCs).

2.3 Media and Other Cell Reagents

1. Dulbecco's PBS (DPBS).
2. Endothelial cell growth medium (EGM-2).
3. Fetal bovine serum (FBS).

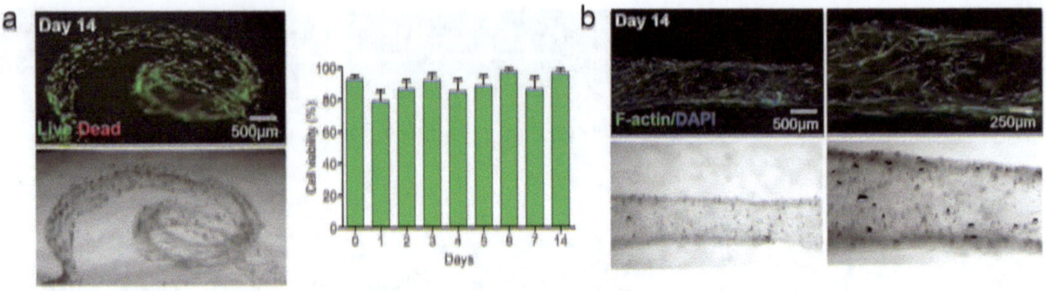

Fig. 5 The bioprinted vascular conduits. (**a**) Fluorescence microscopic images showing the viable vascular cells within different layers of the bioprinted vascular conduits at day 14, followed by the viability of the bioprinted vascular cells from day 0 to day 14. (**b**) Fluorescence microscopic images showing F-actin and nuclei staining of HUVECs and hSMCs within the bioprinted vascular conduits at day 14

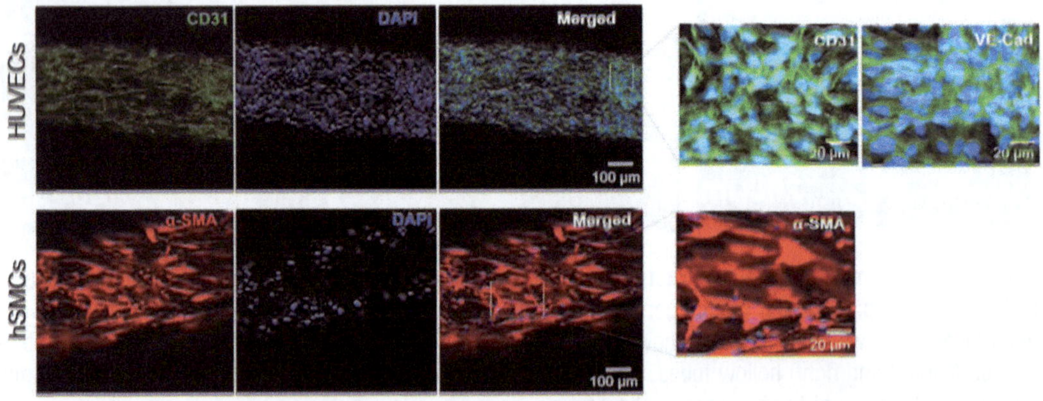

Fig. 6 Functionality of the bioprinted vascular conduits. Confocal microscopy images of the bioprinted vascular conduits showing the expressions of endothelial cell-specific biomarkers, CD31 (green) and VE-cadherin (green) by the inner HUVEC layer, and α-SMA (red) by the outer hSMC layer of the vessel wall, at day 14 of culture

4. Smooth muscle cell growth media-2 (SmGM-2).
5. 0.5% Trypsin (10X).

2.4 MCCES

1. Blunt needles of desired sizes, specifically 27G, 19G, and 14G.
2. 5-min epoxy.
3. Metal connectors.

2.5 Bioprinting of the Vascular Tubes

1. Bioprinter and controller.
2. Heater.
3. Syringe pumps.
4. 3-mL syringes.
5. Tygon tubing (0.020″ ID × 0.060″ OD).

2.6 Structural and Functional Properties of Bioprinted Vascular Tubes

1. Fluorescent microbeads with different colors.
2. Live/Dead® Viability/Cytotoxicity Kit.
3. Paraformaldehyde (PFA).
4. Triton X-100.
5. Bovine serum albumin (BSA).
6. Rabbit anti-human CD31 antibody.
7. Rabbit anti-mouse VE-cadherin antibody.
8. Mouse anti-α-SMA actin antibody.
9. Alexa Fluor® 594- and 488-conjugated goat anti-rabbit or goat anti-mouse antibodies.
10. 4′,6-Diamidino-2-phenylindole (DAPI).
11. Alexa Fluor® 488-phalloidin.

3 Methods

3.1 Preparation of GelMA

1. Prepare 0.01-M PBS in Milli-Q water, and store at room temperature until use.
2. Add 10.0 g of gelatin to 100 mL of PBS in a clean Erlenmeyer flask. Set the undissolved mixture on a heating plate with moderate magnetic stirring at 240 rpm and 50 °C until gelatin is completely dissolved. Cover the Erlenmeyer flask with aluminum foil during the process to prevent evaporation. Monitor temperature regularly with a thermometer.
3. In a chemical fume hood, add 8 mL of methacrylic anhydride to the gelatin solution drop-wisely using a syringe at approximately 200μL min^{-1} under vigorous stirring. Cover the Erlenmeyer flask with aluminum foil when the addition of methacrylic anhydride is finished, and continue vigorous stirring along with heating at 50 °C for 2–3 h. Quench the reaction by a five-fold dilution of the reaction mixture with warm DPBS (*see* **Notes 1–3**).
4. Prepare dialysis membranes by cutting in proper sizes and immerse into Milli-Q water to soften them. Close one side of each by twisting the membrane end and make a knot to obtain the dialysis bags. Examine the membranes for any leakage by adding water. Transfer the lukewarm GelMA solution into each dialysis bag with the help of a glass funnel. Close the other end of the bags with membrane clips and check again for leakage.
5. Place all GelMA-filled dialysis bags into a 5-L glass beaker, and dialyze against warm Milli-Q water at 40 °C under continuous stirring at 500 rpm for 5–7 days in a chemical fume hood to remove unreacted methacrylic anhydride and other possible impurities. Change lukewarm Milli-Q water two times a day,

and flip the dialysis bags upside down 5–6 times during dialysis to homogenize the contents (*see* **Notes 4** and **5**).

6. Filter the dialyzed GelMA solution through 0.22-μm filters. Divide the filtered GelMA solution into 30 mL aliquots in 50 mL conical tubes, and freeze them in a −80 °C refrigerator. Finally, lyophilize the aliquots using a freeze-dryer until they dry completely to obtain GelMA in the form of white porous foams. Seal the 50-mL conical tubes with lids. Store the lyophilized GelMA at −20 °C until use (*see* **Note 6**).

3.2 Preparation of the Blend Bioink

1. Prepare 1-M HEPES in DPBS, and store (*see* **Note 10**) at 4 °C until use.
2. Prepare 100 mL of buffer solution containing 10% (v/v) FBS and 25-mM HEPES in DPBS, and store at 4 °C until use.
3. Prepare 14% (w/v) GelMA precursor stock solution in buffer solution, and store at 4 °C until use (*see* **Notes 7–9**).
4. Prepare 10% (w/v) sodium alginate stock solution in buffer solution, and store at 4 °C until use (*see* **Notes 7–9**).
5. Prepare 10% (w/v) PEGOA stock solution in buffer solution, and store at 4 °C until use (*see* **Notes 8** and **9**).
6. Prepare 3% (w/v) $CaCl_2$ solution in buffer solution and store at 4 °C until use.
7. Prepare 20-mM EDTA solution in DPBS, and store at 4 °C until use.
8. Prepare 5 mL of the bioink by mixing 2.5 mL of the GelMA stock solution, 1 mL of the sodium alginate stock solution, 1 mL of the PEGOA stock solution, and 500μL of the photoinitiator (Irgacure 2959) solution. The final bioink comprises of 7% GelMA, 2% sodium alginate, 2% PEGOA, and 0.25% photoinitiator (Fig. 1) (*see* **Note 10**).

3.3 Maintenance and Preparation of Cells

1. Culture HUVECs and hSMCs in EGM-2 and SmGM-2 media, respectively, in T-75 flasks at 37 °C in a 5% CO_2 cell incubator (*see* **Note 11**).
2. Culture the cells to ~80% confluency. Change media every 2–3 days.
3. Immediately before bioprinting, dissociate the cells using the 0.05% (w/v) trypsin solution for 3–5 min. Resuspend the cells in 1 mL of respective media, and count the cells under microscope using a hemocytometer. Wash the cells twice with DPBS.
4. Resuspend the HUVECs in 2 mL of the bioink at a concentration of 1×10^7 cells mL^{-1}. Prepare the homogenous cell suspension in the bioink by pipetting it gently for 5–10 times.

5. Similarly, resuspend the hSMCs in 2 mL of the bioink at a concentration of 5–6 × 10^6 cells mL^{-1}. Prepare the homogenous cell suspension in the bioink by pipetting it gently for 5–10 times.
6. Transfer HUVEC- and hSMC-embedded bioinks separately into 3-mL sterile syringes. Protect these bioinks from the light.

3.4 Fabrication of the Three-Layered MCCES

1. Connect the barrel of a 27G blunt needle (1.5″-long) with the barrel of a 23G blunt needle (1.0″-long) such that the needles face the opposite directions. Seal the junction with epoxy glue. The 23G blunt needle serves as the inlet for the $CaCl_2$ solution.
2. Insert the 27G blunt needle, from **step 1**, into a 19G blunt needle (1.0″-long) to construct the concentric two-layered microfluidic printhead. Align these needles manually to serve as the core and the middle layers. Seal the junction of the barrels with epoxy glue.
3. Insert the two-layered microfluidic printhead, prepared in **step 2**, into a 14G blunt needle (0.5″-long) to obtain the MCCES with three concentric layers of a 27G needle in the core layer, a 19G needle in the middle layer, and a 14G needle in the outer layer (*see* **Note 12**). Seal the junction of the barrels with epoxy glue.
4. Drill holes on the outside of the barrels of both the middle and the outer needles, and insert metal connectors (23G, 0.5″-long), and seal with epoxy glue. These connectors serve as the inlets for the bioinks for the two layers of the vascular conduits (Fig. 2a, b).

3.5 Bioprinting of Multilayered Vascular Conduits Using the MCCES

1. Fill a 3-mL sterile syringe with 3% $CaCl_2$ solution, and connect it to the core needle of the MCCES through Tygon tubing.
2. Install the MCCES onto the head of the bioprinter with the help of a poly(methyl methacrylate) (PMMA) holder customized to the bioprinter.
3. Connect the inlets of the middle layer and the outer layer of the MCCES to the syringes loaded with HUVECs- and hSMCs-embedded bioinks, respectively.
4. Place the syringes containing the $CaCl_2$ solution, the HUVEC-embedded bioink, and the hSMC-embedded bioink on separate syringe pumps for injection into the respective layers of the MCCES.
5. Turn on the syringe pumps to start injection of the bioinks and the $CaCl_2$ solution simultaneously, at a flow rate of 200μL min^{-1} for the bioinks and 400μL min^{-1} for the $CaCl_2$ solution (*see* **Note 13**). Subsequently, turn on the bioprinter to control the movement of the printhead at a speed of

approximately 4 mm s^{-1}. The bioprinting process is usually performed in the class II biological safety cabinet at room temperature (21–25 °C). The delivery of $CaCl_2$ in the core of the MCCES allows the rapid ionic crosslinking of the alginate component of the bioink and extrusion of continuous hollow tubes (*see* **Note 14**).

6. Collect the bioprinted double-layered hollow tubes on a glass slide (*see* **Note 15**). The diameters of inner lumen and the outer lumen should be 663 ± 52 and 977 ± 63μm, respectively. Similarly, the thicknesses of the inner wall and the outer wall should be 62 ± 10μm and 94 ± 10μm, respectively.
7. Expose the bioprinted tubes at 6.9 mW cm^{-2} of UV light (360–480 nm) for 30 s to chemically crosslink the GelMA component, thus forming the stable and perfusable double-layered vascular conduits containing HUVECs in the inner layer and hSMCs in the outer layer of the walls.
8. Immerse the bioprinted vascular tubes in a 20-mM EDTA solution for 5 min to minimize the sodium alginate concentration. The reduction of alginate enhances the adhesion and spreading of cells within the bioprinted vascular conduits.
9. Rinse the bioprinted vascular conduits with DPBS.
10. Culture the bioprinted vascular conduits in 6- or 12-well plates using the common medium, comprised of both EGM-2 and SmGM-2 media mixed at a 1:1 ratio, at 37 °C and 5% CO_2 for up to 14–21 days (Fig. 2c).
11. Change the medium every 2–3 days (*see* **Note 16**).
12. Monitor the morphologies of the HUVECs and hMSCs under a microscope during the culture.
13. Characterize the bioprinted vascular conduits for viability and functionality using standard protocols such as Live/Dead staining, immunostaining, and molecular biology assays.

3.6 Visualization of the Bioprinted Hollow Tubes Using Fluorescent Microbeads

1. Add 20μL/3 mL of green and red fluorescent microbeads separately in the bioinks for inner layer and the outer layer, respectively.
2. Bioprint the hollow tubes as described above, and image the samples under a fluorescence microscope.
3. The bioprinted hollow tubes should consist of the inner green layer and the outer red layer, demonstrating successful formation of the double-layered tubes (Fig. 3).
4. The perfusability of the bioprinted hollow tubes can be assessed by the active perfusion with fluorescent microbeads along the entire length (Fig. 4).

3.7 Viability of Cells Within Bioprinted Vascular Conduits

1. Bioprint the vascular conduits, and culture them as described above.
2. At designated time points, wash the bioprinted vascular conduits three times with DPBS.
3. Add 100μL well^{-1} of the combined Live/Dead assay reagents (2μM of calcein-AM and 4μM of ethidium homodimer-1), and incubate at 37 °C and 5% CO_2 for 30 min.
4. Wash three times with DPBS.
5. Observe and image the samples under a fluorescence microscope. The live cells appear as green, whereas dead cells appear as red (Fig. 5a).

3.8 Functionality of Cells Within Bioprinted Vascular Conduits

1. Bioprint the vascular conduits, and culture them as described above.
2. At designated time points, fix the bioprinted vascular conduits with 4% PFA after washing three times with DPBS, and permeabilize them with Triton X-100. Then block the samples with 1% (w/v) BSA for 1 h at room temperature.
3. Incubate the samples with primary antibodies, anti-human CD31 antibody or anti-mouse VE-cadherin antibody for the inner endothelial layer and anti-α-smooth muscle actin antibody for the outer smooth muscle layer, at 4 °C overnight with gentle shaking. For F-actin staining, incubate the samples with Alexa Fluor® 488-phalloidin.
4. After washing with blocking buffer, incubate the samples with compatible secondary antibodies (e.g., Alexa Fluor®-conjugated) at room temperature for 2 h with gentle shaking.
5. Add 2.5μg ml^{-1} of DAPI in DPBS for 15 min at room temperature to stain the nuclei.
6. Observe and image the samples with a fluorescence and/or confocal microscope (Figs. 5b and 6).

4 Notes

1. Perform GelMA preparation in a chemical fume hood.
2. The ratio of gelatin and methacrylate anhydride, reaction time, and reaction temperature determine the degree of methacryloyl substitution of gelatin. Calculate the volumes of methacrylate anhydride and gelatin according to degree of methacryloyl substitution requirements. In low-GelMA, 1–2 mL of methacrylate anhydride is added to 10 g of gelatin, while 5 mL is needed for preparing medium-GelMA; the high-GelMA preparation requires 8 mL, whereas very high-GelMA is recommended to use 20 mL of methacrylate anhydride [35].

3. Long-term overheating or boiling of gelatin can cause it to denature, and thus the temperature should be monitored regularly during the GelMA preparation process. The dissolution step can be stopped when the solution is uniform without visible granules.
4. GelMA solution should be filled up to only 1/3 of the entire length of the dialysis bag.
5. The dialysis process of GelMA is to remove the toxic unreacted methacrylate anhydride. Dialysis should be continued until odor of residual methacrylic anhydride disappears completely and the GelMA solution becomes clear. The temperature should be maintained at 40 °C throughout the dialysis period. Alternatively, dialysis can be done at 4 °C for an extended period to minimize degradation. The frequency of changing the dialysis water and duration of dialysis depend on the amount of GelMA dialyzed.
6. Protect GelMA from moisture during storage.
7. The solution may require incubating at 50 °C oven to dissolve completely.
8. Sterilize all solutions by filtering through 0.2-μm syringe filters under a class II biological laminar flow cabinet.
9. The amounts of GelMA, sodium alginate, and PEGOA solutions required depend on the final concentrations of each component and volume of the bioink needed. Yet, the recommended starting concentrations are listed in Subheading 3.1, **step 2**. The amount of individual component can be determined using the equation:

$$\text{Mass} = \text{Concentration} \times \text{Volume}$$

10. Since the GelMA and the PEGOA components are light-sensitive in the presence of the photoinitiator, the bioinks (including those laden with cells) should be covered with aluminum foil to protect from light.
11. The selection of the vascular cells will depend on the target vessel types to be bioprinted. Both ECs and SMCs can be chosen in a vessel type-specific manner as appropriate.
12. Ensure that all three layers of the needles are concentrically aligned in the coaxial form when the epoxy glue starts to solidify, so that the configuration is fixed in the final MCCES.
13. Syringe pump operations allow easy and quick adjustments of the flow parameters in each layer. Alternatively, automatically controlled pneumatic pressures that are programmed into the operations of the bioprinter may also be used, if more convenient.

14. The flow rate of the bioinks and $CaCl_2$ solution during the bioprinting process may require fine-tuning for each new setup to achieve optimal bioprinting.
15. Depending on the need, the bioprinted vascular conduits can be either deposited on a glass slide as the substrate or directly collected in a bath of $CaCl_2$ solution. For the former, rough-surfaced glass slides are usually used to avoid dragging of the deposited hollow fibers with the movement of the printhead. For the latter, there is no need of moving the printhead.
16. Instead of static culture of the bioprinted vascular conduits, they can also be integrated into customized bioreactors for continuous perfusion culture. An example of the bioreactor design can be found in ref. 36.

Acknowledgments

This work was supported by funding from the National Institutes of Health (R00CA201603, R21EB025270, R21EB026175, R01EB028143, R01HL153857, R21EB030257), National Science Foundation (CBET-EBMS-1936105), the Brigham Research Institute, and the American Fund for Alternatives to Animal Research (AFAAR). Y.S.Z. sits on the Scientific Advisory Board of Allevi, Inc., which however, did not support or bias this work. The other authors declare no competing financial/commercial interest.

References

1. Benjamin EJ, Muntner P, Bittencourt MSJC (2019) Heart disease and stroke statistics-2019 update: a report from the American Heart Association. Circulation 139(10): e56–e528
2. Pashneh-Tala S, MacNeil S, Claeyssens F (2015) The tissue-engineered vascular graft—past, present, and future. Tissue Eng Part B Rev 22:68–100
3. Seifu DG, Purnama A, Mequanint K, Mantovani DJNRC (2013) Small-diameter vascular tissue engineering. Nat Rev Cardiol 10(7):410
4. Ma Z, Kotaki M, Yong T, He W, Ramakrishna S (2005) Surface engineering of electrospun polyethylene terephthalate (PET) nanofibers towards development of a new material for blood vessel engineering. Biomaterials 26:2527–2536
5. Lord MS, Yu W, Cheng B, Simmons A, Poole-Warren L, Whitelock JM (2009) The modulation of platelet and endothelial cell adhesion to vascular graft materials by perlecan. Biomaterials 30:4898–4906
6. Taylor LM, Edwards JM, Brant B, Phinney ES, Porter JM (1987) Autogenous reversed vein bypass for lower extremity ischemia in patients with absent or inadequate greater saphenous vein. Am J Surg 153:505–510
7. Kannan RY, Salacinski HJ, Butler PE, Hamilton G, Seifalian AM (2005) Current status of prosthetic bypass grafts: a review. J Biomed Mater Res B Appl Biomater 74 (1):570–581
8. Catto V, Farè S, Freddi G, Tanzi MC (2014) Vascular tissue engineering: recent advances in small diameter blood vessel regeneration. Vasc Med 2014:923030
9. Bordenave L, Menu P, Baquey C (2008) Developments towards tissue-engineered,

small-diameter arterial substitutes. Expert Rev Med Devices 5:337–347

10. Dean EW, Udelsman B, Breuer CK, medicine (2012) Current advances in the translation of vascular tissue engineering to the treatment of pediatric congenital heart disease. Yale J Biol Med 85:229
11. Tara S, Rocco KA, Hibino N, Sugiura T, Kurobe H, Breuer CK, Shinoka T (2014) Vessel bioengineering– development of small-diameter arterial grafts. Circ J 78(1):17–19
12. Schöneberg J, De Lorenzi F, Theek B, Blaeser A, Rommel D, Kuehne AJC, Kießling F, Fischer H (2018) Engineering biofunctional in vitro vessel models using a multilayer bioprinting technique. Sci Rep 8 (1):10430
13. Datta P, Ayan B, Ozbolat IT (2017) Bioprinting for vascular and vascularized tissue biofabrication. Acta Biomater 51:1–20
14. Christensen K, Xu C, Chai W, Zhang Z, Fu J, Huang Y (2015) Freeform inkjet printing of cellular structures with bifurcations. Biotechnol Bioeng 112:1047–1055
15. Bishop ES, Mostafa S, Pakvasa M, Luu HH, Lee MJ, Wolf JM, Ameer GA, He T-C, Reid RR (2017) 3-D bioprinting technologies in tissue engineering and regenerative medicine: current and future trends. Genes Dis 4:185–195
16. Murphy SV, Atala A (2014) 3D bioprinting of tissues and organs. Nat Biotechnol 32:773
17. Yu Y, Zhang Y, Martin JA, Ozbolat IT (2013) Evaluation of cell viability and functionality in vessel-like bioprintable cell-laden tubular channels. J Biomech Eng 135:091011
18. Zhang Y, Yu Y, Ozbolat IT (2013) Direct bioprinting of vessel-like tubular microfluidic channels. J Nanotechnol Eng Med 4:020902
19. Seol Y-J, Kang H-W, Lee SJ, Atala A, Yoo JJ (2014) Bioprinting technology and its applications. Eur J Cardiothorac Surg 46:342–348
20. Xiong R, Zhang Z, Chai W, Huang Y, Chrisey DB (2015) Freeform drop-on-demand laser printing of 3D alginate and cellular constructs. Biofabrication 7:045011
21. Guillotin B, Ali M, Ducom A, Catros S, Keriquel V, Souquet A, Remy M, Fricain J-C, Guillemot F (2013) Laser-assisted bioprinting for tissue engineering. In: Biofabrication. Elsevier, pp 95–118
22. Heinrich MA, Liu W, Jimenez A, Yang J, Akpek A, Liu X, Pi Q, Mu X, Hu N, Schiffelers RM (2019) 3D bioprinting: from benches to translational applications. Small 15(23): e1805510
23. Maina RM, Barahona MJ, Finotti M, Lysyy T, Geibel P, D'Amico F, Mulligan D, Geibel JP (2018) Generating vascular conduits: from tissue engineering to three-dimensional bioprinting. Innov Surg Sci 3:203–213
24. Wang Z, Abdulla R, Parker B, Samanipour R, Ghosh S, Kim K (2015) A simple and high-resolution stereolithography-based 3D bioprinting system using visible light crosslinkable bioinks. Biofabrication 7:045009
25. Ma X, Qu X, Zhu W, Li Y-S, Yuan S, Zhang H, Liu J, Wang P, Lai CSE, Zanella F, Feng G-S, Sheikh F, Chien S, Chen S (2016) Deterministically patterned biomimetic human iPSC-derived hepatic model via rapid 3D bioprinting. Proc Natl Acad Sci U S A 113:2206–2211
26. Pi Q, Maharjan S, Yan X, Liu X, Singh B, van Genderen AM, Robledo-Padilla F, Parra-Saldivar R, Hu N, Jia WJAM (2018) Digitally tunable microfluidic bioprinting of multilayered cannular tissues. Adv Mater 30 (43):1706913
27. Jia W, Gungor-Ozkerim PS, Zhang YS, Yue K, Zhu K, Liu W, Pi Q, Byambaa B, Dokmeci MR, Shin SRJB (2016) Direct 3D bioprinting of perfusable vascular constructs using a blend bioink. Biomaterials 106:58–68
28. Kolesky DB, Homan KA, Skylar-Scott MA, Lewis JA (2016) Three-dimensional bioprinting of thick vascularized tissues. Proc Natl Acad Sci U S A 113(12):3179–3184
29. Jafarkhani M, Salehi Z, Aidun A, Shokrgozar MA (2019) Bioprinting in Vascularization Strategies. Iran Biomed J 23(1):9–20
30. Norotte C, Marga FS, Niklason LE, Forgacs G (2009) Scaffold-free vascular tissue engineering using bioprinting. Biomaterials 30 (30):5910–5917
31. Freeman FE, Kelly DJ (2017) Tuning alginate bioink stiffness and composition for controlled growth factor delivery and to spatially direct MSC fate within bioprinted tissues. Sci Rep 7 (1):17042–17042
32. Ooi HW, Mota C, Ten Cate AT, Calore A, Moroni L, Baker MBJB (2018) Thiol–ene alginate hydrogels as versatile bioinks for bioprinting. Biomacromolecules 19(8):3390–3400
33. Loessner D, Meinert C, Kaemmerer E, Martine LC, Yue K, Levett PA, Klein TJ, Melchels FPW, Khademhosseini A, Hutmacher DW (2016) Functionalization, preparation and use of cell-laden gelatin methacryloyl–based hydrogels as

modular tissue culture platforms. Nat Protoc 11(4):727–746

34. Klotz BJ, Gawlitta D, Rosenberg AJ, Malda J, Melchels FPW (2016) Gelatin-methacryloyl hydrogels: towards biofabrication-based tissue repair. Trends Biotechnol 34(5):394–407
35. Yue K, Li X, Schrobback K, Sheikhi A, Annabi N, Leijten J, Zhang W, Zhang YS, Hutmacher DW, Klein TJ, Khademhosseini A (2017) Structural analysis of photocrosslinkable methacryloyl-modified protein derivatives. Biomaterials 139:163–171
36. Cao X, Ashfaq R, Cheng F, Maharjan S, Li J, Ying G, Hassan S, Xiao H, Yue K, Zhang YS (2019) A tumor-on-a-Chip system with bioprinted blood and lymphatic vessel pair. Adv Funct Mater 29(31):1807173

Chapter 7

In Situ Fabrication and Perfusion of Tissue-Engineered Blood Vessel Microphysiological System

Xu Zhang, Nadia O. Abutaleb, Ellen Salmon, and George A. Truskey

Abstract

Human tissue-engineered blood vessels (TEBVs) that exhibit vasoactivity can be used to test drug toxicity, modulate pro-inflammatory cytokines, and model disease states in vitro. We developed a novel device to fabricate arteriole-scale human endothelialized TEBVs in situ with smaller volumes and higher throughput than previously reported. Both primary and induced pluripotent stem cell (iPSC)-derived cells can be used. Four collagen TEBVs with 600μm inner diameter and 2.9 mm outer diameter are fabricated by pipetting a solution of collagen and medial cells into a three-layer acrylic mold. After gelation, the TEBVs are released from the mold and dehydrated. After suturing the TEBVs in place and changing the mold parts to form a perfusion chamber, the TEBVs are endothelialized in situ, and then media is perfused through the lumen. By removing 90% of the water after gelation, the TEBVs become mechanically strong enough for perfusion at the physiological shear stress of 0.4 Pa within 24 h of fabrication and maintain function for at least 5 weeks.

Key words Tissue engineering, In vitro models, Blood vessels, Biofabrication

1 Introduction

Vascular microphysiological systems (MPS) represent small-scale models of blood vessels in vitro. The MPS are comprised of either human primary cells or cells derived from induced pluripotent stem cells (iPSCs). Human-sourced vascular MPS can serve as models to study vessel disease or tools to evaluate therapeutics [1]. Tissue-engineered blood vessels (TEBVs) are three-dimensional models of blood vessels consisting of a natural or synthetic scaffold with cells added prior to implantation or following ingrowth from blood or surrounding tissue [2, 3]. Human vascular MPS can be used to model diseases in vitro which cannot always be replicated accurately in animals [4–6]. The scale ranges from models of the microvasculature [7–9] to medium-size arterioles [1, 4, 5, 10, 11].

Feng Zhao and Kam W. Leong (eds.), *Vascular Tissue Engineering: Methods and Protocols*,
Methods in Molecular Biology, vol. 2375, https://doi.org/10.1007/978-1-0716-1708-3_7,

Considerations for in vitro human vascular MPS, also known as "vessels on a chip," include the ability to easily fabricate, sufficient mechanical strength to resist physiological pressures, a nonthrombogenic surface, and vasoactivity. A wide range of materials have been used to make arterial- and arteriole-scale tissue-engineered vascular grafts for implantation [2, 3]. While the mechanical strength of these bioengineered grafts is very good, there are challenges matching the stress-strain curve of arteries and ensuring a nonthrombogenic surface that facilitates endothelialization [12–14]. Further, relative to drug screens using cells in two-dimensional culture, TEBVs are limited by low production efficiency and a complex fabrication process. While these TEBVs can model large arteries and arterioles, other systems have been developed to model the microvasculature through either self-assembly of capillary networks [7, 8] or creation of endothelialized channels in compatible biomaterials [8, 15].

To address the challenges limiting TEBV throughput, we developed a novel method to fabricate arteriole-scale endothelialized TEBVs in situ while maximizing throughput and ease of fabrication. The device we describe can be easily assembled first into a seeding chamber, then a perfusion chamber by switching out two components. To begin fabrication, the seeding chamber is assembled and includes molds for four parallel TEBVs. A mixture of collagen gel and medial cells such as neonatal human dermal fibroblasts (hNDFs), primary smooth muscle cells (SMCs), mesenchymal stem cells (MSCs), or iPSC-derived SMCs (viSMCs) is injected into the molds to fabricate four collagen TEBVs in a single chamber. After gelation, the TEBVs undergo a process of plastic compression [16], removing over 90% of the water, and are then sutured in place. Vessels are endothelialized by injecting endothelial cells such as blood-derived endothelial colony-forming cells (ECFCs) [17], primary vascular endothelial cells, or iPSC-derived endothelial cells (viECs). Following endothelialization, the perfusion chamber is assembled, and the four TEBVs are perfused in situ (Fig. 1). These TEBVs can resist physiological pressures, exhibit vasoactivity levels found in arteries, and exhibit endothelial release of nitric oxide (NO) throughout the duration of an experiment.

2 Materials

2.1 Cells

The TEBVs can be fabricated using either primary human cells or cells differentiated from iPSCs.

1. Primary human neonatal dermal fibroblasts (hNDFs, Thermo Fisher #C0045C), passages 5–11.
2. Primary human mesenchymal stem cells (MSC, ATCC #PCS-500-010, or Lonza #PT-2501).

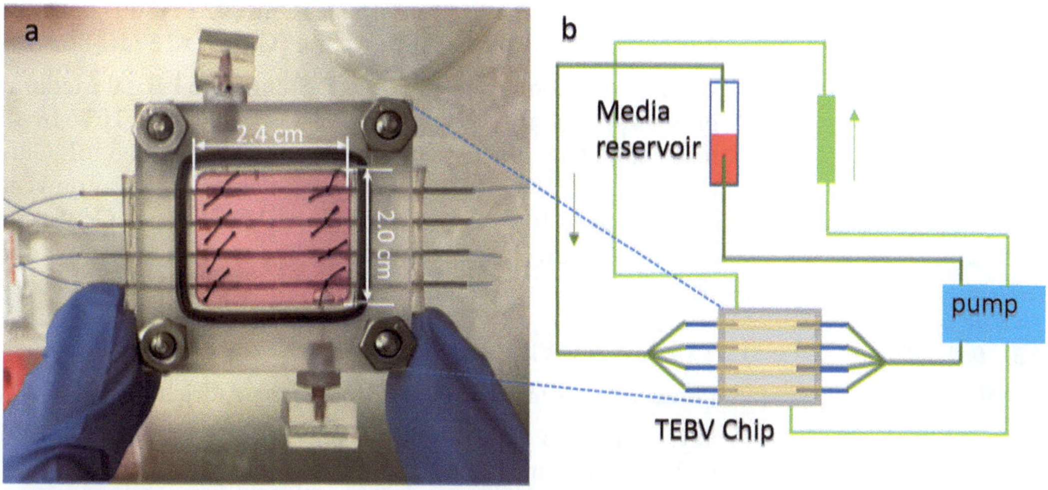

Fig. 1 (**a**) Chip housing 4 TEBVs. (**b**) Flow loops showing position of TEBV chip, pump, media reservoir, and tubing

3. Primary human endothelial cells (ECs) (Thermo Fisher, aortic endothelial cells, #C0065C; coronary artery endothelial cells, Lonza #CC-2585) (*see* **Note 1**).
4. iPSCs can be obtained from the Coriell Institute for Medical Research (www.coriell.org) or WiCell (www.wicell.org) (*see* **Note 2**).

2.2 Cell Culture Media

1. hNDF growth media, comprised of Dulbecco's Modified Eagle's Medium (DMEM) with 4.5 g/L glucose (Gibco, #11960-044) supplemented with 10% HI-FBS (Gibco, #16140-071), 1 × Non-essential amino acids (NEAA, Gibco, #11140–076), 1% Penicillin/Streptomycin (Gibco, #15140-122), 1 × Glutamax (Gibco, #35050–061), 1 × Sodium Pyruvate (Gibco, #11360-070), and 0.1% β-mercaptoethanol (Gibco, #21985023).
2. hMSC media, comprised of 500 mL MEM Alpha Base Media (Gibco, #12561-056), 100 mL heat-inactivated fetal bovine serum (HI-FBS), 10 mL L-Glutamine, and Penicillin/Streptomycin, 5 mL.
3. viSMC media, comprised of 250 mL DMEM/F12, HEPES medium (Gibco, #11330032) combined with 250 mL Neurobasal medium (Gibco, #21103049) and supplemented with 10 mL (2% v/v) B27 without vitamin A (Gibco, #12587010), 5 mL (1% v/v) N2 supplement (Gibco 17502048), 500μL (0.1% v/v) β-mercaptoethanol (Gibco 21985-023), 5 mL Penicillin/Streptomycin (Gibco, #15140-122), 2 ng/mL Activin A (Peprotech, #120-14E), and 2μg/mL heparin (Sigma, #H3149).

4. Endothelial cell media (Cell Applications, #211-500) containing 1% (v/v) Antibiotic-Antimycotic (Gibco, #15240-062).
5. viEC media, comprised of 500 mL StemPro-34 Basal Medium plus supplement (Thermo Fisher 10639011), 10% heat-inactivated fetal bovine serum (Gibco #10082147), 5 mL GlutaMAX (Gibco #35050-061), 5 mL Penicillin/Streptomycin (Gibco 315140-122), 50 ng/mL VEGF165 (GenScript #Z02689), and 2μg/mL heparin (Sigma #H3149) (*see* **Note 3**).

2.3 Cell Culture Supplies

1. T75 flasks (VWR, #BD353136).
2. Phosphate-buffered saline (PBS) (Sigma-Aldrich, #P4593).
3. Trypsin (Gibco #25300062), 0.05% diluted 1:1 in sterile PBS without Ca^{++} and Mg^{++}.
4. Accutase (Innovative Cell Technologies #AT104).

2.4 Materials for Mold Assembly

1. The molds are created using acrylic or polycarbonate and composed of five parts: parts A–E (Fig. 2).
2. Part A (Fig. 2) (40 mm × 42 mm × 8 mm) houses the TEBVs and connects them to the perfusion loop via metal tubing with inner diameter of 0.33 mm and outer diameter of 0.63 mm (New England Small Tube).
3. Parts A, B, and C are assembled for the seeding chip, and parts A, D, and E can be assembled for the perfusion chip (*see* **Note 4**).
4. The holes for the tubing and TEBVs are spaced 4.84 mm apart. The inner opening is 26.5 mm × 26.5 mm.
5. There are four holes placed at the corners of parts A–E for 6/32 screws and nuts. O-rings (70 BUNA, Size 24) are needed to provide a good seal. The side ports fit #10–32 UNF straight hole adapters to connect to tubing.
6. Parts B and C are 40 mm × 42 mm × 6 mm (the raised area is 3.9 mm), and the opening dimensions are the same as for part A.
7. The channels in parts B and C (each forms a hemicylindrical region of diameter 2.2 mm, length 24 mm) serve as molds for the TEBVs.
8. Parts D and E are 40 mm × 42 mm × 4.5 mm (the raised area is 2 mm), and the opening dimensions are the same as for part A.
9. Eight stainless steel mandrels with outer diameter of 0.63 mm and inner diameter of 0.33 mm (New England Small Tube).
10. PDMS (10:1 mixed, Sylgard 184 PDMS Kit, Dow, Inc.).

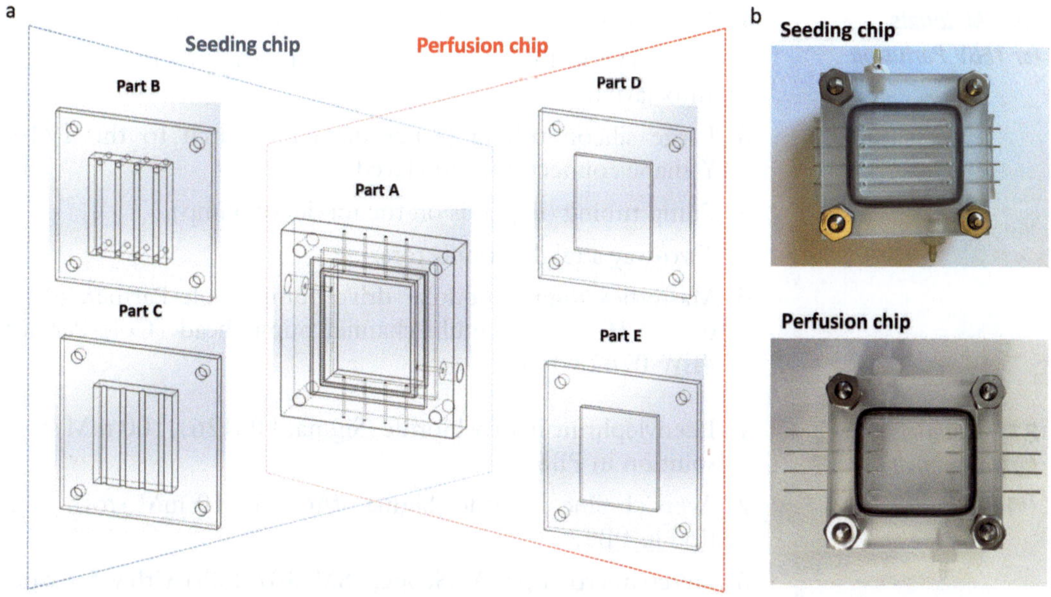

Fig. 2 The design of the mold. (**a**) The drafts of all the parts to assemble the mold. (**b**) The assembled seeding chip (TEBV fabrication mold) and perfusion chip (TEBV culture chamber)

11. Two barbed tubes with screws (match the size on part A, McMaster Carr, Elmhurst, IL).
12. Two O-rings (match the sealing grooves on part A).

2.5 Materials for TEBV Fabrication

1. Rat-tail collagen I, high concentration (8–10 mg/mL), 100 mg (VWR/Corning, #47747-218).
2. Acetic acid (0.6%), diluted from glacial acetic acid in deionized water and sterile filtered.
3. Dulbecco's Modified Eagle's Medium, Low Glucose, 10X (Sigma, #D2429).
4. 1 M NaOH, sterile, prepared by dissolving NaOH pellets in deionized water and sterile filtering after cooling.
5. 50 mL Falcon tube.
6. Two 1 mL sterile disposable syringe, BD Biosciences.
7. One 3 mL sterile disposable syringe, BD Biosciences.
8. Sterile sutures (REDISILK, #SK683-SI).
9. 5–6 Kimwipes, autoclaved.
10. 15–16, 2 cm × 2 cm Kimwipes (cut), autoclaved.
11. Two BD Adapters, LUER-STUB 23GA (BD, #427565).
12. Two stainless steel tweezers, sharp tip, autoclaved.
13. One small scissor, autoclaved.
14. Four screws with four nuts (6/32 Screw 3/4″, McMaster-Carr, #99607A124), autoclaved.

2.6 Materials for TEBV Perfusion

1. 1/16″ barb fitting (Y-shape connector of syringe to tubing).
2. Small Tygon tubing (inner diameter match to the mandrel), autoclaved.
3. Large silicon tubing (inner diameter match to the 1/16″ Y-shape connector), autoclaved.
4. Pump tubing (depends on the model of pump).
5. Two-way straight connectors.
6. Masterflex digital modular drive pump (Cole-Parmer, #EW-07557-10) with a multi-channel pump head (Cole-Parmer, #EW-07623-10).

2.7 Materials for Vasoactivity Testing

1. Phenylephrine hydrochloride (Sigma, #P6126), 100 mM stock solution in PBS.
2. Acetylcholine chloride (Sigma, #A6625), 10 mM stock solution in PBS.
3. Stereo microscope (AmScope, #SM-3BZ-80S) with video camera 10 MP USB 2.0 Microscope (AmScope, #MA1000).

3 Methods

3.1 Set up Mandrels and Side Connectors

1. Fix the mandrels on part A of the mold using PDMS (Fig. 3).
2. Screw the two barbed tubes into part A of the mold.
3. Set the O-rings in the sealing grooves on the part A of the mold.

3.2 Passage Primary Cells (ECs, hNDFs, and hMSCs)

1. Aspirate the media from the culture flask.
2. Wash the flask with PBS 1–2 times.
3. Add 5 mL 0.025% trypsin (0.05% for hNDFs) into the T-75 flask, and incubate at 37 °C for 5 min.
4. Add 5 mL of culture media to inhibit trypsin activity.
5. Transfer the cell suspension into centrifuge tubes.
6. Centrifuge cells at 1000 rpm (150 × *g*) for 5 min.
7. After centrifugation, remove media, and resuspend 10^6 cells in 60μL fresh culture media.

3.3 Passage viSMCs and viECs

1. Aspirate the media from the culture flask.
2. Wash the flask once with warm PBS.
3. Add 5 mL Accutase to T-75 flask, and incubate at 37 °C for 3 min.
4. Dilute Accutase with 5 mL culture media.

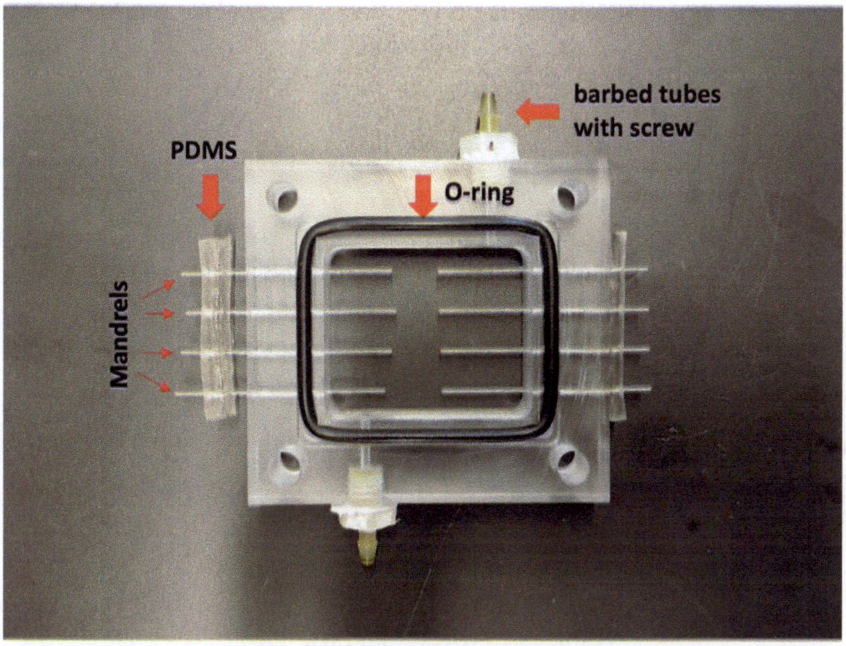

Fig. 3 Set up the mandrels and the side connectors on the part A of the chip

5. Transfer cell suspension to centrifuge tube.
6. Centrifuge cells at 1000 rpm (150 × *g*) for 5 min.
7. Remove supernatant, and resuspend cells in desired amount of fresh culture media.

3.4 TEBV Fabrication

1. Insert four mandrels into the chamber at each side of part A. Make sure each pair of mandrels at opposite positions of the chamber is touching each other at the middle to form one long tube.
2. Assemble part A, part B, and part C to form the seeding chip (Fig. 3).
3. Passage and resuspend 10^6 hNDFs in 60μL hNDF media (or 3×10^6 viSMCs in 60μL viSMC media).
4. Make 7 mg/mL solution of rat-tail collagen I in 1.5 mL Eppendorf tube, total volume 1 mL (*see* **Note 5**).
5. The formula for 1 mL collagen and cell mixture for the four TEBV system is (*see* **Note 6**):

   Volume of 10× DMEM = 94μL

   Volume of collagen solution (V_c) = 1 mL × (7 mg/mL)/ collagen stock concentration).

   Volume 1 M NaOH (V_{NaOH}) = Volume of collagen solution × 0.023.

   Volume of hDNF suspension = 60μL.

Volume of hNDF media (V_{hNDF}) = 1 mL − 94μL − V_c − V_{NaOH} − 60μL (if negative then do not add this extra media)

6. Add the solutions in the following sequence: 10 × DMEM→NaOH→Collagen→ hDNF suspension→ hNDF media.
7. Using a 200μL pipette, add collagen mixture into each mold chamber (four chambers per chip, the volume of each chamber depends on design).
8. Allow the collagen solution with cells to gel in the chip for 30 min in a cell culture incubator at 37 °C.

3.5 Plastic Compression and Dehydration of TEBVs In Situ

1. Move the mold from the incubator to the biosafety cabinet.
2. Remove part B and part C of the mold (Fig. 4a).
3. Dehydrate the TEBVs with 2 cm × 2 cm wipes five times on each side (Fig. 4b–d).
4. Allow Kimwipes to remain in place until water absorption is visible (*see* **Notes** 7 and **8**).
5. Attach part D to the bottom of part A.
6. Add 2 mL warm hNDF media (or viSMC media for TEBVs made with iPS-derived cells) into the chamber using a serological pipette.
7. Soak the dehydrated TEBVs for at least 1 min.
8. Remove the media from chamber.
9. Draw out all the mandrels such that there is 5 mm remaining in the interior of the chip on each side.
10. Firmly attach TEBVs to the mandrels with sutures, tying in double knots if needed (Fig. 5a).
11. Cover the top layer of the chamber with part E, screw in place, and fill the chamber with warm EGM media (or viSMC media for iPSC-derived TEBVs) using a syringe and tubing through the barbed side port tube (Fig. 5b).

3.6 Endothelial Seeding of TEBVs

1. Passage EPCs (or viECs).
2. Prepare a suspension of 6 × 10^6 cells/mL in 0.45 mL EC media (or 4 × 10^6 cells/mL in 0.45 mL viEC media for iPSC-derived TEBVs).
3. Connect one empty syringe (1 mL syringe with BD adapter needle, 22 gauge or greater) to one mandrel (link the adapter needle and mandrel using Tygon tubing).
4. Fill another 1 mL syringe with EC suspension, and connect it to the mandrel at opposite position of the chamber.

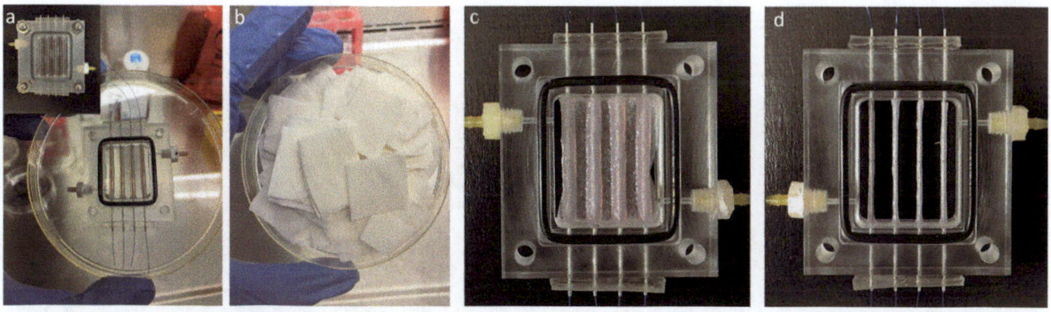

Fig. 4 Dehydration of the TEBVs. (**a**) Release the TEBVs from the mold. (**b**) The cut and autoclaved wipes. (**c**) Dehydrate the TEBVs using wipes. (**d**) The dehydrated TEBVs

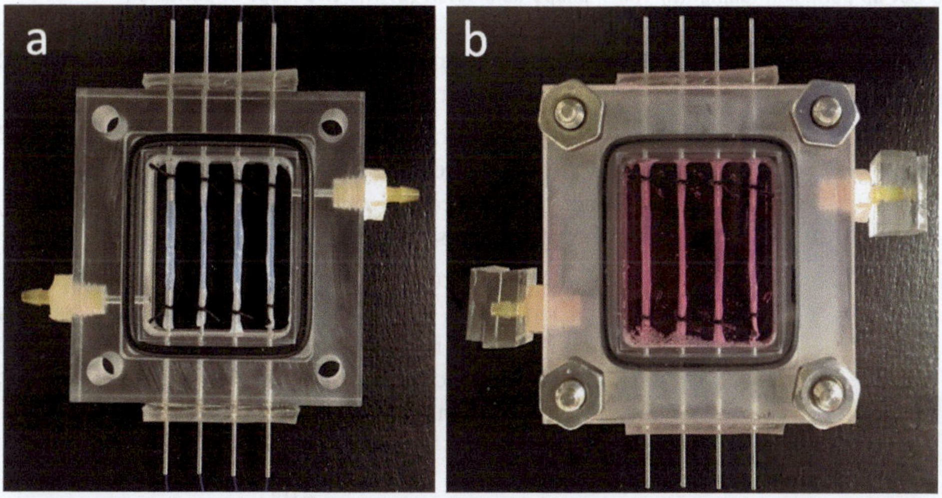

Fig. 5 Suture the vessels and set up the culture chamber. (**a**) Part A of vessel chamber with drawn-out mandrels and sutured TEBVs. (**b**) Fully assembled TEBV chip filled with media

5. Slowly push suspension through, counteracting by pulling on the other syringe to act as a pressure balance. Ensure that suspension is going through the TEBV and filling the whole lumen.
6. Move the syringes to the next vessel, and repeat the filling step. Seed the four vessels one by one.
7. Place the perfusion chip with ECs for at least 30 min in incubator, and flip it after 15 min.
8. Keep the perfusion chip in the incubator overnight. Set up the perfusion 24 h after fabrication.

3.7 TEBV Perfusion

There are two perfusion loops for the chip: the vessel loop and the side loop (Fig. 1). The perfusion tubing is composed of silicon tubing and pump tubing. The tubing is linked by two-way straight connectors. The length of each loop depends on the experiment.

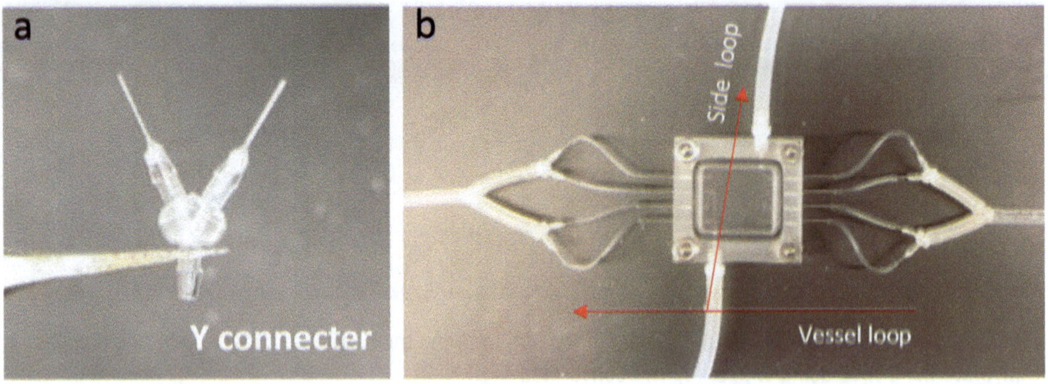

Fig. 6 Set up perfusion tubing. (**a**) The Y-shape connector. (**b**) The two loops of perfusion

1. Set up the side loop, and fill the loop using EC media (viSMC media for iPSC-derived TEBVs).
2. Set up the vessel loop, and fill the loop using EC media (viSMC media for iPSC-derived TEBVs). The four TEBVs in one chamber are linked to one perfusion tubing using 1/16″ Y-shape connectors (Fig. 6).
3. Set up the pump tubing on the pump and perfuse. The flow rate depends on the desired wall shear stress (usually 0.1–1.5 Pa). The relationship between wall shear stress (τ_w, Pa) and volumetric flow rate (Q, m^3/s) for a tube of radius R (m) and Newtonian fluid of viscosity μ (Pa s) is:

$$\tau_w = \frac{4\mu Q}{\pi R^3} \quad (1)$$

For culture media at 37 °C, $\mu = 0.087$ Pa s (*see* **Note 9**).

3.8 Vasoactivity Testing

1. Place perfusion chamber with TEBVs beneath stereo microscope, and image TEBVs at 9× magnification.
2. Start video recording of the TEBVs being perfused at flow rate that meets the desired shear stress (Eq. 1) to acquire vessel baseline diameter.
3. After at least 30 s of baseline, add phenylephrine at 1μM (final concentration) to the media stock vial of the vessel loop.
4. Record phenylephrine response for 6 min.
5. Add acetylcholine at 1μM (final concentration) to the media stock vial of the vessel loop.
6. Record acetylcholine response for 6 min.
7. Obtain three frames of images from the video: (1) before adding phenylephrine as base line, (2) after phenylephrine treatment for 6 min, and (3) after acetylcholine treatment for 6 min.

8. Measure diameter of each vessel at three to five different locations under baseline conditions. For example, for each TEBV divide the vessel into four equal parts, and measure the diameter in the middle of each region.
9. Average the values for each vessel to obtain the average TEBV diameter before treatment. Then average the diameter of each vessel.
10. Repeat the process for the images after 6 min of treatment with 2μM phenylephrine (PE) and 6 min of treatment with 2μM acetylcholine (ACh). Then average the diameter of each vessel.
11. Divide the average diameter of phenylephrine-treated TEBV by the diameter of TEBV at baseline, subtract from 1, and multiply by 100 to obtain percent vasoconstriction: (1-ave Diameter after PE/Initial Diameter) × 100% (*see* **Note 10**).
12. Use the average diameter of acetylcholine-treated TEBV divided by the diameter of phenylephrine-treated TEBV, and subtract 1 to get vasodilation: (Diameter after ACh)/Diameter after PE -1) × 100%.

3.9 En Face Immunofluorescence Imaging of TEBV Endothelium

1. At the end of an experiment, perfuse the TEBVs with 4% paraformaldehyde (PFA) solution for 1 h at room temperature.
2. Following fixation, remove the sample from the samples from the chamber, and transfer to a well plate containing PBS.
3. Wash samples in PBS three times to remove excess PFA.
4. If imaging intracellular proteins (e.g., von Willebrand Factor, vWF), samples should be permeabilized with 0.1% Triton-X for 5 min.
5. After rinsing three times with Dulbecco's phosphate-buffered saline (DPBS) (Sigma Aldrich, D8662) (following permeabilization), incubate in blocking buffer (PBS + 10% goat serum +5% BSA) for 2 h at room temperature.
6. Incubate the lumen with one of the following antibodies overnight at 4 °C: CD31 (BD #550389, 1:200 dilution), VCAM-1 (Santa Cruz Biotechnology (SCBT), #sc-13160, 1:200 dilution), ICAM-1 (Santa Cruz Biotechnology, #sc-107,1:200 dilution), E-selectin (SCBT, #sc-14011, 1:200 dilution), VE-Cadherin (Thermo Fisher, Clone 16B1, #14-1449-82), vWF (Thermo Fisher, Clone F8/86, #MA5-14029, 1:100 dilution).
7. Rinse the samples three times with DPBS.
8. Incubate the lumen with the following secondary antibodies for 160 min at room temperature: Alexa Fluor 488-goat anti-mouse IgG (Thermo Fisher, #A-32723), Alexa Fluor 488-goat anti-rabbit IgG (Thermo Fisher, #A-32731), Alexa Fluor

546-goat anti-mouse IgG (Thermo Fisher, #A-11030), Alexa Fluor 546-goat anti-rabbit IgG (Thermo Fisher, #A-11035), Alexa Fluor 633-goat anti-mouse (Thermo Fisher, #A21052) each at 1:200 dilution, and Hoechst dye (Thermo Fisher, #H1398, 1:1000 dilution) to stain nuclei.

9. Cut the vessel longitudinally to open, with the endothelial surface facing up. Avoid contact with endothelial layer, and place flat between two coverslips. Acquire images on a confocal microscope at 20× or 40× magnification (*see* **Note 11**).

4 Notes

1. Alternatively, blood-derived endothelial cells (also known as endothelial colony-forming cells, ECFCs) derived as described by Ingram et al. [18] and used through passages 4–6.
2. Protocols for differentiating iPS cells to vascular smooth muscle cells (viSMCs) and endothelial cells (viECs) [4, 19] were adapted from Patsch et al. [20].
3. Warm media to 37 °C prior to use.
4. Acrylic components can be sterilized with ethylene oxide (ETO) treatment. Alternatively, clear polycarbonate can be autoclaved.
5. All solutions need to be placed on ice. The operation of mixing collagen needs to be performed very carefully to avoid air bubbles.
6. Use NaOH and acetic acid to regulate the collagen pH at 8.5. Solution should change from yellow to pink.
7. Kimwipes applications 1–2: double layer wipe on each side.

   Kimwipes applications 3–5: single layer wipe on each side.
8. To ensure that at least 90% of the water is removed, weigh the TEBV cell suspension mixture before adding Kimwipes, and weigh each Kimwipes before and after contacting TEBV. Calculate weight change, and divide by the initial amount of water in TEBV to determine water lost from TEBV.
9. Viscosity is measured at 37 °C by a fall ball viscometer (Gilmont, Cole-Parmer, #EW-08701-00) or a rheometer that can provide results in shear rate ranges of 150–2500 s^{-1} (e.g., Brookfield Engineering Laboratories DV-III rheometer).
10. Vasoactivity measurements are performed to identify the functional behavior of the TEBVs over time and after treatment with drugs, cytokines, or other molecules that may modulate function.

11. Portions of the vessels overlapping with the chamber grips should not be imaged, as endothelial cells are unable to adhere well in these areas, and shear stresses will be inconsistent.

Acknowledgments

This work was supported by NIH grants HL138252 and UH3TR0002142.

References

1. Fernandez CE, Yen RW, Perez SM, Bedell HW, Povsic TJ, Reichert WM, Truskey GA (2016) Human vascular microphysiological system for in vitro drug screening. Sci Rep 6:21579
2. Fernandez CE, Achneck HE, Reichert WM, Truskey GA (2014) Biological and engineering design considerations for vascular tissue engineered blood vessels (TEBVs). Curr Opin Chem Eng 3:83–90
3. Liu RH, Ong CS, Fukunishi T, Ong K, Hibino N (2017) Review of vascular graft studies in large animal models. Tissue Eng Part B Rev 24(2):133–143
4. Atchison L, Abutaleb N, Snyder-Mounts E, Gete Y, Ladha A, Ribar T, Cao K, Truskey GA (2020) iPSC-derived endothelial cells affect vascular function in a tissue engineered blood vessel model of Hutchinson-Gilford progeria syndrome. Stem Cell Reports 14 (2):325–337
5. Atchison L, Zhang H, Cao K, Truskey GA (2017) A tissue engineered blood vessel model of Hutchinson-Gilford Progeria syndrome using human iPSC-derived smooth muscle cells. Sci Rep 7(1):8168
6. Zhang B, Montgomery M, Chamberlain MD, Ogawa S, Korolj A, Pahnke A, Wells LA, Masse S, Kim J, Reis L, Momen A, Nunes SS, Wheeler AR, Nanthakumar K, Keller G, Sefton MV, Radisic M (2016) Biodegradable scaffold with built-in vasculature for organ-on-a-chip engineering and direct surgical anastomosis. Nat Mater 15(6):669–678
7. Jeon JS, Bersini S, Whisler JA, Chen MB, Dubini G, Charest JL, Moretti M, Kamm RD (2014) Generation of 3D functional microvascular networks with human mesenchymal stem cells in microfluidic systems. Integr Biol 6 (5):555–563
8. Wang X, Phan DTT, George SC, Hughes CCW, Lee AP (2017) 3D anastomosed microvascular network model with living capillary networks and endothelial cell-lined microfluidic channels. Methods Mol Biol 1612:325–344
9. Wang YI, Abaci HE, Shuler ML (2017) Microfluidic blood–brain barrier model provides in vivo-like barrier properties for drug permeability screening. Biotechnol Bioeng 114 (1):184–194
10. Ji H, Atchison L, Chen Z, Chakraborty S, Jung Y, Truskey GA, Christoforou N, Leong KW (2016) Transdifferentiation of human endothelial progenitors into smooth muscle cells. Biomaterials 85:180–194
11. Jung Y, Ji H, Chen Z, Fai Chan H, Atchison L, Klitzman B, Truskey G, Leong KW (2015) Scaffold-free, human mesenchymal stem cell-based tissue engineered blood vessels. Sci Rep 5:15116
12. Ding N, Dou C, Wang Y, Liu F, Guan G, Huo D, Li Y, Yang J, Wei K, Yang M, Tan J, Zeng W, Zhu C (2018) Antishear stress bionic carbon nanotube mesh coating with intracellular controlled drug delivery constructing small-diameter tissue-engineered vascular grafts. Adv Healthc Mater 7(11):e1800026
13. Huo D, Liu G, Li Y, Wang Y, Guan G, Yang M, Wei K, Yang J, Zeng L, Li G, Zeng W, Zhu C (2017) Construction of antithrombotic tissue-engineered blood vessel via reduced graphene oxide based dual-enzyme biomimetic cascade. ACS Nano 11(11):10964–10973
14. Li Y, Wan S, Liu G, Cai W, Huo D, Li G, Yang M, Wang Y, Guan G, Ding N, Liu F, Zeng W, Zhu C (2017) Netrin-1 promotes inflammation resolution to achieve endothelialization of small-diameter tissue engineering blood vessels by improving endothelial progenitor cells function in situ. Adv Sci (Weinh) 4 (12):1700278
15. Zheng W, Huang R, Jiang B, Zhao Y, Zhao W, Jiang X (2016) An early-stage atherosclerosis research model based on microfluidics. Small 12:2022–2034

16. Ghezzi CE, Risse P-A, Marelli B, Muja N, Barralet JE, Martin JG, Nazhat SN (2013) An airway smooth muscle cell niche under physiological pulsatile flow culture using a tubular dense collagen construct. Biomaterials 34:1954–1966
17. Yoder MC, Mead LE, Prater D (2007) Redefining endothelial progenitor cells via clonal analysis and hematopoietic stem/progenitor cell principals. Blood 109:1801–1809
18. Ingram DA, Mead LE, Tanaka H, Meade V, Fenoglio A, Mortell K, Pollok J, Ferkowicz MJ, Gilley D, Yoder MC (2004) Identification of a novel hierarchy of endothelial progenitor cells using human peripheral and umbilical cord blood. Blood 104(9):2752–2760
19. Abutaleb NO, Truskey GA. (2021) Differentiation and characterization of human iPSC-derived vascular endothelial cells under physiological shear stress STAR Protoc. 2(2):100394
20. Patsch C, Challet-Meylan L, Thoma EC, Urich E, Heckel T, O'Sullivan JF, Grainger SJ, Kapp FG, Sun L, Christensen K, Xia Y, Florido MH, He W, Pan W, Prummer M, Warren CR, Jakob-Roetne R, Certa U, Jagasia R, Freskgard PO, Adatto I, Kling D, Huang P, Zon LI, Chaikof EL, Gerszten RE, Graf M, Iacone R, Cowan CA (2015) Generation of vascular endothelial and smooth muscle cells from human pluripotent stem cells. Nat Cell Biol 17(8):994–1003

Chapter 8

Peritoneal Pre-conditioning Method for In Vivo Vascular Graft Maturation Utilizing a Porous Pouch

Mahyar Sameti and Chris A. Bashur

Abstract

Tissue-engineered vascular grafts (TEVGs) require strategies to allow graft remodeling but avoid stenosis and loss of graft mechanics. A variety of promising biomaterials and methods to incorporate cells have been tested, but intimal hyperplasia and graft thrombosis are still concerning when grafting in small-diameter arteries. Here, we describe a strategy using the peritoneal cavity as an "in vivo" bioreactor to recruit autologous cells to electrospun conduits, which can improve the in vivo response after aortic grafting. We focus on the methods for a novel hydrogel pouch design to enclose the electrospun conduits that can avoid peritoneal adhesion but still allow infiltration of peritoneal fluid and cells needed to provide benefits when subsequently grafting in the aorta.

Key words Small-diameter arteries, Hydrogel, Vascular graft, Electrospun conduits, Peritoneal cavity, In situ tissue engineering

1 Introduction

Coronary artery disease is one of the leading causes of death worldwide. The most promising treatment is bypassing with autologous vessels, such as the saphenous vein, although this is not a viable option in more than 30% of the patients due to systematic vascular disease or previous usage [1, 2]. There are also limitations with the other current grafting options. For example, non-degradable polyester or Dacron grafts have significant concerns with stenosis and occlusion for high-pressure, small-diameter (<4 mm) vessels [3, 4]. Tissue-engineered vascular grafts (TEVGs) are a promising alternative, but new methods still need to be developed for these grafts to address the clinical needs, which include graft stenosis but also aneurysmal dilation [5–8]. The material composition of the scaffold (e.g., natural and/or synthetic) and their properties are an important consideration, and a large variety of scaffold materials

Feng Zhao and Kam W. Leong (eds.), *Vascular Tissue Engineering: Methods and Protocols*,
Methods in Molecular Biology, vol. 2375, https://doi.org/10.1007/978-1-0716-1708-3_8,

have been used [9, 10]. Developing an ideal biomaterial for a vascular graft has been challenging.

In addition to the biomaterial composition and processing techniques, the method to incorporate cells is also very important and is the focus of this chapter. Many TEVG conduits under investigation are acellular (i.e., they will recruit cells in the body). This can have several benefits, such as reduced processing time, but it also still has limitations that still need to be addressed. These limitations include insufficient biomaterial hemocompatibility and the promotion of an extended inflammatory response that can lead to intimal hyperplasia. Alternatively, multiple studies have shown the benefits of the pre-seeding of cells on the TEVGs [11–15]. However, there are also limitations with this approach, including questions about cell source. The approach that our group, as well as other groups, have used involves utilizing the body itself to recruit autologous cells and extracellular matrix proteins before grafting into the artery. We have shown that there are benefits of this approach with peritoneal pre-implantation over a comparable acellular conduit (e.g., a reduction in long-term inflammation) [16].

There are variations on this in vivo bioreactor strategy that have been used, including implantation of a silicone tube [17], silicon cage [18], and mesh conduit [19, 20]. However, one general concern with these approaches is the potential for adhesions to form to the implanted conduit. For example, many of the conduits that have been implanted in prior studies could not be retrieved due to adhesion formation [21]. In this chapter, we describe the generation of a macroporous poly(ethylene glycol) diacrylate (PEGDA) pouch to enclose the conduits during pre-implantation in the peritoneal cavity. This is flexible for different types of enclosed conduits and can prevent adhesion formation to the conduits themselves. Further, it can uniquely prevent adhesions to the pouch as well as further adhesion formation in the peritoneal cavity after the pouch is removed [22].

2 Materials

2.1 Electrospinning

1. Poly(ε-caprolactone) (PCL) ester terminated with the inherent viscosity of 1.0–1.3 dl/g in chloroform (*see* **Note 1**).
2. Collagen type I, lyophilized, acid-solubilized from calf skin, and salt-free (*see* **Note 2**).
3. 1,1,1,3,3,3-Hexafluoro-2-propanol (HFIP) with the purity of >99% as a solvent.
4. 1.6 mm outer diameter stainless steel mandrel.

5. High-voltage power supply up to 15 kV.
6. 22-gauge needles.

2.2 Hydrogel Pouch Fabrication

1. PEGDA, Mw = 3350, for hydrogel fabrication.
2. 2-Hydroxy-4′-(2-hydroxyethoxy)-2-methylpropiophenone (Irgacure 2959) with the purity of 98% as a polymerization initiator.
3. 0.22 μm syringe filter.
4. 184 silicone elastomer base and 184 silicone elastomer curing agent for preparing the molds.
5. A 100 W long-wave UV lamp (365 nm).
6. Sterile deionized water.
7. Glass aspirating pipette with an outer diameter of 1.4 mm.
8. Transparent glass microscope slide.
9. Vortex mixer and hot water bath.

2.3 Pouch Implantation in the Peritoneal Cavity and Aortal Grafting

1. Outbred male Sprague-Dawley rats (SD) (200–250 g) (*see* **Note 3**).
2. Anesthesia agent, isoflurane as an inhalation anesthetic.
3. Polyglactin 910, 4-0 suture for muscle closure.
4. 9-0 nylon suture with 0.10 mm needle diameter for aortal grafting.
5. 4.8 mm closed position skin staplers.
6. Injectable heparin sodium, 1000 μ/ml.
7. Analgesic, buprenorphine HCl (Buprenex®), 0.3 mg/ml injection solution.
8. Warm saline solution.
9. Surgical tools.

3 Methods

3.1 Electrospinning

1. Blends of collagen/PCL (w/w) are prepared and loaded in a syringe. For example, pure PCL as well as 10% and 25% blends (collagen/PCL) were prepared in HFIP and required concentrations in solvent ranging from 12 to 14 wt.% (*see* **Note 4**).
2. The electrospinning parameters should be: high voltage supply 15 kV, fixed throw distance of 10 cm, syringe flow rate of 0.8 ml/h, and a constant duration of the electrospinning on each mandrel (*see* **Note 5**). The electrospinning process was performed at room temperature in a chemical fume hood.

3. Controllable lateral movement and slow rotation (<100 rpm) should be applied for randomly oriented conduits (*see* **Note 5**).
4. Ethanol is used to remove the conduits from the mandrels.
5. Conduits are cut into the same segment size and stored in a desiccator until use. Ends of the conduits will be closed using solvent-bonding prior to implantation to prevent cell and tissue deposition inside the lumen.
6. Conduits need to be sterilized with ethylene oxide prior to usage.
7. The resulting conduits are characterized as described previously [19].

3.2 Hydrogel Pouch Fabrication

1. Measure PEGDA powder, and dissolve in sterile DI water at 37 °C to make 25% (w/v) solutions (*see* **Note 6**).
2. Add the Irgacure 2959 initiator at a ratio of 2.5 mg/75 mg of PEGDA and protect from light (*see* **Note** 7).
3. The solution needs to be sterilized with a 0.22 μm syringe filter.
4. Mix the components of the 184 Silicone elastomers at a ratio of 10:1 (base: curing agent), and then cure in an oven at 60 °C overnight (*see* **Note 8**).
5. The silicon mold is easiest to prepare with three stacked layers. Each layer is cut in the shape of a hollow rectangle, to result in a 1.9 × 1.4 × 0.5 cm pouch. This mold is sterilized using the autoclave (*see* **Note 9**).
6. A transparent glass piece with dimensions of 1.4 × 0.9 × 0.25 cm is placed in between the silicon stacks to retain the hollow area inside of the pouch after curing.
7. The sterile PEGDA solution is added to the molds in a sterile laminar hood and placed under the UV lamp with a distance of 8.5 cm for 10 min until the gels solidify (Fig. 1).
8. After solidifying, a sterile glass rod will be used to generate four pores on both sides of the pouch (*see* **Note 10**). The glass needs to be removed carefully without breaking the pouch, which results in a hollow space in the PEGDA hydrogels.
9. Two electrospun conduits are placed into the pouch.
10. The open side of the pouch, from where conduits are inserted, is then closed using additional PEGDA solution and a second UV cross-linking step (same timing) (Fig. 2).
11. The pouches are soaked in DI water until use to prevent any shrinkage and to remove any excess PEGDA and Irgacure 2959. They are typically used within 24 h.

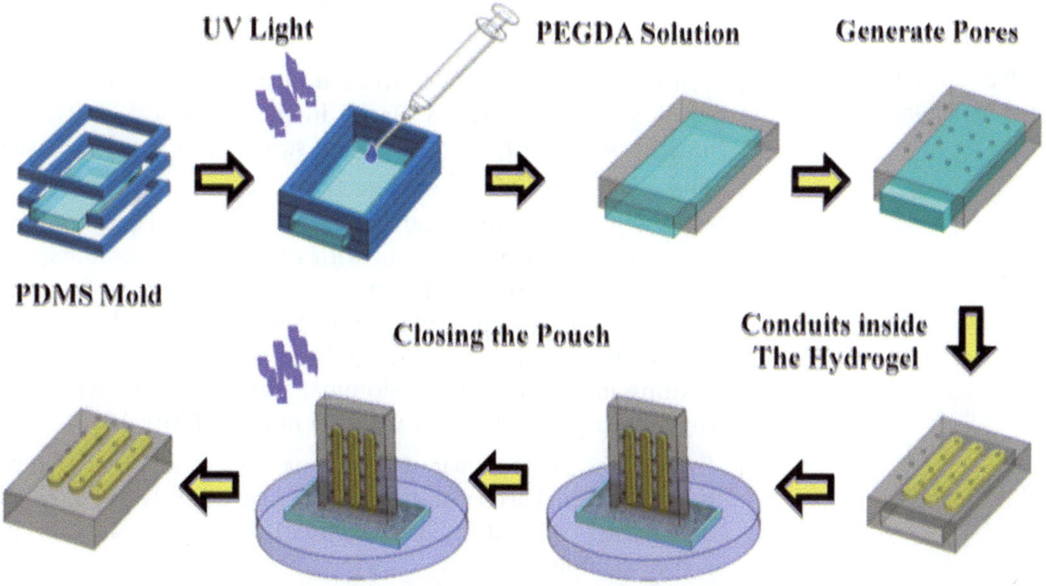

Fig. 1 Schematic showing PEGDA pouch fabrication with enclosed electrospun conduits. (Reproduced from ref. 22 with permission from Elsevier)

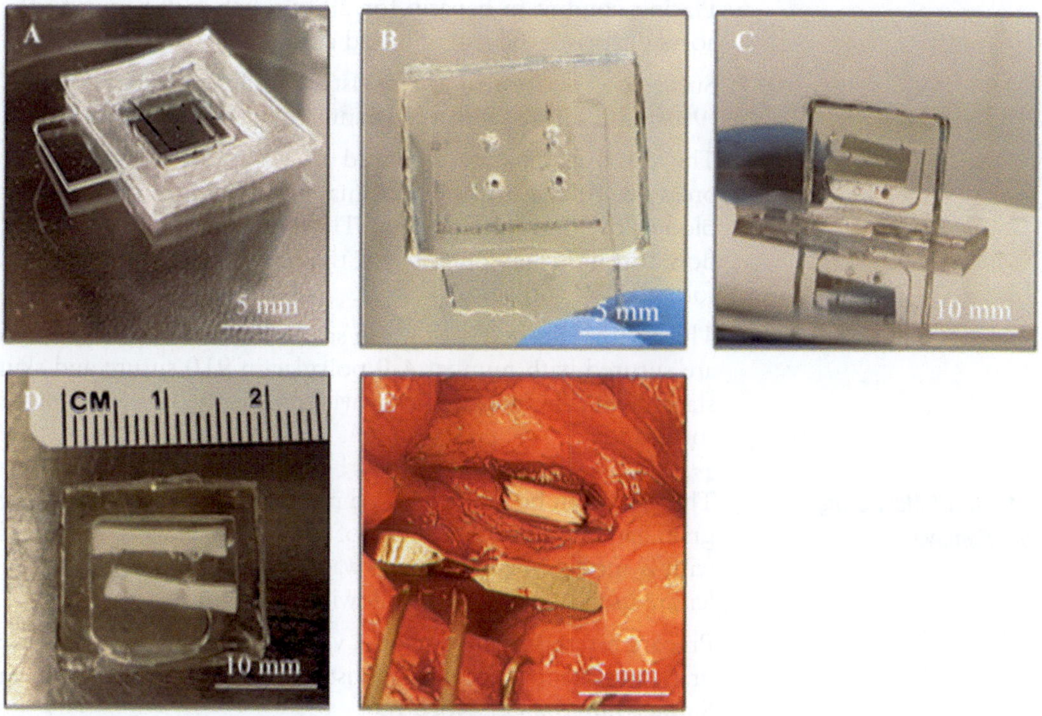

Fig. 2 PDMS mold with microscope slide (**a**). PEGDA pouch generated with the PDMS mold before (**b**) and after (**c**) adding the conduits inside. Representative PEGDA pouch generated with the PDMS mold (**d**). Representative image of a conduit from the hydrogel grafted in a rat aorta (**e**)

3.3 Pouch Implantation in the Peritoneal Cavity and Aortal Grafting

1. Rats are put under anesthesia with inhaled isoflurane at 2–3% (*see* **Note 11**).
2. Ventral laparotomy is performed to expose the peritoneal cavity. A prepared pouch is placed in the peritoneal cavity, the muscle closed with 4-0 polyglactin suture, and the skin closed with staples (*see* **Note 12**).
3. After 4 weeks of implantation, another ventral laparotomy is performed. Hydrogels, along with the enclosed conduits, are harvested from the peritoneal cavity to perform intra-aortal grafting. One conduit, which was harvested from the animal, will be chosen for intra-aortal grafting on the same animal. For grafting in the infrarenal abdominal aorta, the diameter of the surrounding artery is typically ~1.5 mm (*see* **Note 13**).
4. Expose the aorta, separate the vena cava, and place clamps below the renal artery and above the iliac bifurcation. 0.05 ml of heparin (100 unit/kg of the animal) is injected in a local vein before clamping.
5. Cut the aorta to result in an approximately 10 mm gap after elastic recoil. Small volumes of heparin added to the proximal and distal sides of the aorta.
6. Placing conduit in heparin for 20 min before suturing to the aorta decreases the risk of blood clots.
7. Suturing is performed at the distal and proximal anastomoses (9-0 nylon suture) with 12–16 interrupted sutures at each side.
8. The distal clamps are removed to verify no leaking at low pressure. Afterward, the proximal clamp is removed to allow blood flow through the graft. The confirmation of the blood flow through the conduit and the absence of leaking is performed after suturing.
9. The cavity is flushed with warm saline and closed. The muscles are sutured with number 4-0 polyglactin 910 suture and skin stapled. Finally, analgesia is provided, and the rats are allowed to recover.

3.4 Graft Monitoring and Removal

1. The grafts can be monitored via non-invasive imaging to verify graft patency and for analysis of graft properties. Ultrasound with B-mode, M-mode, pulse-wave Doppler, and color Doppler are used in our studies following standard procedures.
2. Prepare for graft removal with a ventral laparotomy, expose the graft, and clamp proximal and distal to the graft, as described in Subheading 3.2 (*see* **Note 14**).
3. Excise the graft with the surrounding artery at the anastomoses.
4. Process the graft for ex vivo analysis. This will typically include sectioning and imaging but will vary depending on the specific goals of the study.

4 Notes

1. Due to the difference of inherent viscosity of PCL, be consistent with the batch you choose at the beginning of the experiment. Differences in polydispersity can also have an impact [23].
2. This product is commercially available collagen type I prepared according to the method of Gallop and Seifter [24]. In the method, fresh calf skin is extracted with 0.5 M sodium acetate to remove non-collagen proteins. The soluble collagen is extracted with 0.075 M sodium citrate pH 3.7 and precipitated as fibrils by dialysis against 0.02 M disodium phosphate. Collagen type I extracted by this method is soluble in 0.01–0.5 M acetic acid.
3. Sprague-Dawley rats are the most common rats used for aortal grafting and vascular surgeries. The use of an outbred strain is not a concern for grafting of acellular conduits or conduits with only autologous cells.
4. To dissolve the materials, PCL will first be added to a container with HFIP and will be on a shaker overnight at room temperature. To add collagen, it will be added 20–30 min before electrospinning to avoid changes in viscosity. The concentration in HFIP needs to be adjusted when collagen is added to keep the fiber diameters consistent.
5. These parameters can impact fiber diameter, orientation, conduit wall thickness, and other properties, so it is best to keep these parameters consistent. Instead, adjusting solution concentration is one of the best ways to vary fiber diameter.
6. PEGDA dissolves in DI water fastest with heat and vortexing. PEGDA pouches are made with 25% (w/v) concentration, instead of lower concentrations, due to its mechanical properties which were explained in our previous study [22].
7. Irgacure 2959 is very sensitive to light, especially to UV light. To prevent unwanted polymerization to the polymer, this step needs to be done in darkness.
8. Mixing the silicone elastomers will make many bubbles which affect the surface smoothness and mechanical properties. After combining in a collection tube, the prepared solution needs to be transferred to a petri dish and left overnight at room temperature so the bubbles will be removed. After bubble removal, it can be transferred to the oven to cure.
9. The size of the pouch can be varied to incorporate either more conduits or longer conduits that would be needed for a larger animal model. For recent studies, the size was minimized to reduce the length of the abdominal incision.

10. The purpose of the pores is to enable conduit contact with the peritoneal fluid to allow for cell entry and tissue generation. The number of pores can be adjusted, but the impacts on the number of peritoneal adhesions that form and mechanical integrity of the pouch should be considered.
11. The level of isoflurane should be monitored throughout this surgery and adjusted to prevent the rat from being in too deep of a plane of anesthesia.
12. One hydrogel pouch per rat is used in this study (containing two conduits), but this could be increased depending on the size of the pouch and the animal model used.
13. The diameter of the conduit could be adjusted depending on the size of the rat and the section of the aorta used for grafting. Mismatch in diameter between the graft and aorta will impact the hemodynamics and also the grafting results.
14. It is often more challenging to separate the aorta and vena cava on graft removal due to the formation of additional tissue on the graft adventitia.

References

1. Shojaee M, Bashur CA (2017) Compositions including synthetic and natural blends for integration and structural integrity: engineered for different vascular graft applications. Adv Healthc Mater 6:1700001
2. Nieponice A et al (2010) *In Vivo* assessment of a tissue-engineered vascular graft combining a biodegradable elastomeric scaffold and muscle-derived stem cells in a rat model. Tissue Eng Part A 16:1215–1223
3. Singh C, Wong C, Wang X (2015) Medical textiles as vascular implants and their success to mimic natural arteries. J Funct Biomater 6:500–525
4. Ribas LM et al (2017) Experimental implantation of an arterial substitute made of silicone reinforced with polyester fabric in rabbits. Clinics 72:780–784
5. Pashneh-Tala S, MacNeil S, Claeyssens F (2016) The tissue-engineered vascular graft—past, present, and future. Tissue Eng B Rev 22:68–100
6. Syedain Z et al (2016) Tissue engineering of acellular vascular grafts capable of somatic growth in young lambs. Nat Commun 7:12951
7. Klinkert P, Post PN, Breslau PJ, van Bockel JH (2004) Saphenous vein versus PTFE for above-knee femoropopliteal bypass. A review of the literature. Eur J Vasc Endovasc Surg 27:357–362
8. Roger VL et al (2012) Heart disease and stroke statistics-2012 update: a report from the American heart association. Circulation 125: e2–e220
9. Williams DF (2019) Challenges with the development of biomaterials for sustainable tissue engineering. Front Bioeng Biotechnol 7:127
10. Thottappillil N, Nair PD (2015) Scaffolds in vascular regeneration: current status. Vasc Health Risk Manag 11:79–91
11. van Wachem PB, Stronck JWS, Koers-Zuideveld R, Dijk F, Wildevuur CRH (1990) Vacuum cell seeding: a new method for the fast application of an evenly distributed cell layer on porous vascular grafts. Biomaterials 11:602–606
12. Pawlowski KJ, Rittgers SE, Schmidt SP, Bowlin GL (2004) Endothelial cell seeding of polymeric vascular grafts. Front Biosci 9:1412–1421
13. Roh JD et al (2008) Small-diameter biodegradable scaffolds for functional vascular tissue engineering in the mouse model. Biomaterials 29:1454–1463
14. Perea H, Aigner J, Hopfner U, Wintermantel E (2006) Direct magnetic tubular cell seeding: a novel approach for vascular tissue engineering. Cells Tissues Organs 183:156–165

15. Tiwari A, Punshon G, Kidane A, Hamilton G, Seifalian AM (2003) Magnetic beads (Dynabead™) toxicity to endothelial cells at high bead concentration: implication for tissue engineering of vascular prosthesis. Cell Biol Toxicol 19:265–272
16. Shojaee M, Wood KB, Moore LK, Bashur CA (2017) Peritoneal pre-conditioning reduces macrophage marker expression in collagen-containing engineered vascular grafts. Acta Biomater 64:80–93
17. Song L, Wang L, Shah PK, Chaux A, Sharifi BG (2010) Bioengineered vascular graft grown in the mouse peritoneal cavity. J Vasc Surg 52:994–1002
18. Furukoshi M, Moriwaki T, Nakayama Y (2016) Development of an *in vivo* tissue-engineered vascular graft with designed wall thickness (biotube type C) based on a novel caged mold. J Artif Organs 19:54–61
19. Birthare K, Shojaee M, Jones CG, Brenner JR, Bashur CA (2016) Collagen incorporation within electrospun conduits reduces lipid oxidation and impacts conduit mechanics. Biomed Mater 11:025019
20. Bashur CA, Eagleton MJ, Ramamurthi A (2013) Impact of electrospun conduit fiber diameter and enclosing pouch pore size on vascular constructs grown within rat peritoneal cavities. Tissue Eng Part A 19:809–823
21. Chue WL et al (2004) Dog peritoneal and pleural cavities as bioreactors to grow autologous vascular grafts. J Vasc Surg 39:859–867
22. Shojaee M et al (2020) Design and characterization of a porous pouch to prevent peritoneal adhesions during *in vivo* vascular graft maturation. J Mech Behav Biomed Mater 102:103461
23. Nezarati RM, Eifert MB, Cosgriff-Hernandez E (2013) Effects of humidity and solution viscosity on electrospun fiber morphology. Tissue Eng Part C Methods 19:810–819
24. Hu CL, Crombie G, Franzblau C (1978) A new assay for collagenolytic activity. Anal Biochem 88:638–643

Chapter 9

Fabrication of a Completely Biological and Anisotropic Human Mesenchymal Stem Cell-Based Vascular Graft

Weilue He, Dhavan Sharma, Wenkai Jia, and Feng Zhao

Abstract

Tissue-engineered small-diameter vascular grafts are required to match mechanical properties as well as cellular and extracellular architecture of native blood vessels. Although various engineering technologies have been developed, the most reliable strategy highlights the needs for incorporating completely biological components and anisotropic cellular and biomolecular organization into the tissue-engineered vascular graft (TEVG). Based on the antithrombogenic, immunoregulatory, and regenerative properties of human mesenchymal stem cells (hMSCs), this chapter provides a step-by-step protocol for generating a completely biological and anisotropic TEVG that comprises of hMSCs and highly aligned extracellular matrix (ECM) nanofibers. The hMSCs were grown on an aligned nanofibrous ECM scaffold derived from an oriented human dermal fibroblast (hDF) sheet and then wrapped around a temporary mandrel to form a tubular assembly, followed by a maturation process in a rotating wall vessel (RWV) bioreactor. The resulting TEVG demonstrates anisotropic structural and mechanical properties similar to that of native blood vessels. A completely biological, anisotropic, and mechanically strong TEVG that incorporates immunoregulatory hMSCs is promising to meet the urgent needs of a surgical intervention for bypass grafting.

Key words Tissue-engineered vascular graft, Human mesenchymal stem cells, Extracellular matrix nanofibers, Anisotropic structure

1 Introduction

Due to the high occurrence and mortality of cardiovascular diseases worldwide, there is a huge demand for small-diameter blood vessels (inner diameter (ID) < 6 mm) to perform coronary bypass surgeries following myocardial infarction. Although autografts harvested from saphenous vein, arm vein, or radial artery remain the clinical standard for bypassing occluded coronary arteries, severe drawbacks still exist, including limited graft availability and highly invasive harvesting procedures. As a result, cellular and acellular tissue-engineered vascular grafts (TEVGs) have emerged and demonstrated propitious results.

Feng Zhao and Kam W. Leong (eds.), *Vascular Tissue Engineering: Methods and Protocols*, Methods in Molecular Biology, vol. 2375, https://doi.org/10.1007/978-1-0716-1708-3_9,

While designing a TEVG, major considerations include the construct's mechanical properties, cellular structure, biocompatibility, and fabrication processability. Since synthetic materials excel at accessibility, reproducibility, controllability, mechanical properties, and fabrication feasibility, they have been utilized to construct TEVGs, especially for large-diameter blood vessels (ID > 6 mm). Electrospun nanofibrous scaffolds assembled from biodegradable polymers, such as polycaprolactone-polylactic acid copolymer (PCL-PLA), polyglycolic acid (PGA), and polyurethane (PU), mimic the anisotropic organization of native extracellular matrix (ECM) of blood vessels [1–3]. However, the lack of bioactivity and the potential adverse immune responses have dramatically limited their applications in vascular tissue engineering. Alternatively, highly biocompatible natural ECM biomolecules, including collagen [4, 5], fibrin [6, 7], fibronectin [8], elastin [5, 8], chondroitin sulfate [9], and hyaluronic acid [10], have been used to engineer TEVGs. Nevertheless, their applications are normally restricted by their poor mechanical properties [11]. To resolve the problem, these TEVGs can be toughened by introducing mechanically strong components. For example, PCL-PLA was blended into the collagen solution to improve the mechanical strength of electrospun nanofibers for TEVG fabrication, which has also manifested high biocompatibility in supporting human coronary artery EC growth [2].

Decellularized animal tissues from porcine blood vessels and small intestine submucosa are also attractive candidates for TEVG fabrication. Decellularization removes cell components, typically cell membrane and genetic materials, while retaining ECM components, thereby minimizing the potential immune responses. Since the physical and biochemical cues of the native ECM are largely preserved, a decellularized tissue normally demonstrates proper mechanical property and excellent biocompatibility that promote cell adhesion and remodeling. However, the quality of the decellularized tissue varies with its source and the decellularization procedure, which severely compromises its reproducibility.

Human mesenchymal stem cells (hMSCs) possess excellent antithrombogenic, immunoregulatory, and regenerative properties that are especially appealing for cardiovascular tissue engineering [12–15]. Moreover, autologous hMSCs offer extra opportunities for engineering patient-specific TEVGs. Our lab has previously developed a technology to fabricate an anisotropic hMSC-based TEVG which exhibits proper mechanical properties and highly organized smooth muscle cell (SMC)-like cell layers [16]. In this book chapter, a step-by-step protocol is provided to create a completely biological TEVG by utilizing aligned ECM nanofibers, hMSCs, and a rotating wall vessel (RWV) bioreactor (Fig. 1). Highly aligned human dermal fibroblast (hDF) sheets were decellularized to obtain aligned ECM nanofibrous scaffolds, which were

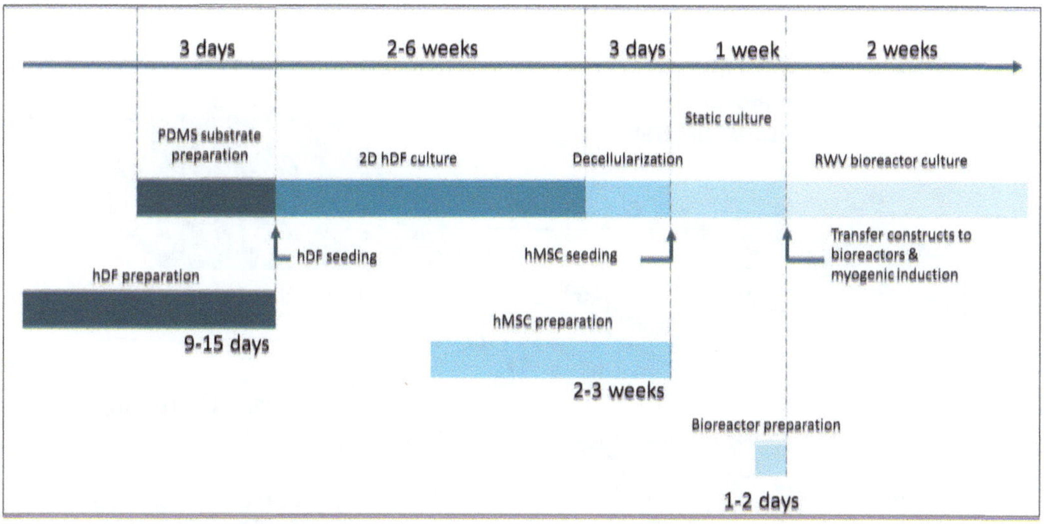

Fig. 1 Timeline of TEVG fabrication

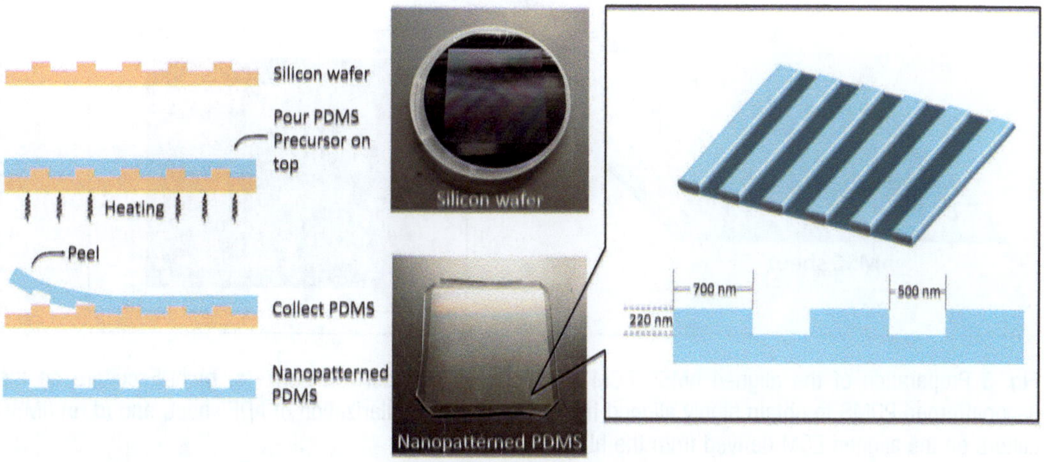

Fig. 2 Fabrication of aligned nanopatterned PDMS substrates using silicon wafer

then populated with hMSCs (Fig. 2). A 3D tubular hMSC assembly was obtained by rolling hMSC-ECM sheets around a temporal mandrel (Fig. 3). The maturation of this hMSC-ECM tube was achieved in a dynamic culture using a RVW bioreactor, where both chemical factors and physical stimuli were introduced for myogenic induction of hMSCs towards SMC-like cells (Fig. 4). The produced hMSC-based TEVG demonstrated anisotropic structural and mechanical properties that were very similar to that of native vessels (Fig. 5). Starting from the pre-made ECM, it only requires three weeks to fabricate a mature TEVG.

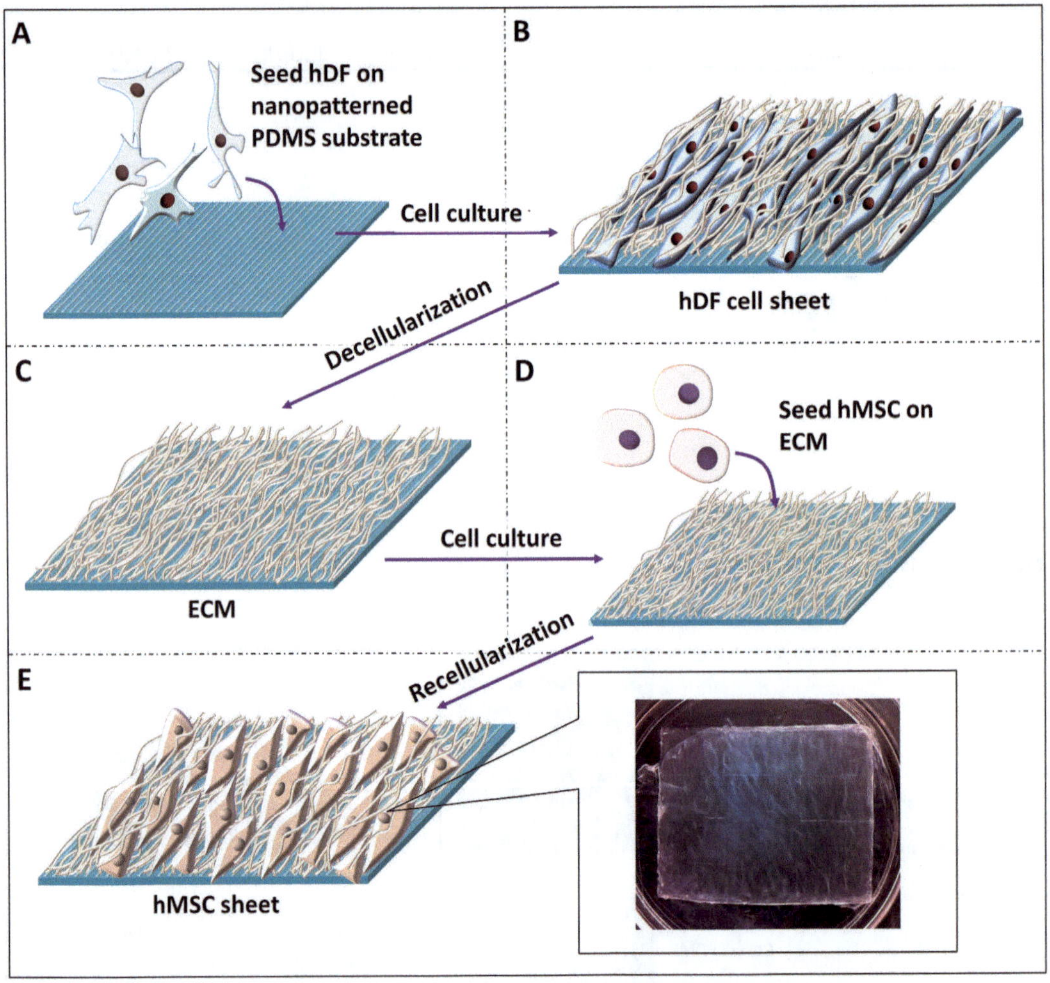

Fig. 3 Preparation of the aligned hMSC-ECM sheet on the hDF-derived ECM. (**a**, **b**) hDF culture on the nanopatterned PDMS to obtain highly aligned hDF sheet, (**c**) decellularization of hDF sheet, and (**d**, **e**) hMSC culture on the aligned ECM derived from the hDF sheet

2 Materials

2.1 Aligned Polydimethylsiloxane (PDMS) Substrate Preparation

1. PDMS precursor solutions: Sylgard™ 184 silicone elastomer kit containing PDMS base and curing agent.
2. Nanopatterned silicon wafer with aligned grooves (220 nm groove depth, 500 nm groove width, and 1200 nm period) (Fig. 2).
3. Vacuum pump, vacuum chamber, and hotplate.
4. Detergent: Micro-90® concentrated alkaline cleaning solution.
5. Dry etch system: Trion ICP/RIE Etch PHTII-4301.
6. 70% (v/v) ethanol solution prepared in deionized (DI) water.

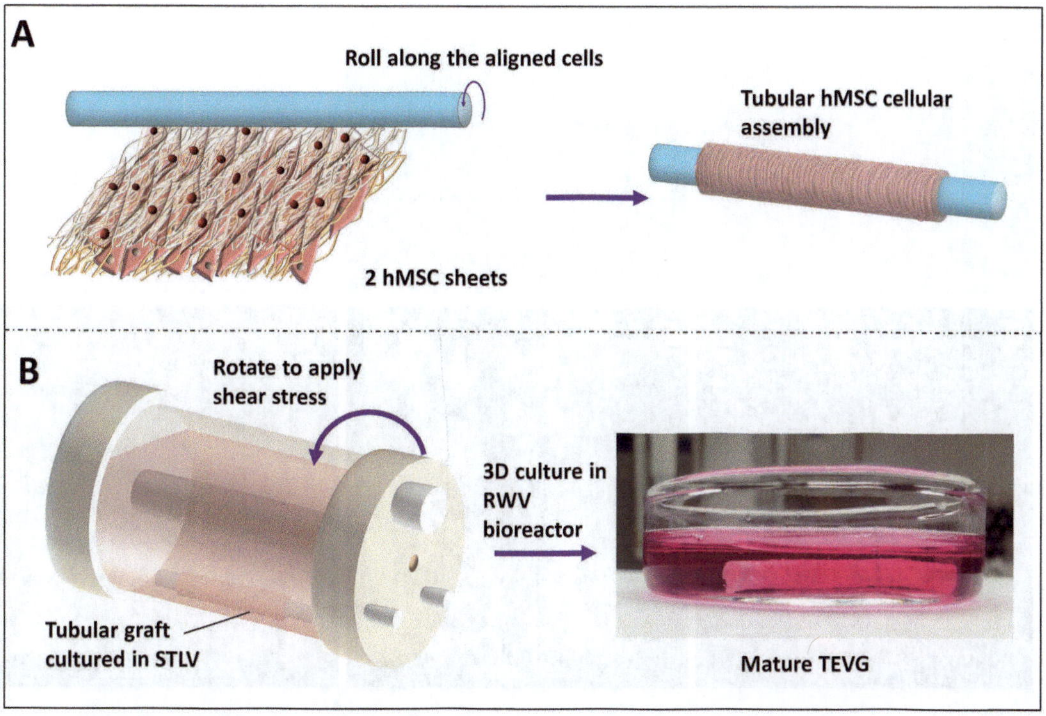

Fig. 4 Fabrication of TEVG. (**a**) Generation of the 3D construct from 2D hMSC-ECM sheets. (**b**) 3D culture of the tubular hMSC-ECM construct to generate a mature TEVG by using a RWV bioreactor

2.2 Cell Culture Reagents

1. Bovine collagen type I working solution*: 20μg/mL in 1 M hydrochloric acid (HCl) (*see* **Note 1**).
2. 0.25% trypsin-EDTA solution.
3. Fibroblast culture media$^{\Delta}$: Mix 600 mL Dulbecco's Modified Eagle's Medium (DMEM), 200 mL fetal bovine serum (FBS), 200 mL Ham F12 medium, and 10 mL penicillin/streptomycin (10,000 units penicillin and 10 mg streptomycin/mL) (*see* **Note 2**).
4. Fibroblast cell-sheet culture media$^{\Delta}$: Add 0.0991 g sodium ascorbate into 1 L fibroblast cell culture medium.
5. hMSC culture medium$^{\Delta}$: Mix 780 mL minimum essential medium α (α-MEM), 200 mL FBS, 10 mL L-glutamine, and 10 mL penicillin/streptomycin.
6. Smooth muscle myogenic induction medium$^{\Delta}$: Mix 790 mL α-MEM, 200 mL FBS, 10 mL penicillin/streptomycin, 1μg transforming growth factor beta 1 (TGF-β1), and 5.9 mg ascorbic acid.
7. Phosphate-buffered saline (PBS): Add 8 g of sodium chloride, 0.2 g of potassium chloride, 1.44 g of sodium phosphate dibasic, and 0.24 g of monopotassium phosphate into 1 L DI water. Adjust pH to 7.4 with 1 M HCl.

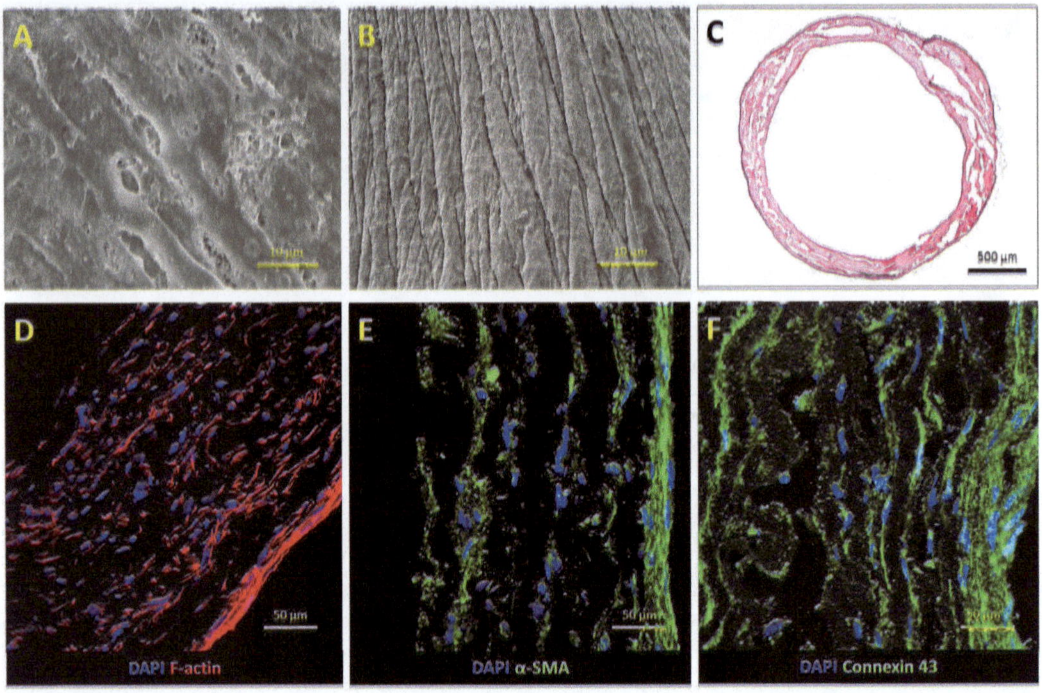

Fig. 5 Structure and differentiation characterization of TEVG. (**a, b**) SEM images showing the internal and external surfaces of the TEVG, respectively, (**c**) H&E staining image of the TEVG cross section, and (**d–f**) immunofluorescence images of the TEVG cross sections, showing the differentiation of hMSC towards SMC-like phenotype (**d, f**)

2.3 Reagents for Sample Preparation

1. Decellularization solution I: 1 M sodium chloride, 10 mM Tris, and 5 mM ethylenediaminetetraacetic acid (EDTA) in DI water.
2. Decellularization solution II: 0.1% (w/w) sodium dodecyl sulfate (SDS), 10 mM Tris, and 5 mM EDTA in DI water.
3. Equilibrium medium$^{\Delta}$: 80% (v/v) DMEM and 20% (v/v) FBS.
4. Glutaraldehyde solution*: 2.5% glutaraldehyde prepared in PBS.
5. Gradient ethanol solutions*: 50, 60, 70, 80, 90, and 100% (v/v) ethanol prepared in PBS.
6. Hexamethyldisilazane.
7. Optimum cutting temperature (OCT) compound.
8. Liquid nitrogen.
9. Formaldehyde working solution: 4% formaldehyde prepared in DI water.
10. Hematoxylin solution: Hematoxylin (Gill NO. 3) 6 g/L hematoxylin (*see* **Note 3**).

11. Acidified eosin Y solution: 0.5% (w/v) in acidified 90% ethanol (*see* **Note 3**).
12. Xylene.
13. Permount mounting medium.
14. Transparent nail polish.
15. Bovine serum albumin (BSA) blocking buffer*: 1% (w/v) BSA prepared in 0.2% Triton.
16. Primary antibody working solutions*: Dilute antibody (alpha smooth muscle actin (α-SMA), calponin, and connexin 43) stocks in the BSA blocking buffer using dilution ratios provided by manufacturers' product sheets, normally 1:200 (*see* **Note 4**).
17. Secondary antibody working solutions*: Dilute the corresponding fluorescent-dye conjugated antibody stocks in the BSA blocking buffer using dilution ratios provided by manufacturers' product sheets, normally 1:400 (*see* **Note 4**).
18. 4′,6-Diamidino-2-phenylindole (DAPI) working solution*: Dilute DAPI stock (5 mg/mL) in PBS using 1:1000 ratio.
19. Mounting medium: PermaFluor™ aqueous mounting medium.
20. mRNA extraction kit: RNeasy Mini Kit (Qiagen Germantown, MD).
21. cDNA preparation kit: RevertAid RT Reverse Transcription Kit (Thermo Fisher, Waltham WA).
22. Primer KiCqStart® SYBR® Green Primers (Sigma, St. Louis, MO).
23. SYBR™ Green PCR Master Mix (Thermo Fisher).

2.4 Other Equipment

1. RWV bioreactor system: Synthecon 3D culture system with the slow turning lateral vessels (STLV) (Synthecon Houston, TX).
2. CO_2 cell culture incubator.
3. Cressington 208HR high-resolution sputter coater.
4. Scanning electron microscope (SEM): Hitachi S-4700 field emission SEM.
5. Cryostat Microtome.
6. Optical microscopes: Olympus BX51-DP70.
7. Fluorescent microscope: Olympus FV-1000 confocal microscope.
8. StepOnePlus RT-PCR system.
9. Mechanical testing machine: 8872 Instron system.

3 Methods

3.1 Development of Aligned PDMS

1. Prepare PDMS Sylgard™ 184 (Dow Corning, Midland, MI) according to manufacturer's instruction. Thoroughly mix PDMS base and curing agent with a clean stirring rod using 10:1 (w/w) ratio.
2. Pour the PDMS precursor onto the nanopatterned silicon wafer (Fig. 2).
3. Transfer the PDMS precursor into a vacuum chamber to remove bubbles.
4. Cure PDMS by heating the precursor through a pre-leveled hotplate at 65 °C for 4 h.
5. Peel off the nanopatterned PDMS sheet from the wafer, thoroughly clean it with the detergent solution, and air-dry under the chemical fume hood (*see* **Note 5**).
6. Plasma etch the PDMS sheets with 200 mTorr oxygen plasma for 60 s at the radio-frequency power of 100 W by using Trion ICP/RIE Etch PHTII-4301 system.
7. Immerse plasma-etched PDMS sheets in 70% ethanol for 30 min immediately after plasma etching (*see* **Note 6**).
8. Remove ethanol, and air-dry the etched PDMS sheets in the biosafety cabinet.
9. Surface coat PDMS sheets with sterile 20μg/mL bovine collagen type I solution for 2 h.
10. Remove excess collagen solution by aspiration, and air-dry the nanopatterned PDMS sheets in the biosafety cabinet for at least 20 min (*see* **Note 7**).
11. Immerse collagen-coated PDMS sheets into fibroblast culture medium, and incubate overnight at 37 °C prior to hDF seeding (Fig. 2).

3.2 hDF Sheet Culture

1. Expand hDFs (Passage 3 to 5) using fibroblast culture medium in standard cell culture plates.
2. When cells reach 80–90% confluence, trypsinize and seed them onto the nanopatterned PDMS substrate with seeding density of 10,000 cells/cm^2 (Fig. 3a).
3. Culture hDFs on the PDMS using fibroblast culture medium for 1 week at 37 °C with 5% CO_2 supply till cells are confluent. Change media every 48 h.
4. After 1 week, switch to fibroblast sheet medium and continue culturing hDFs on the PDMS for another 5 weeks at 37 °C with 5% CO_2 supply to generate aligned hDF sheets. Change medium every 48 h (Fig. 3b).

3.3 hDF Sheet Decellularization

1. Remove the fibroblast sheet medium, and thoroughly wash the hDF sheets with PBS three times.
2. Immerse hDF sheets in the decellularization solution I, and incubate on the shaker for 1 h at room temperature (RT).
3. Wash hDF sheets using PBS for 5 min on the shaker. Repeat this step three times.
4. Immerse hDF sheets in decellularization solution II, and incubate for 10 min at RT with gentle shaking.
5. Wash hDF sheet-derived ECM using PBS for 5 min on the shaker. Repeat this step three times.
6. Incubate the ECM in the equilibrium medium for 48 h at RT with gentle shaking. Change the medium at 24 h point.
7. Rinse ECM thoroughly with PBS three times.
8. Store the hDF-ECM sheets at −80 °C until further use (*see* **Note 8**).
9. Immerse the hDF-ECM sheets into hMSC culture medium for 20 min at 37 °C prior to hMSC seeding (Fig. 3c).

3.4 hMSC-ECM Sheet Culture

1. Expand hMSCs (Passage 4 to 6) using hMSC culture medium in standard culture plates.
2. When hMSCs reach 80–90% confluence, trypsinize and seed them onto the hDF-ECM sheets at a density of 10,000 cells/ cm^2 (Fig. 3d).
3. Continue culturing hMSCs on the hDF-ECM sheets using hMSC culture medium for 1 week at 37 °C with 5% CO_2 supply. Change medium at 72 h of interval (Fig. 3e).

3.5 TEVG Assembly

1. Thoroughly clean and autoclave the 250 mL STLVs (*see* **Note 9**).
2. Set up the power supply and the rotary base. Preset the rotating speed to 20 rmp (*see* **Note 10**).
3. Assemble the autoclaved units in the biosafety cabinet. Add 200 mL smooth muscle myogenic induction medium into the STLV chamber to equilibrate the system. Tighten all parts of the STLV in the biosafety cabinet.
4. Move the STLV out from the biosafety cabinet, and mount it on to the rotary base in the 37 °C incubator. Turn on the power supply, and allow the STLV to rotate overnight (*see* **Note 11**).
5. Stack two hMSC-ECM sheets, and roll them carefully around a 2-mm-diameter and 9-cm-long mandrel to form the 3D tubular hMSC assembly (*see* **Note 12**) (Fig. 4a).
6. Transfer the 3D tubular hMSC assembly along with the mandrel into the STLV for 1-day static culture in the 37 °C

incubator with 5% CO_2 to allow for tight attachment between the cell layers.

7. Turn on the power supply, and make sure that the rotating speed is stabilized at 20 rpm (*see* **Note 13**).
8. Culture the 3D construct under shear stress for 13 days to form the hMSC-based TEVG. Change medium after 7 days (*see* **Note 14**).
9. Collect the resulting construct after 2 weeks by slowly and gently removing the TEVG from the mandrel (Fig. 4b).

3.6 ECM Alignment Examination by SEM

1. Cut the resulting TEVG into 0.4–0.5-mm-long samples.
2. Fix samples in 2.5% glutaraldehyde solution for 30 min at RT.
3. Wash samples with PBS for 5 min. Repeat this step three times.
4. Dehydrate glutaraldehyde-fixed samples in the gradient ethanol solutions in ascending order. Incubate samples in each solution for 5 min. Repeat the incubation step in 100% ethanol one more time.
5. Immerse samples in hexamethyldisilazane for 30 min at RT for final dehydration.
6. Open up samples using scissors along the longitudinal direction to expose both external and internal surfaces.
7. Mount samples onto SEM holders, and coat samples with 10-nm-thick platinum by using a sputter coater.
8. Observe the TEVG surfaces under SEM (*see* **Note 15**) (Fig. 5a, b).

3.7 H&E Staining

1. TEVG samples (0.4–0.5 mm long) are embedded in the OCT compound and frozen in liquid nitrogen for 2 min.
2. Set the working temperature of cryostat to −20 °C, and use the cryostat microtome to obtain 5–10-μm-thick tissue sections.
3. Thaw tissue sections on the histological slides.
4. Fix the samples in 4% formaldehyde for 30 min at RT.
5. Wash samples using PBS for 2 min. Repeat this step for three times.
6. Immerse tissue sections in hematoxylin solution Gill NO. 3 for 90 s.
7. Remove hematoxylin and wash the samples carefully with tap water until water is clear (*see* **Note 16**).
8. Immerse tissue sections in acidified eosin Y solution for 30 s.
9. Remove eosin solution and wash samples carefully with tap water (*see* **Note 16**).
10. Gently dip slides into 95% ethanol to remove the excessive eosin dye (*see* **Note 17**).

11. Dehydrate the samples by slowly dipping the slides into 100% ethanol twice.
12. Dip samples in xylene in the chemical hood.
13. Mount samples using Permount and coverslips. Seal edges of the coverslips with transparent nail polish, and observe under optical microscope (*see* **Note 18**) (Fig. 5c).

3.8 Differentiation Examination

1. Fix tissue sections using 4% formaldehyde for 30 min.
2. Wash samples with PBS for 5 min at RT. Repeat this step three times.
3. Add BSA blocking buffer, and incubate for 30 min at RT.
4. Remove blocking buffer, and add working solutions of primary antibodies, i.e., α-SMA, calponin, and connexin 43. Incubate samples for 1 h at RT or overnight at 4 °C.
5. Remove excess antibody working solutions, and wash the samples using the blocking buffer for 5 min. Repeat this step three times.
6. Add working solutions of secondary antibodies. Incubate samples for 1 h in the dark at RT.
7. Wash samples using the BSA blocking buffer for 5 min. Repeat this step three times.
8. Add 4′,6-diamidino-2-phenylindole (DAPI, Sigma-Aldrich, St. Louis, MO) solution, and incubate for 10 min at RT.
9. Wash samples thoroughly with PBS for 5 min. Repeat this step three times.
10. Mount samples using mounting medium and coverslips. Seal edges of the coverslips using transparent nail polish, and observe under confocal microscope (*see* **Note 19**) (Fig. 5d–f).
11. Extract and purify mRNA using mRNA extraction kit.
12. Reverse transcribe extracted mRNA into cDNA using cDNA preparation kit.
13. Perform real-time qPCR using customized KiCqStart® SYBR® Green Primers, SYBR™ Green PCR Master Mix, and StepOnePlus RT-PCR system (*see* **Note 20**).

3.9 Mechanical Strength Test

1. Cut open a 12-mm-long TEVG segment along the longitudinal direction, so that rectangular strips can be obtained.
2. Trim the samples into 6 mm by 12 mm strips (circumferential by longitudinal directions).
3. Gently wrap specimens by sandpapers, and carefully mount onto the sample grips.
4. Set pulling force to 0.5 N/min, and maintain until sample failure. Record the load and displacement by using an 8872 Instron system.

5. The thickness of the specimen is obtained from measuring the thickness of H&E-stained TEVG samples.
6. Calculate ultimate tensile strength (UTS), ultimate strain, and the elastic modulus of TEVG in both longitudinal and circumferential directions (*see* **Note 21**).

4 Notes

1. * indicates the reagents that need to be prepared freshly before use.
2. Δ indicates the reagents that can be prepared ahead of time and stored in the 4 °C refrigerator until use. Reagents should be warmed up to RT before use. The recommended maximum storage time is 1 month.
3. Hematoxylin and eosin solutions purchased from different suppliers can be different. Check the specific product information for optimal staining protocols.
4. Antibody's properties can vary significantly. Make sure to check specific product information for optimal protocols before use.
5. The prepared PDMS sheets can be cut into desired dimensions (e.g., 6 cm × 5 cm), and the direction of the alignment should be marked for ease of use for the following experiments.
6. This step is to sterilize and prevent the plasma-etched PDMS surface from being oxidized. Starting from this step, following procedures should be performed in the sterile environment until the TEVG is matured, and all the reagents should be sterilized by 0.2μm filtration before use.
7. Surface treatment by using plasma etching and collagen solution is to enhance cell adhesion. Cells normally do not attach to a pristine PDMS surface. If the plasma etching machine is not available, other surface modification methods that achieve good cell attachment can also be used [17].
8. Perform the entire decellularization process very gently in order to prevent the detachment of ECM sheet from the PDMS substrate as hMSCs need be seeded and cultured on a flatten ECM sheet.
9. The RWV bioreactor system may be prepared one or two days before the end of hMSC-ECM sheet culture depending on specific devices chosen. For example, when the autoclavable Synthecon 3D STLV system is used, cleaning, autoclave, and leak testing are recommended to start 2 days before the 3D culture. If other types of RWV bioreactor are used, preparation protocols should be optimized according to manufacturer's guidance.

10. In our design, the rotary base is set inside a 37 °C 5% CO_2 incubator, and the power supply is hooked up outside for the ease of control.
11. This step is critical to equilibrating the bioreactor chamber, testing the experimental setup, and ensuring there is no leakage.
12. The diameter and length of the mandrel are adjustable, to customize the diameter and length of the TEVG.
13. Rotating speed can be adjusted to provide a specific shear stress depending on specific requirements. If the same RVW bioreactor is used, refer to [18] for shear force calculation.
14. Air present in the incubator can directly supply oxygen and CO_2 to the culture medium through the gas exchanging membrane at the center of the STLV. Therefore, no more gas line is needed in our system. However, if oxygen concentration is a factor, gas lines can be connected to supply specific types of gas.
15. In this study, SEM images were taken by Hitachi S-4700 field emission SEM (Fig. 5a, b). The results show hMSCs and the ECM remained aligned on internal and external surfaces, respectively.
16. To prevent tissue sections from being washed away, one alternative method would be slowly dipping the slides into water repetitively. However, this method may take longer time.
17. It is recommended to examine the red stain under the microscope after each dipping to find out when to stop washing. This procedure can be used to generate more consistent staining results.
18. H&E staining results are shown in Fig. 5c. Samples were observed under Olympus BX51-DP70 microscope.
19. Fluorescent images are shown in Fig. 5d–f. Images were taken by using Olympus FV-1000 confocal microscope. Induced by shear stress, F-actin signal was slightly higher on the outermost surface of TEVG (Fig. 5d). Myogenic differentiation of hMSCs was verified by the presence of SMC markers, i.e., α-SMA and connexin 43 (Fig. 5e, f).
20. Refer to [16] for detailed results where both α-SMA and calponin expressions were enhanced by shear stress, further indicating shear stress enhanced hMSC differentiation towards SMC-like cells in this system.
21. Tensile tests can be performed based on the standard test protocol ASTM D638. Refer to [16] for detailed results where UTS, ultimate strain, and the elastic modulus of TEVG were slightly smaller but within the same order of magnitude as those of the native blood vessels.

Acknowledgments

This study was supported by the National Institutes of Health (R01HL146652 and R15HL145654) and the National Science Foundation (2106048) to F.Z.

References

1. Williamson MR, Black R, Kielty C (2006) PCL–PU composite vascular scaffold production for vascular tissue engineering: attachment, proliferation and bioactivity of human vascular endothelial cells. Biomaterials 27 (19):3608–3616
2. He W et al (2005) Fabrication and endothelialization of collagen-blended biodegradable polymer nanofibers: potential vascular graft for blood vessel tissue engineering. Tissue Eng 11(9–10):1574–1588
3. Vaz C et al (2005) Design of scaffolds for blood vessel tissue engineering using a multilayering electrospinning technique. Acta Biomater 1(5):575–582
4. Weinberg CB, Bell E (1986) A blood vessel model constructed from collagen and cultured vascular cells. Science 231(4736):397–400
5. Buijtenhuijs P et al (2004) Tissue engineering of blood vessels: characterization of smooth-muscle cells for culturing on collagen-and-elastin-based scaffolds. Biotechnol Appl Biochem 39(2):141–149
6. Grassl E, Oegema T, Tranquillo RT (2002) Fibrin as an alternative biopolymer to type-I collagen for the fabrication of a media equivalent. J Biomed Mater Res 60(4):607–612
7. Swartz DD, Russell JA, Andreadis ST (2005) Engineering of fibrin-based functional and implantable small-diameter blood vessels. Am J Phys Heart Circ Phys 288(3):H1451–H1460
8. Heilshorn SC et al (2003) Endothelial cell adhesion to the fibronectin CS5 domain in artificial extracellular matrix proteins. Biomaterials 24(23):4245–4252
9. Thalla PK et al (2014) Chondroitin sulfate coatings display low platelet but high endothelial cell adhesive properties favorable for vascular implants. Biomacromolecules 15 (7):2512–2520
10. Zavan B et al (2008) Neoarteries grown in vivo using a tissue-engineered hyaluronan-based scaffold. FASEB J 22(8):2853–2861
11. Pashneh-Tala S, MacNeil S, Claeyssens F (2016) The tissue-engineered vascular graft—past, present, and future. Tissue Eng Part B Rev 22(1):68–100
12. Wang Y et al (2013) Long-term cultured mesenchymal stem cells frequently develop genomic mutations but do not undergo malignant transformation. Cell Death Dis 4(12): e950–e950
13. Kim HJ, Park J-S (2017) Usage of human mesenchymal stem cells in cell-based therapy: advantages and disadvantages. Dev Reprod 21 (1):1
14. Fu Y et al (2017) Trophic effects of mesenchymal stem cells in tissue regeneration. Tissue Eng Part B Rev 23(6):515–528
15. Williams C et al (2011) Stacking of aligned cell sheets for layer-by-layer control of complex tissue structure. Biomaterials 32(24):5625–5632
16. Xing Q et al (2017) Aligned nanofibrous cell-derived extracellular matrix for anisotropic vascular graft construction. Adv Healthc Mater 6 (10):1601333
17. Sharma D et al (2019) Polydopamine and collagen coated micro-grated polydimethylsiloxane for human mesenchymal stem cell culture. Bioactive Materials 4:142–150
18. Hammond T, Hammond J (2001) Optimized suspension culture: the rotating-wall vessel. Am J Physiol Renal Physiol 281(1):F12–F25

Chapter 10

Engineering Vascular Grafts with Multiphase Structures

Bryan D. James and Josephine B. Allen

Abstract

Blood vessels in the body are multiphasic organs with microenvironmental niches specific to the cells that inhabit each section. Electrospinning is a fabrication technique used to produce nano- to microfibrous architectures capable of mimicking native extracellular matrix structure. Likewise, polycitrate elastomers are favorable luminal materials for vascular applications because of their hemocompatibility and mechanical properties. Here we describe the procedure for fabricating a biphasic polycitrate elastomer, collagen, and elastin electrospun composite to spatially tailor both composition and architecture for recapitulating the intimal and medial layers of the blood vessel in a vascular graft.

Key words Vascular graft, Electrospinning, Biphasic, Polycitrate elastomers, Tissue engineering, Collagen, Elastin, Hemocompatible, Blood vessel

1 Introduction

One of the namesake goals of tissue engineering is recapitulating the native structure of tissue in an engineered scaffold to provide the necessary cues for stimulating healthy tissue formation. Engineering vascular graft scaffolds remains challenging because these tissues require a hemocompatible lumen, matched mechanical properties with the surrounding native vessel, and a fibrous, multiphasic architecture to mirror the structural and compositional complexity of the blood vessel consisting of the intimal, medial, and adventitial layers [1, 2]. One approach is to use electrospinning, which is a fabrication strategy capable of forming tubular fibrous structures [3]. The process works by extruding a polymer-laden solution through a needle across a high voltage potential onto a rotating mandrel. The needle and mandrel are connected to form an open circuit. By tuning several processing parameters, the polymer solution can be pulled across from the needle tip to the mandrel target resulting in polymer fibers collecting at the target. Several processing parameters that govern the organization and size

Feng Zhao and Kam W. Leong (eds.), *Vascular Tissue Engineering: Methods and Protocols*,
Methods in Molecular Biology, vol. 2375, https://doi.org/10.1007/978-1-0716-1708-3_10,

of the electrospun fibers include solution viscosity, working distance between the needle and the target, voltage potential, solvent dielectric and volatile properties, mandrel rotation rate, flow rate, and temperature [3]. Additionally, by using polycitrate elastomers the hemocompatibility and mechanical property requirements for a vascular graft can be addressed [4–7]. Specifically, poly (1,8-octanediol citrate) (POC) has an elastic modulus similar to that of native veins and arteries as well as being both biocompatible and hemocompatible [4–7]. By combining electrospinning and polycitrate elastomers, we are able to generate extracellular matrix topographies and sharp interfaces to spatially control graft structure for recapitulating native vessel architecture [8].

Herein we describe the procedure for fabricating a tunable biphasic vascular graft with a microfibrous embossed poly (1,8-octanediol citrate) intimal layer and an electrospun microfibrous collagen, POC, elastin medial layer.

2 Materials

Where indicated, prepare all solutions using ultrapure water (prepared by purifying deionized water, to attain a sensitivity of 18 MΩ-cm at 25 °C) and high-quality grade reagents. Prepare and store all reagents at room temperature (unless indicated otherwise). Always wear the proper personal protective equipment specified in the safety data sheets for each reagent, and follow all local waste disposal regulations when disposing waste materials.

2.1 Required Reagents

1. Citric acid powder, stored at room temperature.
2. 1,8-Octanediol powder, stored at room temperature.
3. Ethanol, 200 proof, stored at room temperature in a flammables cabinet.
4. Hexafluoroisopropanol (HFIP) liquid, stored at room temperature in a flammables cabinet.
5. Lyophilized type I collagen, bovine, stored at 4 °C, Elastin Products Company, Inc.
6. Lyophilized elastin, bovine, stored at 4 °C, Elastin Products Company, Inc.
7. 5% (w/v) polyethylene oxide (PEO) solution in water, stored at room temperature.
8. Glutaraldehyde (GTA) liquid, stored at 4 °C.
9. Triton X-100 detergent, stored at room temperature.

2.2 Electrospinning Equipment

1. Tunable high voltage power supply.
2. Anodized multipurpose 6061 aluminum rod with a 3/16″ diameter.
3. Drill.
4. Nonconductive box for electrospinning.
5. Cable with alligator clips.
6. Programmable syringe pump.
7. 10 mL polyethylene syringe with female luer fitting.
8. 25-gauge blunted syringe needle with male luer fitting.

2.3 Mechanical Characterization Equipment

1. Digital calipers.
2. Razor blade.
3. Uniaxial mechanical load frame, Instron 5942.
4. Two stainless steel hooks compatible with the uniaxial mechanical load frame.

3 Methods

Carry out all procedures at ambient conditions on the benchtop unless otherwise specified (Fig. 1).

3.1 Synthesis of POC Pre-polymer

1. Add 0.5 moles of 1,8-octanediol powder and 0.5 moles of citric acid powder to a 250 mL three-neck round-bottom flask with inlet and outlet adapters to support the flow of nitrogen gas (*see* **Note 1**).
2. Heat the mixture at 165 °C using a hot oil bath under magnetic stirring and light nitrogen gas flow to melt the 1,8-octanediol and citric acid powders.

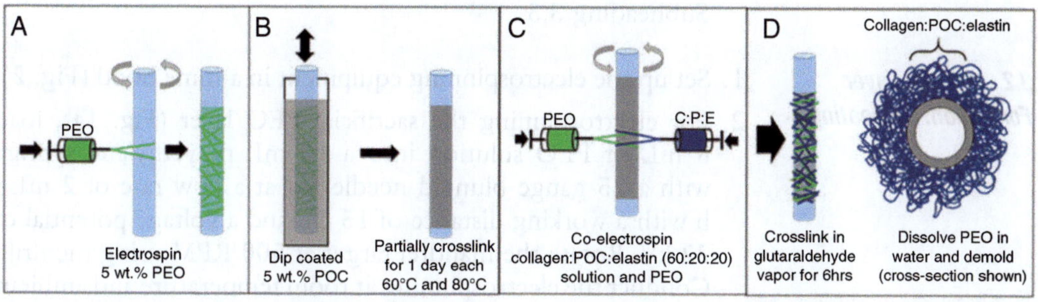

Fig. 1 Process schematic of bilayer graft fabrication. (**a**) Electrospinning of sacrificial PEO layer. (**b**) Dipcoating and cross-linking of POC intimal layer. (**c**) Co-electrospinning of PEO and collagen/POC/elastin medial layer. (**d**) Chemical cross-linking and collection of the fabricated bilayer scaffold. Reproduced from ref. 8 with permission from John Wiley & Sons

3. Upon complete melting of the constituents, reduce the temperature to 140 °C for 30 min to form the POC pre-polymer (*see* **Note 2**).
4. Quench the POC pre-polymer reaction by immersing the reaction vessel in an ice bath.
5. Dilute the cooled POC pre-polymer with a volume of ethanol three times the weight of the combined reactants used in **step 1** (*see* **Note 3**).
6. Agitate the vessel on an orbital shaker overnight to facilitate complete dissolution of the POC pre-polymer.
7. Dropwise add the POC pre-polymer in ethanol solution to a volume of water three times that of the POC pre-polymer in ethanol solution to purify the POC pre-polymer by recrystallization (*see* **Note 4**).
8. Gently decant the volume of liquid above the precipitated POC pre-polymer.
9. Repeat the recrystallization purification step by re-dissolving the precipitated POC pre-polymer in ethanol and adding the solution dropwise to a volume of water three times that of the POC pre-polymer in ethanol solution. Mass the dry, empty container used for precipitating the POC pre-polymer before starting.
10. Remove the volume of liquid above the reprecipitated POC pre-polymer.
11. Freeze dry the purified POC pre-polymer to remove any remaining water and ethanol.
12. Mass the dry purified POC pre-polymer.
13. Dissolve the dry purified POC pre-polymer in ethanol to form a 50% (w/v) solution for use in Subheading 3.2.
14. Dissolve a separate amount of the dry purified POC pre-polymer in HFIP to form a 10% (w/v) solution for use in Subheading 3.3.

3.2 Intimal Layer Formation: Dipcoating

1. Set up the electrospinning equipment in a fume hood (Fig. 2).
2. For electrospinning the sacrificial PEO layer (Fig. 1a), load 8 mL of PEO solution into a 10 mL polyethylene syringe with a 25-gauge blunted needle fed at a flow rate of 2 mL/h with a working distance of 15 cm and a voltage potential of 22 kV. Rotate the mandrel target at 500 RPM using the drill. Conduct the electrospinning at room temperature and ambient humidity conditions (*see* **Note 5**).
3. Dip the PEO electrospun covered mandrel once in a 50% (w/v) solution of POC pre-polymer in ethanol (Fig. 1b) (*see* **Note 6**).

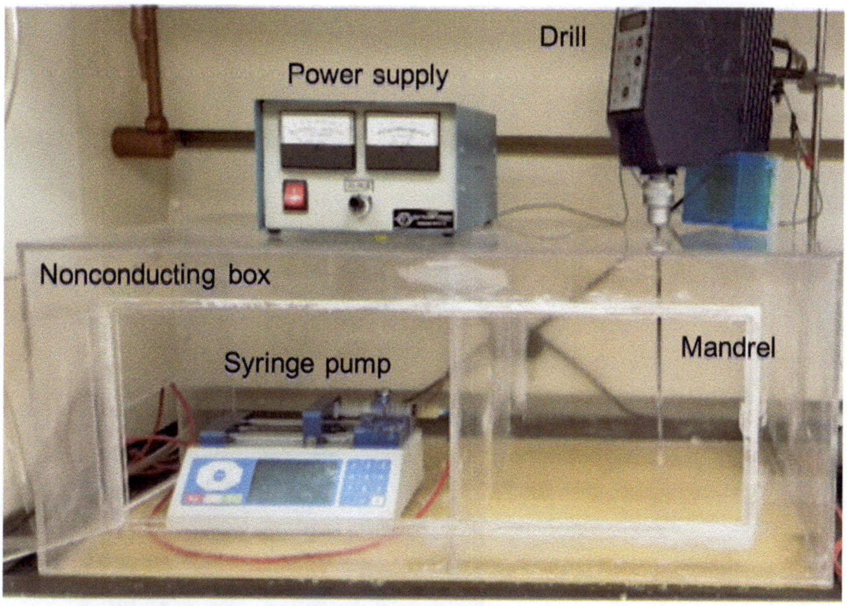

Fig. 2 Example electrospinning setup in the fume hood

4. Place the mandrel standing upright in an oven at 60 °C for 1 day and then at 80 °C for an additional 1 day to cross-link the POC layer (Fig. 1b) (*see* **Note 7**).

3.3 Medial Layer Formation: Electrospinning

1. Prepare a 60:20:20% (w/w) solution of collagen/POC/elastin in HFIP. To do this, first dissolve the POC pre-polymer in HFIP to form a 10% (w/v) solution. Then dissolve the lyophilized collagen and lyophilized elastin in the POC pre-polymer in HFIP solution.
2. Set up the electrospinning equipment in a fume hood.
3. For co-electrospinning PEO and the collagen/POC/elastin solution, one syringe is aimed from the left and the other from the right (Fig. 1c). Load 8 mL of PEO solution into a 10 mL polyethylene syringe with a 25-gauge blunted needle fed at a flow rate of 2 mL/h with a working distance of 15 cm and a voltage potential of 22 kV. Rotate the mandrel target at 500 RPM using the drill. For the other syringe, load 5 mL of collagen/POC/elastin solution into a 10 mL polyethylene syringe with a 25-gauge blunted needle fed at a flow rate of 1.25 mL/h with a working distance of 15 cm and a voltage potential of 22 kV.
4. In a fume hood, chemically cross-link the electrospun scaffold in GTA vapor for 6 h by suspending the mandrel ~1 cm above a volume of 20 mL GTA in a 100 mL glass graduated cylinder capped with paraffin wax (Figs. 1d and 3).

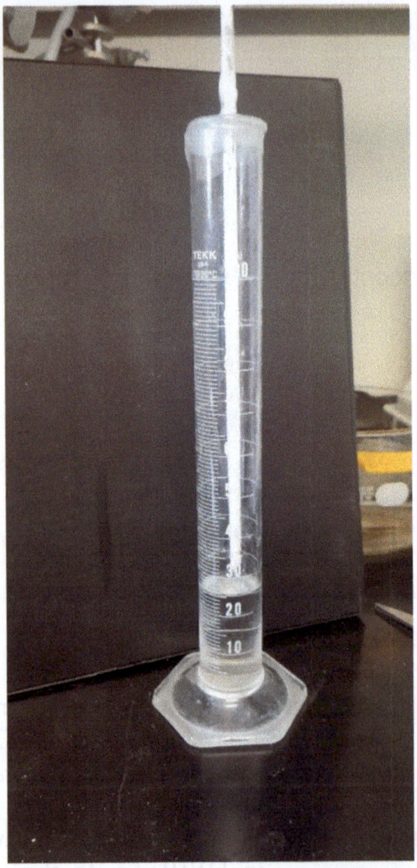

Fig. 3 Setup for chemically cross-linking of the scaffold with GTA

5. Place the mandrel after cross-linking in 1 L of water for 2 days to dissolve the PEO sacrificial layer and the co-spun PEO fibers (Fig. 1d).
6. Remove the mandrel using a small amount of Triton X-100 (Fig. 4) (*see* **Note 8**).

3.4 Mechanical Characterization of the Graft

1. Cut grafts into 5-mm-width ringlets using a razor blade.
2. Hydrate the ringlets for 1 h in water.
3. Measure the ringlet wall thickness (t), ringlet width (l), and ringlet outer diameter (d) with the calipers each five times at separate positions around the ringlet. Use the median value for each parameter in later calculations.
4. Secure the two stainless steel hooks in the upper and lower crossheads of the uniaxial mechanical load frame. Then load a ringlet by hanging it on the upper hook and lowering the crosshead of the load frame until the ringlet is held loosely by both the upper and lower hooks of the uniaxial mechanical load frame (Fig. 5a).

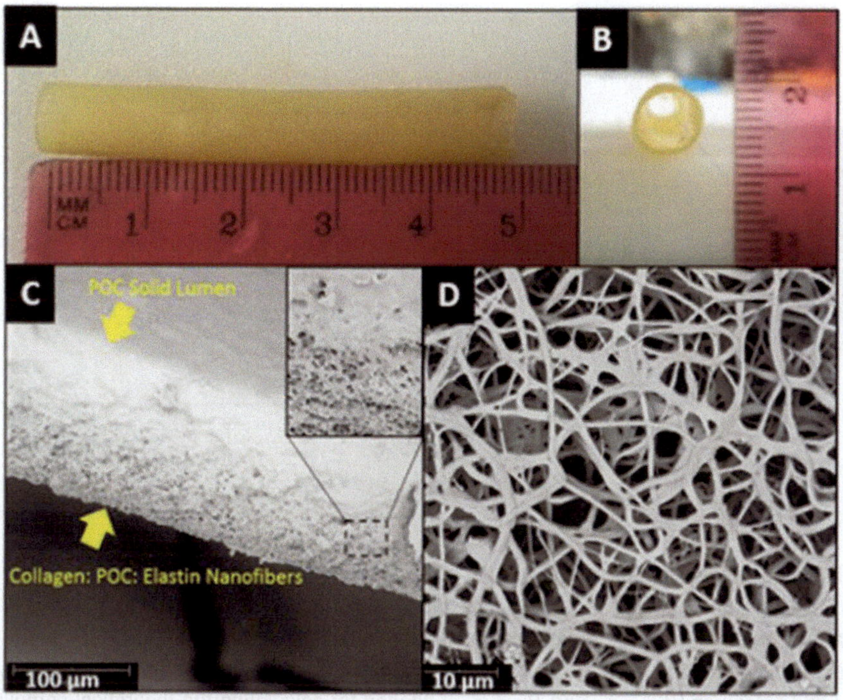

Fig. 4 Digital images of the scaffold in the transverse (**a**) and the cross section (**b**). SEM images of a cross section of the biphasic scaffold taken at 760× (**c**), fibers of the abluminal collagen/POC/elastin layer taken at 5000× (**d**). Reproduced from ref. 8 with permission from John Wiley & Sons

5. Test the ringlet in tension until failure using a stroke rate of 5 mm/min as determined for rigid and semi-rigid tubes from ASTM D638-14. Collect load (F) and displacement (e) data from the tensile test. Test at least five ringlets.
6. Convert the load and displacement data from each tensile test to stress (σ) and strain (ε) using the following equations.

$$\sigma = \frac{F}{A_0} \quad (1)$$

$$A_0 = lt \quad (2)$$

$$\varepsilon = \frac{200e}{2\pi d} \quad (3)$$

7. Plot stress versus strain, and calculate Young's modulus (E) for the ringlet by fitting a straight line to the linear portion of the stress versus strain curve. Young's modulus is equal to $E = \frac{\sigma}{\varepsilon}$ (4) in the linear portion (Fig. 5b). This value is Young's modulus for the graft.

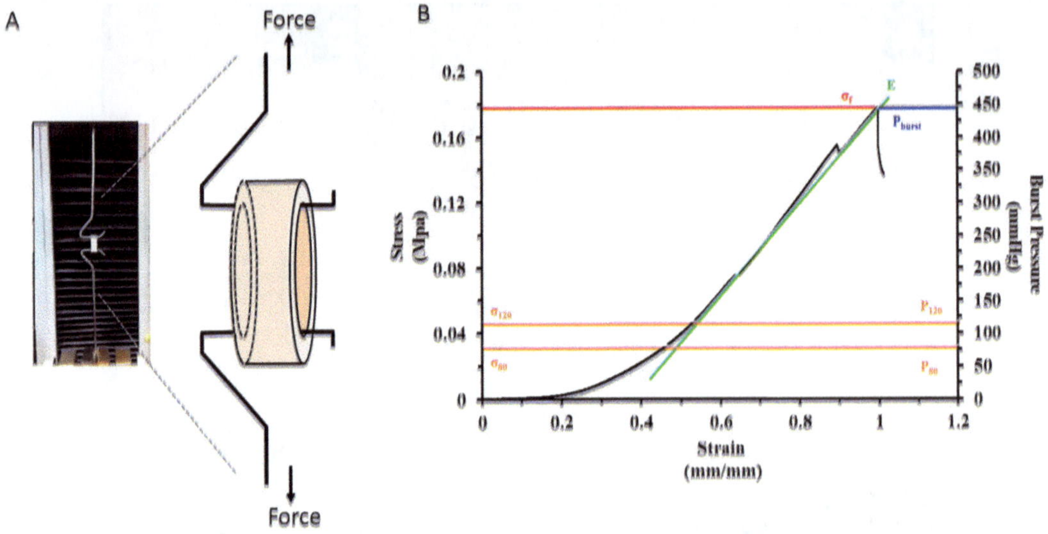

Fig. 5 Digital image and schematic of the sample setup for the ringlet or hoop mechanical testing of the graft (**a**) and an example uniaxial tensile testing data for a graft ringlet (**b**). The plot in **b** shows the different mechanical properties that can be determined including Young's modulus of the graft (E), the failure stress (σ_f), the extrapolated burst pressure of the graft (P_{burst}), and the extrapolated stresses ($\sigma_{80\ mmHg}$ and $\sigma_{120\ mmHg}$) and pressures ($P_{80\ mmHg}$ and $P_{120\ mmHg}$) needed for calculating the theoretical compliance of the graft

8. Extrapolate the burst pressure (P_{burst}) for the graft from the measured failure stress (σ_f) of the ringlet using Eq. 5 (Fig. 5b).

$$P_{burst} = \frac{2\sigma_f t}{d} \tag{5}$$

9. Replot the tensile test data with two vertical axes, one being stress (σ) and the other being pressure (P), and the horizontal axis being strain (ε) (Fig. 5b). Extrapolate the theoretical compliance ($C_{theoretical}$) for the graft by extrapolating the volume change of the graft underloading of 80 mmHg and 120 mmHg using the measured stress supported by the ringlet at each of those pressures (σ_{P_x}) according to the following equations:

$$d_{P_x} = \frac{2\sigma_{P_x} t}{d} \tag{6}$$

$$V_{P_x} = \pi l \left(\frac{d_{P_x}}{2}\right)^2 \tag{7}$$

$$C_{theoretical} = \frac{V_{120} - V_{80}}{P_{120} - P_{80}} \times 100\% \tag{8}$$

4 Notes

1. Adjust both the reactant masses and reaction vessel size as appropriate for the amount of POC pre-polymer being synthesized.
2. Successful progression of the polymer synthesis can be confirmed via a qualitative increase in solution viscosity.
3. POC pre-polymer is soluble in ethanol and insoluble in water.
4. The reactants (1,8-octanediol and citric acid) are soluble in water.
5. Maintaining minimal humidity is necessary for achieving a stable electrospun PEO layer. High humidity will cause the PEO to flake off the mandrel because it is soluble in water. If flaking occurs at the laboratory's ambient conditions, flowing dry air through the nonconducting electrospinning box can help to control humidity.
6. The POC pre-polymer will flow down the mandrel giving a slight non-uniform thickness along the length of the mandrel. To mitigate this, the mandrel can be placed upright and inverted every 5 min while the ethanol evaporates to give a more uniform thickness along the length of the mandrel. At this point, the experimenter can proceed to the next step. If the experimenter desires a thicker lumen layer, the mandrel may be repeatedly dipped in the POC pre-polymer following evaporation of the solvent before proceeding to the next step.
7. These heating conditions will partially cross-link the POC pre-polymer. This is necessary to ensure sufficient lamination between the POC intimal layer and the electrospun collagen/POC/elastin medial layer. Complete cross-linking by heating at elevated temperatures (120 °C) or for longer times (3–4 days) will reduce the tackiness of the POC leading to delamination between the intimal and medial layers. Under cross-linking will lead to the POC intimal layer being unable to maintain its shape as well as being unable to be removed from the mandrel intact.
8. Triton X-100 is used as a surfactant to assist in demolding the scaffold from the mandrel. Applying ethanol will also work.

Acknowledgments

This work was funded by a US Army Medical Research and Material Command Grant (W81XWH-15-1-0066) awarded to J.A. Opinions, interpretations, conclusions, and recommendations in this work are those of the authors and are not necessarily endorsed by the Department of Defense.

References

1. Goins A, Webb AR, Allen JB (2019) Multi-layer approaches to scaffold-based small diameter vessel engineering: a review. Mater Sci Eng C 97:896–912
2. James BD, Allen JB (2018) Vascular endothelial cell behavior in complex mechanical microenvironments. ACS Biomater Sci Eng 4:3818–3842
3. Kny E, Ghosal K, Thomas S (eds) (2018) Electrospinning. Royal Society of Chemistry, Cambridge
4. Yang J, Webb AR, Pickerill SJ, Hageman G, Ameer GA (2006) Synthesis and evaluation of poly(diol citrate) biodegradable elastomers. Biomaterials 27:1889–1898
5. Yang J, Webb AR, Ameer GA (2004) Novel citric acid-based biodegradable elastomers for tissue engineering. Adv Mater 16:511–516
6. Motlagh D et al (2007) Hemocompatibility evaluation of poly(diol citrate) in vitro for vascular tissue engineering. J Biomed Mater Res Part A 82A:907–916
7. Yang J, Motlagh D, Webb AR, Ameer GA (2005) Novel biphasic elastomeric scaffold for small-diameter blood vessel tissue engineering. Tissue Eng 11:1876–1886
8. Goins A, Ramaswamy V, Lichlyter D, Webb AR, Allen JB (2018) Fabrication of a bilayer scaffold for small diameter vascular applications. J Biomed Mater Res Part A 106:2850–2862

Chapter 11

Fabrication of Small-Diameter Tubular Grafts for Vascular Tissue Engineering Applications Using Mulberry and Non-mulberry Silk Proteins

Prerak Gupta and Biman B. Mandal

Abstract

Silk fibroin (SF) is a natural well-known biomaterial that has widely been explored for various tissue engineering applications with great success. Herein, we describe the methodology for fabricating two different types of tubular silk scaffolds aimed for vascular grafting. The first method emphasizes the use of very thin (10–15μm) silk films with unidirectional longitudinal micro-patterns, followed by their sequential rolling, which results in a multilayered tubular graft mimicking native-like cellular composition. The second method describes the fabrication of a bi-layered tubular scaffold comprising of a highly porous inner layer covered with an outer nanofibrous electrospun layer.

Key words Vascular grafts, Tissue engineering, Mulberry and non-mulberry silk, Biomaterials, Patterned films, Bi-layered tubular scaffolds

1 Introduction

Cardiovascular diseases remain the primary cause of death globally, with coronary artery occlusion contributing utmost [1]. Currently, the most reliable treatment option includes the replacement of a diseased or occluded blood vessel with a healthy counterpart via surgical intervention, which necessitates the availability of a suitable graft. For this purpose, autografts are the preferred choice for clinicians; however, their implementation is sometimes restricted because of their limited availability due to secondary diseased conditions and prior harvesting. In addition, obtaining an autograft causes donor site morbidity [2]. Vascular grafts made of synthetic polymers like polytetrafluoroethylene (PTFE) are a suitable choice for replacing larger-diameter (>6 mm) vessels, but due to compliance mismatch, they are not suited for smaller vessels in high-pressure regions [3]. In this scenario, the majority of research is now focused on creating vascular grafts using biodegradable

Feng Zhao and Kam W. Leong (eds.), *Vascular Tissue Engineering: Methods and Protocols*,
Methods in Molecular Biology, vol. 2375, https://doi.org/10.1007/978-1-0716-1708-3_11,

polymers (natural and synthetic), which have a better chance to integrate with the native tissue leading to neo-vessel formation and vascularization [4].

Among other biomaterials, silk, a natural structural protein-polymer obtained from silkworms, has shown promising aspects towards various tissue engineering applications [5, 6]. In general, silkworms are classified into two categories, mulberry (e.g., *Bombyx mori*) and non-mulberry (e.g., *Antheraea assama* or Indian Muga silk and *Philosamia ricini* or Eri silk), based on their feeding habits. The aqueous silk fibroin (SF) can be obtained from cocoons and glands, which could further be fabricated into myriad formats, e.g., micro-patterned thin films, porous scaffolds, nanofibers, hydrogels, etc. [7]. In addition, ideal properties of SF protein for tissue engineering applications include biocompatibility, tunable biodegradability, and excellent mechanical strength. We have reviewed different silk varieties and their properties in our previous reports [5, 6, 8]. Vascular grafting with silk fibroin-based tubular scaffolds have shown promising outcomes using both native (silk yarns) [9] and regenerated forms [10]; however, the majority of these studies implemented mulberry *Bombyx mori* silk.

In this chapter, we describe two methodologies for the fabrication of small-diameter tissue-engineered vascular grafts using mulberry (*Bombyx mori*) and Indian endemic non-mulberry (*Antheraea assama* and *Philosamia ricini*) silk. The first method implements micro-patterned silk films aimed to induce the native-like cellular alignment [11], whereas the second method uses a molding/freeze-drying approach, creating a porous tubular scaffold, followed by coating with electrospun nanofibrous layer resulting in a bi-layered graft [12].

2 Materials

Use ultrapure water (sensitivity 18 MΩ-cm at 25 °C) to prepare all the solutions. Take appropriate safety measurements while handling hazardous chemicals. All the reagents are procured from Sigma-Aldrich, USA, unless otherwise noted.

2.1 Silk Fibroin (SF) Source

1. *Bombyx mori* (BM) cocoons (mulberry silk) (*see* **Note 1**).
2. *Antheraea assama* (AA, Muga) and *Philosamia ricini* (PR, Eri) silk glands isolated from mature 5th instar silkworm larvae (non-mulberry silk) (*see* **Note 2**).

2.2 Silk Fibroin (SF) Extraction

1. 0.02 M sodium carbonate: Dissolve 4.24 g of Na_2CO_3 in 2 L of water.
2. 9.3 M lithium bromide (LiBr): Dissolve 15.96 g of LiBr in 20 mL of water (*see* **Note 3**).

3. 1% (w/v) sodium dodecyl sulfate (SDS): Dissolve 0.5 g of SDS in 50 mL of water.
4. Dialysis membrane (12 kDa molecular weight cutoff).
5. Dialysis clips.
6. Plastic beakers (5 L, 2 L) and glass beaker (100 mL).

2.3 Micro-patterned Polydimethylsiloxane (PDMS) Molds and Silk Film Fabrication

1. PDMS molds (4×4 cm^2) with micro-gratings (depth, 200 nm; ridge-to-ridge distance, 3μm) can be obtained by following the detailed methodology described by Rockwood et al. [7].
2. Desiccator and vacuum pump.
3. 80% (v/v) ethanol in water.

2.4 Isolation of Vascular Cells from Porcine Aorta

1. Descending porcine aorta: Collect the aorta within 1 h of animal's sacrifice. Use 200 mL ice-cold sterile phosphate-buffered saline (PBS) to transport 5-cm-long aortic section.
2. Complete cell culture media: Prepare high-glucose Dulbecco's Modified Eagle's Medium (DMEM) in sterile water following the recommended manufacturer's instructions. Add 50 mL of sterile fetal bovine serum (FBS) and 5 mL of the antibiotic-antimycotic solution to 445 mL of basal DMEM media under sterile conditions. All cell culture reagents are procured from Gibco, USA.
3. 0.018% (w/v) collagenase type 1A (from *Clostridium histolyticum*, lyophilized powder 125 $\geq$ CDU/mg): Dissolve 9 mg collagenase in 50 mL of basal DMEM media (without FBS). Filter sterilize the working collagenase solution using a 0.22μm syringe filter (*see* **Note 4**).
4. 0.1% (w/v) collagenase type 1A (from *Clostridium histolyticum*, lyophilized powder 125 $\geq$ CDU/mg): Dissolve 10 mg collagenase in 10 mL of basal DMEM media (without FBS) followed by filter sterilization.
5. 0.5% (w/v) gelatin: Dissolve 0.5 g gelatin in 100 mL PBS, and autoclave the solution.
6. Gelatin-coated 6-well tissue culture plate: Add 2 mL of sterile gelatin in each well, incubate for 2 h in a cell culture incubator, discard the gelatin solution, and allow the plates to dry for 2 h in laminar flow hood. Store the plates at 4 °C under a sterile environment until further use. Coated plates should be prepared 1 day before cell isolation.
7. Standard mammalian cell culture facility.
8. Sterile cell scraper (Himedia, India).
9. Sterile scalpel blade and holder.
10. Sterile 70μm cell strainer.

11. Sterile forceps (pointed and blunt end).
12. 90 mm and 60 mm sterile Petri dishes.

2.5 Multilayered Silk Film-Based Tubular Graft Fabrication and Maturation

1. Sterile inert polypropylene mandrel of required diameter, nearly 6 cm long.
2. Sterile 3-0 silk suture.
3. 90 mm sterile Petri dishes.
4. Sterile scissors.
5. Trypsin-EDTA solution.
6. Cell counter.

2.6 Mold Parts (for Bi-layered Silk Scaffold-Based Vascular Graft)

1. Prepare mold base and mold cap using 3D printer.
2. Polypropylene tube (1.9 mm internal diameter, 3 cm length).
3. Stainless steel rod (1.1 mm diameter, 4 cm length).
4. Stainless steel cylinder (2 mm internal diameter, 3 cm length).

2.7 Bi-layered Tubular Graft Fabrication

1. 2 mL and 5 mL BD Luer-Lok syringes.
2. Lyophilized SF tablets: Prepare 2% SF solution in water, and pour 2 mL solution in each well of the standard 12-well tissue culture plate. Freeze the solution overnight (nearly 12 h) at −20 °C, and lyophilize for 48 h. Take out the tablets, and store them in airtight moisture-free containers at 4 °C. These tablets can be stored for long term (up to 6 months).
3. SF aqueous solution from lyophilized tablets: Dissolve 1.1 g of BM silk tablets in 10 mL of water, keep overnight at 4 °C, centrifuge (8228 × *g*, 10 min), and transfer supernatant in a fresh tube. Make up the volume to obtain 10% (w/v) BM silk. Similarly, dissolve 300 mg AA silk tablets in 10 mL water, and make up the volume to obtain 2% (w/v) AA silk solution (*see* **Note 5**).
4. Standard electrospinning setup with a rotational and linear movable collector. The required collector (stainless steel rod) diameter is 0.8–0.9 mm.
5. Electrospinning solution: Dissolve 500 mg of BM tablets in 5 mL of 1,1,1,3,3,3-hexafluoro-2-propanol (HFP) resulting in 10% (w/v) solution. Dissolve 500 mg of polycaprolactone (PCL, average M_n 80,000) in 5 mL of HFP to obtain 10% (w/v) solution. Final electrospinning solution is blend of 10% (w/v) BM silk and 10% (w/v) PCL solutions in 1:1 (v/v) ratio (*see* **Note 6**).

3 Methods

3.1 GRAFT I (Multilayered Silk Films-Based Vascular Graft)

3.1.1 Fabrication of Silk Films (See Note 7)

1. Rinse the PDMS molds (required number) with ultrapure water, and dry in a hot air oven (60 °C) for 2 h.
2. Keep the dried PDMS molds in the laminar airflow hood.
3. Prepare 15 mL of 1% (w/v) SF solution (for preparing 10 films). Dilute the concentrated stock silk solution to obtain 1% (w/v) silk. For example, the addition of 12.5 mL of water into 2.5 mL of 6% (w/v) stock SF solution would yield 1% (w/v) 15 mL SF solution.
4. Pour 1.5 mL of SF solution on each square-shaped PDMS mold (4×4 cm^2), and spread evenly using the tip of 0.5 mL tube (*see* **Note 8**).
5. Allow the SF solution to dry in the laminar airflow hood for 12–16 h at room temperature.
6. Keep the dried SF-PDMS molds in a vacuum desiccator, and pour 100 mL water at the bottom to create water vapors. Create ~40 psi pressure using a vacuum pump, and allow water vapor annealing for 6 h at room temperature (~25 °C) (*see* **Note 9**).
7. Release the pressure at a minimal rate, and remove the desiccator lid. Detach the films from molds using fine forceps, and transfer them to fresh Petri dishes for further use (*see* **Note 10**).

3.1.2 Isolation and Culture of Vascular Cells (See Note 11)

1. Representative protocol for vascular cell isolation from the porcine descending aorta is shown in Fig. 1.
2. Process the tissue sample in BSL class 2 biosafety cabinet (cell culture hood). Cut 1 cm section of the aortic tube. Wash properly with ice-cold sterile PBS.
3. Chop off the connective tissue from the outer surface using fine, sterile forceps.
4. Incubate the tissue section with 5% (v/v) antibiotic-antimycotic solution prepared in incomplete (without FBS) DMEM media in mammalian cell culture incubator (5% CO_2, 37 °C, 95% humidity) for 30 min.
5. Make approximately 0.5-mm-deep cut along the longitudinal axis using a sterile scalpel blade, and peel off the adventitial layer along the circumference of the tube using blunt end forceps.
6. Cut the adventitial layer (containing fibroblasts) into 1×1 mm^2 pieces, and keep them in 10 mL 0.018% (w/v) collagenase solution in a 50 mL tube for 12 h in the cell culture incubator under constant shaking.

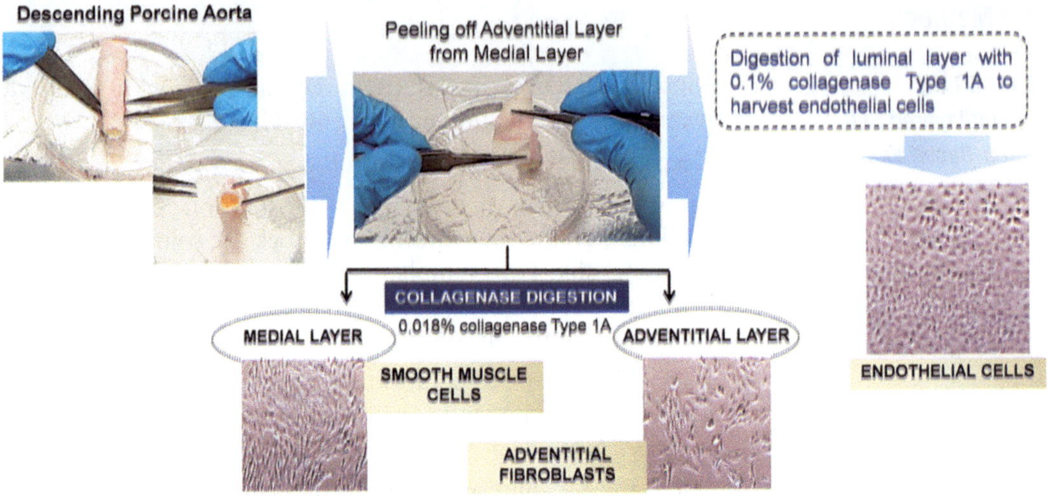

Fig. 1 Schematic representation of isolation of vascular cells (endothelial cells, smooth muscle cells, and fibroblasts) from descending porcine aorta

7. Cut open the remaining part of the aorta, exposing the lumen, resulting in a rectangular flap containing intimal and medial layers. Take 10 mL of 0.1% (w/v) collagenase solution in a 60 mm Petri dish, and keep the lumen in contact with the surface of the Petri dish. Incubate for 30 min in cell culture incubator. Scrap the luminal surface using a cell scraper to isolate the endothelial cells (EC). Add 10 mL of fresh DMEM, and collect the solution in a 15 mL tube.
8. Centrifuge at 804 × *g*, 5 min, and re-suspend the pellet in complete DMEM media. Plate these cells in the gelatin-coated 6-well plate (2 mL media in each well).
9. The remaining media layer from **step** 7 contains smooth muscle cells (SMCs). Cut the medial layer into 1 × 1 mm^2 pieces, and follow **step 6**.
10. Filter (using 70 μm cell strainer) and centrifuge (804 × *g*, 5 min) the digested tissue obtained from **steps 6** and **9**, containing fibroblasts and SMCs, respectively. Re-suspend the pellet in complete DMEM media, and plate the cells in the gelatin-coated 6-well plate following **step 8** (*see* **Note 12**).
11. Leave the plated vascular cells undisturbed in cell culture incubator for 24 h (from **steps 8** to **10**). Wash each well to remove the tissue debris, and replenish the fresh media every alternate day (*see* **Note 13**).
12. Sub-culture cells in tissue culture T75 flasks following standard culture methodology. Use the primary porcine vascular cells at early passages (P2-P6) for developing tubular graft. Use complete DMEM media to culture all types of vascular cells (ECs, SMCs, and fibroblasts).

3.1.3 Culture of Vascular Cells onto Patterned Silk Films

1. Transfer the water vapor annealed films in 10 mL of sterile PBS in 90 mm Petri dishes. Sterilize the films under ultraviolet (UV) light for 30 min in standard cell culture hood (*see* **Note 14**).
2. Remove the PBS using a sterile pipette, and add 10 mL of complete DMEM media submerging the film. Keep the plates in the cell culture incubator overnight for conditioning (*see* **Note 15**).
3. For seeding one silk film (4×4 cm^2), trypsinize the vascular cells, and prepare a cell suspension in complete DMEM media with a cell density of 64×10^4 cells/100μL. Same cell density was used for all vascular cell types.
4. Dispense the conditioning media from Petri dishes (from **step 2**), and add 100μL of cell suspension on patterned silk film surface taking precaution to avoid any spillage. Keep the cell-seeded silk film (with minimal media) in the cell culture incubator, and keep adding 100μL fresh media after each 30 min until 4 h. Finally, add 10 mL fresh media in each Petri dish submerging the cell-seeded silk film, and keep in the incubator (*see* **Note 16**).
5. Use cell-seeded silk films with 70% confluency for tube formation.

3.1.4 Fabrication and Maturation of Multilayered Vascular Graft

1. Keep the EC-, SMC-, and fibroblast-seeded films, sterile silk sutures, and inert mandrel in the cell culture hood. The number of cell-seeded films required for tubular construct fabrication depends on the diameter of inert mandrel and layers per graft (*see* **Note 17**).
2. Using two blunt end forceps, start rolling the EC-seeded film around the mandrel, followed by SMC- and fibroblast-seeded films. After obtaining the required number of layers, cut the remaining film section using sterile scissors. Consider the native cellular alignment while rolling the films, i.e., EC along the longitudinal axis and SMC/fibroblasts along the radial direction, as shown in Fig. 2a. After each successive film wrapping, secure both the ends using silk suture (while on mandrel) to prevent the unfolding. Rolling assembly is shown in Fig. 2b.
3. Add complete DMEM media into the rolling assembly to submerge the rolled films, and place it in a cell culture incubator for maturation over 14 days. Replenish with fresh media every 2 days until termination of the experiment.
4. After 14 days of culture, bring out the rolling assembly into the culture hood. Using blunt end forceps, slide out the multilayered constructs slowly from the mandrel. The matured tube is shown in Fig. 2c.

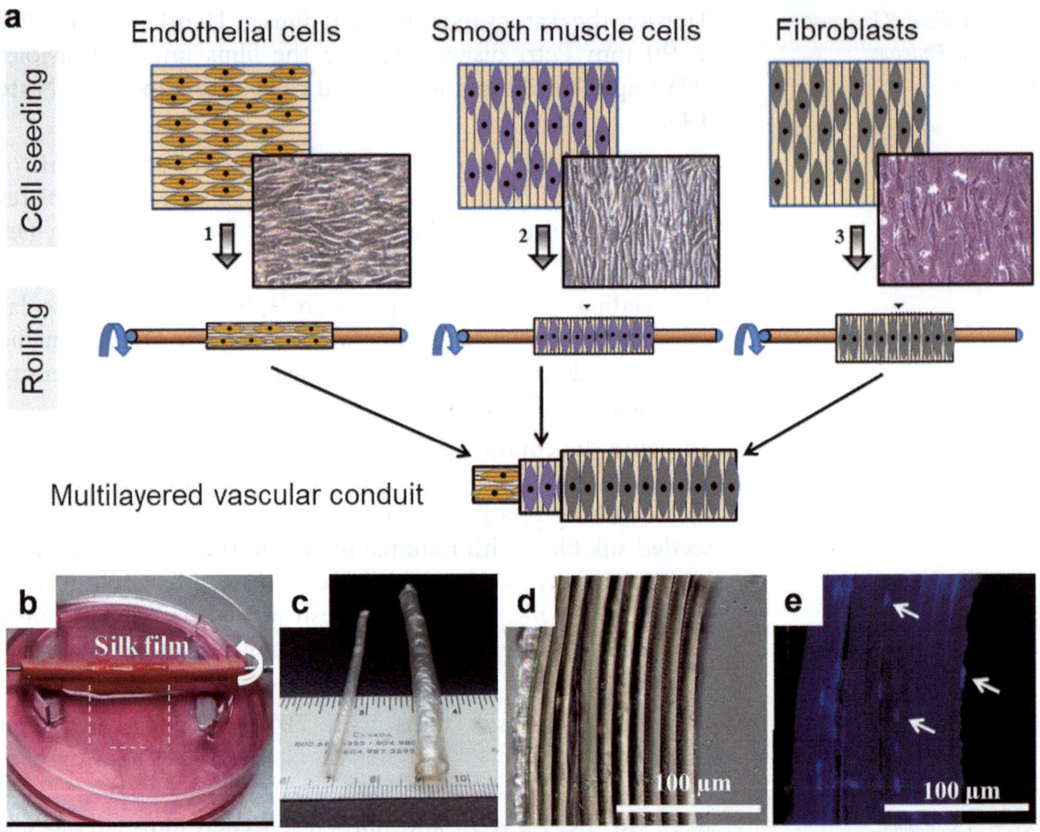

Fig. 2 (**a**) Schematic representation of fabrication of biomimetic multilayered silk film-based vascular conduit. Vascular cells are aligned unidirectionally induced by silk film pattern. The fabrication process involves the seeding of vascular cells on silk films followed by their sequential rolling onto an inert mandrel to obtain a multilayered vascular conduit. (**b**) Assembly used for rolling of vascular cell-seeded silk films. (**c**) Matured cellular vascular graft after 14 days of static culture. (**d**) Phase-contrast image of graft cross section corroborating the intactness of multiple layers. (**e**) Fluorescence imaging showing the presence of Hoechst 33342-stained vascular cells. White arrows are marking the cell nucleus. (Reproduced with permission from ref. [11] from American Chemical Society)

5. Chop off the suture secured ends. At this stage, all layers of silk films should be integrated owing to the secretion of extracellular matrix (ECM) proteins during maturation (Fig. 2d, e). This method provides ~3-cm-long, multilayered cellular tubular functional graft, which could potentially be used for vascular tissue engineering applications.

3.2 GRAFT II (Bi-layered Silk Scaffold-Based Vascular Graft)

1. A custom-made mold is used. As shown in Fig. 3, the mold parts include base, cap, stainless steel rod, polypropylene tube, and stainless steel tube. A representative image of the assembled mold is also shown in the figure (*see* **Note 18**).
2. Rinse all the mold parts in water, and allow for air-drying. Insert the polypropylene tube in the stainless steel tube, so that it covers the lumen (*see* **Note 19**). Insert one end of the

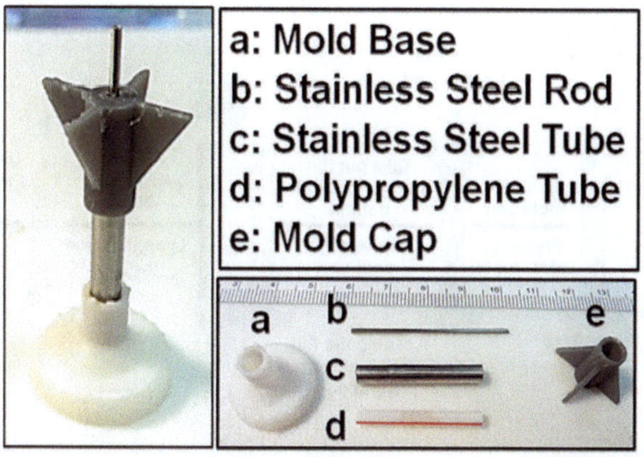

Fig. 3 Custom-designed mold used to fabricate the inner porous layer of bi-layered silk scaffolds. The mold consists of a 3D printed base and cap, polypropylene tube, stainless steel rod, and tube. The image on the left is representing the assembled mold. (Reproduced with permission from ref. [12] from Elsevier)

stainless steel/polypropylene tube in the mold base, and secure the joint using parafilm. Insert the stainless steel rod in the center from the same, end and secure using parafilm. The overall process of bi-layered tubular silk scaffold formation is shown in Fig. 4a.

3. Inject 6% (w/v) silk solution into the void space of the assembled mold slowly, and close the open end of the mold with a cap keeping the stainless steel rod in the center. For each of our designed mold, 250μL of silk solution is required producing four tubular scaffolds per 1 mL of silk solution (*see* **Note 20**).
4. Freeze the assembled mold at −20 °C for 2 h.
5. Take out the central rod by pushing it from the top, and remove the mold cap. It will create a void space along the tube, which forms the lumen of the scaffold (*see* **Note 21**).
6. Keep the remaining mold assembly again for freezing at −20 °C for overnight and lyophilize for 24 h.
7. Disassemble the remaining mold by taking out the base first, followed by pushing the tubular scaffold from one end. Scaffold comes out easily.
8. Soak the tubular porous silk scaffolds (inner layer) with 80% v/v ethanol in water for 15 min. at room temperature, and use the ethanol-soaked scaffold for coating the outer electrospun layer (*see* **Note 22**).
9. Fill in the electrospinning solution in a 5 mL syringe, and set in the syringe pump. Insert three ethanol-soaked scaffolds in the electrospinning collector through the lumen, and use 300μL of electrospinning solution for each cycle maintaining the

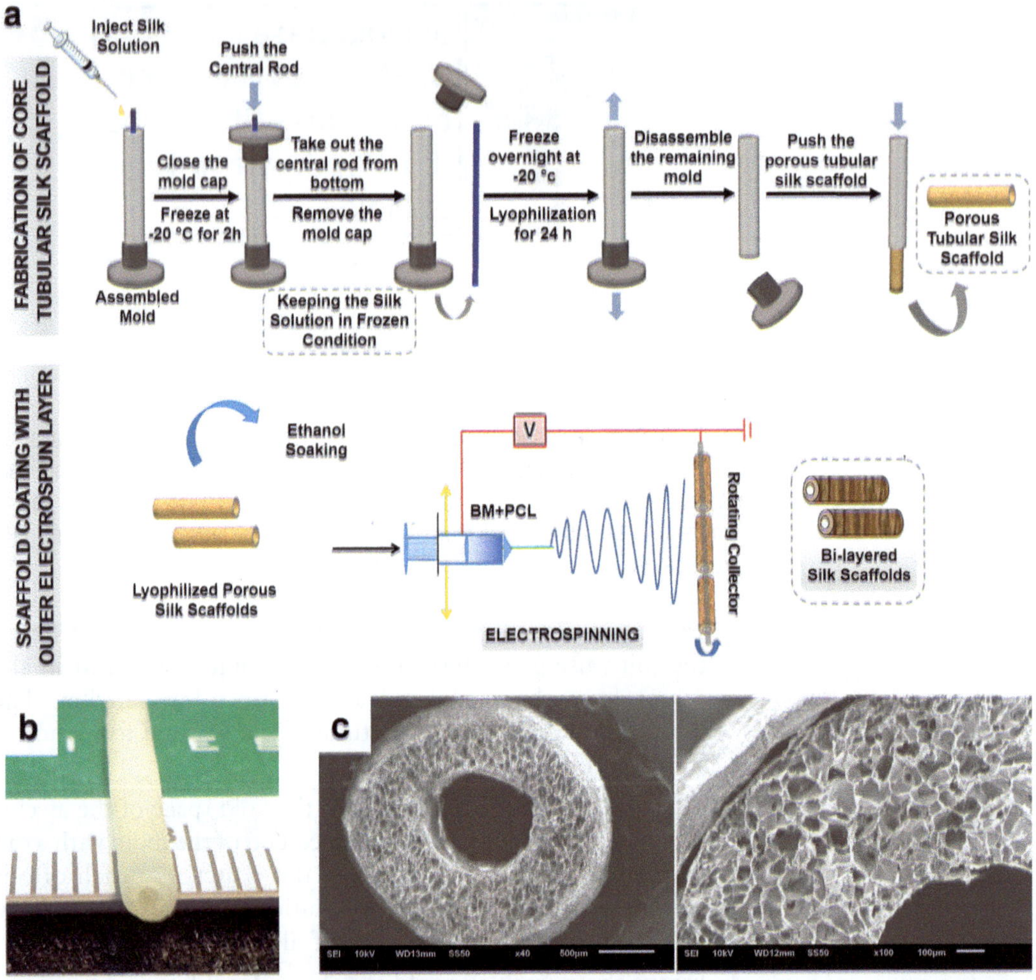

Fig. 4 (**a**) Schematic representation of fabrication of bi-layered tubular silk scaffolds. The inner porous silk layer is fabricated using custom-designed 3D printed molds, whereas the outer layer is coated using an electrospinning machine and depositing silk-PCL blend nanofibers. (**b**) Representative macroscopic image of bi-layered tubular silk vascular graft. (**c**) Representative scanning electron microscope (SEM) images of scaffold cross section are showing the bi-layered architecture of scaffold and the porosity of inner and outer layers. (Reproduced with permission from ref. [12] from Elsevier)

following parameters: tip-to-collector distance (10 cm), flow rate (100μL/min), rotational speed (200 rpm), voltage difference (15 kV), translation speed (50 mm/s), humidity (~47%), and temperature (~25 °C).

10. Cut the ends of each scaffold with a sharp blade without damaging the scaffolds, and take out the bi-layered scaffolds from the collector using blunt end forceps.
11. The bi-layered tubular scaffold can be stored in 80% ethanol at 4 °C up to 1 month.

12. The macro- and microscopic images of developed bi-layered tubular scaffolds are shown in Fig. 4b, c, respectively. The inner layer of the scaffold is highly porous which is mechanically supported by dense outer electrospun layer. The resulting scaffold could be used as acellular graft, or the inner layer of the scaffold can be incorporated with various cell types and bioactive molecules.

4 Notes

1. We procure the *Bombyx mori* cocoons from local silk farms. Only undamaged cocoons should be used to extract the silk fibroin aqueous solution. For a detailed extraction protocol, the reader is suggested to follow the previously reported protocol by Rockwood et al. [7]. Aqueous SF solution should be stored at 4 °C.
2. We procure the *Antheraea assama* (Muga) and *Philosamia ricini* (Eri) silkworms from local silk farms at the matured stage to isolate the silk glands. Silk glands isolated from premature larvae would result in a minimal yield of silk protein. For a detailed extraction protocol, the reader is suggested to follow the previously reported protocol by Mandal et al. [13]. Non-mulberry aqueous SF extracted from silkworm glands is unstable at higher concentrations (>3–4% w/v) and forms hydrogel. Therefore, in both of the described methods for vascular grafts fabrication, we used either 1% or 2% of non-mulberry silk. Aqueous SF solution should be stored at 4 °C; long-term storage (>15 days) is not recommended.
3. LiBr is hygroscopic and produces heat when dissolved in water. Appropriate chemical hazard precautions and heat safety measures should be taken during the handling of LiBr. While preparing LiBr solution (20 mL), dissolve the required amount of LiBr powder in 15 mL water at 60 °C, and then make up the volume to 20 mL.
4. Do not dissolve the collagenase in complete DMEM medium (with FBS); this will neutralize the enzyme activity and result in inadequate tissue digestion.
5. The aqueous SF solution could either be obtained fresh or by reconstituting the lyophilized silk tablets in water. Do not use a high percentage of SF solution (>2% w/v) to prepare the lyophilized tablets. If reconstituting the SF tablets in water to obtain an aqueous SF solution, the high percentage of BM lyophilized tablets will not dissolve completely. However, a lower percentage of lyophilized silk is easy to dissolve. While dissolving the tablets in water, they should not float on the

surface and be completely submerged. Dissolve one extra tablet, and make up the volume to obtain the required percentage of silk solution.

6. Use the airtight glass bottle to prepare BM and PCL solutions in HFP considering its volatile nature. Open the HFP bottle in fume hood. For dissolving BM silk, weigh the tablets, keep them in 20 mL airtight glass bottle, and add HFP from the top. At the beginning, HFP will be soaked by silk tablets. Close the cap, and keep for stirring overnight at room temperature. In contrast to the aqueous solution, BM silk tablets dissolve completely in HFP, so weigh the exact required amount. Similarly, prepare the PCL solution. BM and PCL solutions (10% w/v) are highly viscous. Take precautions to dispense a similar volume while mixing them for preparing the electrospinning solution. Prepare the electrospinning solutions fresh; it may show signs of phase separation if kept for a longer duration (10 days).
7. For a detailed methodology of patterned silk film fabrication, the reader is suggested to refer to earlier reported protocol by Rockwood et al. [7]. A similar methodology is adopted for fabricating both mulberry and non-mulberry silk films.
8. The surface of PDMS molds is hydrophobic, which results in shrinkage of SF solution. Supervise the solution till 2–3 h, and spread the SF solution completely over the molds to prevent the shrinkage. Owing to hydrophobicity and hydrophilicity differences between different silk varieties, the shrinkage pattern varies.
9. The water vapor annealing results in the water stability of silk films by inducing β-sheet transformation.
10. After releasing the pressure, allow the films to air-dry for 10 min. Detaching the films in moist conditions may damage them. Once air-dried, the films will start detaching from molds at the corners; at this stage use fine forceps, hold the film from one corner, and lift completely from the mold. Cut nearly 1–2 mm from all sides to keep them flat. While storing, label the pattern side of the films to avoid any confusion at a later stage. Always handle the films with forceps, and avoid touching with bare hands. Store the films at room temperature in closed Petri dishes no longer than 10–15 days.
11. For isolation of vascular cells (endothelial cells, smooth muscle cells, and fibroblasts) from the porcine descending aorta, we followed previously described protocols [14, 15] with slight modifications.
12. While plating the vascular cells, in order to avoid any possibility of microbial contamination, 3% (v/v) antibiotic-antimycotic solution should be added into culture media initially.

13. Vascular cells isolated from porcine aorta generally attach within 24 h. While washing the wells, ensure complete removal of tissue debris as it may affect the cell growth. After second media change, reduce the antibiotic-antimycotic concentration to 1% (v/v).
14. Handle silk films using blunt end forceps for a better grip. When transferring into PBS for sterilization, use two forceps at both opposite ends of single side, and keep the patterned side on top. Using single forceps would cause folding of films, which would further need to be flattened after transfer to the Petri dish.
15. Use one film in each Petri dish. Take the required number of films for fabricating tubes of different diameters. Conditioning in cell culture media would also confirm if the films are properly sterilized. After overnight incubation, if any microbial contamination is observed, discard the sample.
16. While seeding the vascular cells onto silk films, add 10μL aliquots of cell suspension at ten different locations onto each silk film. It will prevent the cell spillage and allow uniform distribution of cells.
17. For instance, if the diameter of the inert mandrel is 4 mm, the required length of the film to obtain a single layer around the mandrel would be equivalent to the circumference (12.56 mm). Consider the film size (4×4 cm^2) and the desirable number of layers for each cell type (EC, 02; SMC, 06; and fibroblasts, 06); the user would require 01 EC-seeded film, 02 SMC-seeded film, and 02 fibroblast-seeded silk films. All three silk varieties can be used for this purpose. We have shown that vascular cells cultured on BM films tend to form aggregates resulting in cellular detachment while rolling process, which is not the case with vascular cells cultured on *Antheraea assama* (AA) films [16]. Owing to this observation, we suggest using AA films, preferably for fabricating multilayered vascular grafts.
18. Mold cap and base can be printed using any commercially available 3D printer. The key function of 3D printed parts is to hold the stainless steel rod at the center. The design can be modified if required. For cutting the stainless steel rod and tube, use either a laser cutter or electric discharge machine (EDM) with wire cutter setup. It will provide very sharp cut edges. If ends of the tube and rod are not sharp, the user will face difficulty while taking out the lyophilized tubular scaffold from the mold and may damage it.
19. The polypropylene tube inserted in the stainless steel tube allows smooth removal of tubular scaffold after lyophilization. In case if the inner luminal surface is not covered with the

polypropylene tube, the silk solution will stick on the wall, and it will damage the scaffold while taking it out from the mold. Incorporating polypropylene tube increases the efficiency of tube retrieval from the mold to 100% with reproducibility.

20. The inner layer of the scaffold can be fabricated using BM and AA silk. In the first case, 6% BM silk solution can directly be used. However, obtaining a higher percentage (6% w/v) of AA silk is not feasible [12, 13]. In our published methodology in ref. [12], we have shown the use of a combination of BM and AA silk to obtain a final 6% (w/v) concentration. For preparing 1 mL of the combined solution, blend 500μL of BM silk (10%, w/v) with 500μL of AA silk (2% w/v). The BM and AA silk blend form hydrogel if kept at 37 °C for 15 min [17–19]. In order to avoid hydrogel formation, process the silk blend at lower temperatures (~4 °C). While injecting the silk solution into the mold, inject the needle at the bottom of the mold touching base and then dispense the silk solution. Once the mold is filled, move the needle along the central rod to ensure the homogenous distribution of silk solution in the mold. Avoid any bubble formation during this process.
21. Removal of central rod and mold cap should be done as fast as possible, avoiding melting of inner silk solution. If the silk solution melts at this stage, the resulting scaffolds will be defected and non-uniform. Handle the mold by holding the base part at this step. Avoid touching the stainless steel part with bare hands as it might result in the melting of the frozen silk solution.
22. Ethanol-soaked scaffolds allow better adhesion of the outer electrospun layer with an inner porous scaffold layer. We have observed minimal delamination of both scaffold layers using this methodology.

Acknowledgments

We acknowledge the laboratory of Prof. David A. Vorp (University of Pittsburgh, USA) for suggestions during the development of Graft II described in this chapter. P.G. acknowledges the Ministry of Education (MoE), Government of India; the Institute of International Education (IIE), New York; and the United States-India Educational Foundation (USIEF) for research fellowship. B.B.M. acknowledges funding support from the Department of Biotechnology (DBT) and Department of Science and Technology (DST), Government of India. We have filed Indian patents for Graft I (application number 1246/KOL/2015) and Graft II (with or without modifications, application number 201931024432) fabrication methodologies.

References

1. Mathers CD, Loncar D (2006) Projections of global mortality and burden of disease from 2002 to 2030. PLoS Med 3(11):e442
2. Veith FJ, Moss CM, Sprayregen S, Montefusco C (1979) Preoperative saphenous venography in arterial reconstructive surgery of the lower extremity. Surgery 85(3):253–256
3. Lovett M, Cannizzaro C, Daheron L, Messmer B, Vunjak-Novakovic G, Kaplan DL (2007) Silk fibroin microtubes for blood vessel engineering. Biomaterials 28(35):5271–5279
4. Shin'oka T, Imai Y, Ikada Y (2001) Transplantation of a tissue-engineered pulmonary artery. N Engl J Med 344(7):532–533
5. Janani G, Kumar M, Chouhan D, Moses JC, Gangrade A, Bhattacharjee S, Mandal BB (2019) Insight into silk-based biomaterials: from physicochemical attributes to recent biomedical applications. ACS Appl Bio Mater 2 (12):5460–5491
6. Bandyopadhyay A, Chowdhury SK, Dey S, Moses JC, Mandal BB (2019) Silk: a promising biomaterial opening new vistas towards affordable healthcare solutions. J Indian Inst Sci:1–43
7. Rockwood DN, Preda RC, Yücel T, Wang X, Lovett ML, Kaplan DL (2011) Materials fabrication from Bombyx mori silk fibroin. Nat Protoc 6(10):1612
8. Konwarh R, Bhunia BK, Mandal BB (2017) Opportunities and challenges in exploring Indian nonmulberry silk for biomedical applications. Proc Indian Natl Sci Acad 83(1)
9. Enomoto S, Sumi M, Kajimoto K, Nakazawa Y, Takahashi R, Takabayashi C, Asakura T, Sata M (2010) Long-term patency of small-diameter vascular graft made from fibroin, a silk-based biodegradable material. J Vasc Surg 51(1):155–164
10. Rodriguez M, Kluge JA, Smoot D, Kluge MA, Schmidt DF, Paetsch CR, Kim PS, Kaplan DL (2020) Fabricating mechanically improved silk-based vascular grafts by solution control of the gel-spinning process. Biomaterials 230:119567
11. Gupta P, Kumar M, Bhardwaj N, Kumar JP, Krishnamurthy C, Nandi SK, Mandal BB (2016) Mimicking form and function of native small diameter vascular conduits using mulberry and non-mulberry patterned silk films. ACS Appl Mater Interfaces 8 (25):15874–15888
12. Gupta P, Lorentz KL, Haskett DG, Cunnane EM, Ramaswamy AK, Weinbaum JS, Vorp DA, Mandal BB (2020) Bioresorbable silk grafts for small diameter vascular tissue engineering applications: In vitro and in vivo functional analysis. Acta Biomater 105:146–158
13. Mandal BB, Kundu S (2008) A novel method for dissolution and stabilization of non-mulberry silk gland protein fibroin using anionic surfactant sodium dodecyl sulfate. Biotechnol Bioeng 99(6):1482–1489
14. Hao H, Ropraz P, Verin V, Camenzind E, Geinoz A, Pepper MS, Gabbiani G, Bochaton-Piallat M-L (2002) Heterogeneity of smooth muscle cell populations cultured from pig coronary artery. Arterioscler Thromb Vasc Biol 22(7):1093–1099
15. Butcher JT, Nerem RM (2004) Porcine aortic valve interstitial cells in three-dimensional culture: comparison of phenotype with aortic smooth muscle cells. J Heart Valve Dis 13 (3):478–486
16. Gupta P, Moses JC, Mandal BB (2019) Surface patterning and innate physicochemical attributes of silk films concomitantly govern vascular cell dynamics. ACS Biomater Sci Eng 5 (2):933–949
17. Bhunia BK, Mandal BB (2019) Exploring gelation and physicochemical behavior of in situ bioresponsive silk hydrogels for disc degeneration therapy. ACS Biomater Sci Eng 5 (2):870–886
18. Kumar M, Gupta P, Bhattacharjee S, Nandi SK, Mandal BB (2018) Immunomodulatory injectable silk hydrogels maintaining functional islets and promoting anti-inflammatory M2 macrophage polarization. Biomaterials 187:1–17
19. Chouhan D, Lohe T, Samudrala PK, Mandal BB (2018) In situ forming injectable silk fibroin hydrogel promotes skin regeneration in full thickness burn wounds. Adv Healthc Mater 7(24):1801092

Chapter 12

Fabrication and Evaluation of Tissue-Engineered Vascular Grafts with Hybrid Fibrous Structure

He Wang and Qiang Zhao

Abstract

Bypass surgery has proven to be an effective strategy for the treatment of cardiovascular diseases. In this chapter, we describe the preparation of a type of small-diameter vascular grafts with hybrid fibrous structure by co-electrospinning. Of the two components, slow-degrading poly(ε-caprolactone) (PCL) fibers maintain the structural integrity of the graft and provide sufficient mechanical strength. The other component is rapidly degradable polymer polydioxanone (PDS) or type-I collagen that can be used as a carrier to deliver vasoactive molecules at the same time. The in vivo performance of as-prepared vascular grafts was further evaluated in a rat abdominal aorta replacement model.

Key words Tissue-engineered vascular grafts (TEVGs), Co-electrospinning, Rapid degradation, Vasoactive molecules, Rat abdominal aorta replacement model

1 Introduction

Cardiovascular disease (CVD) remains a leading cause for the high morbidity and mortality globally. Bypass surgery has proven to be an effective strategy for CVD treatment. Although commercialized vascular grafts, such as expanded polytetrafluoroethylene (ePTFE) and poly(ethylene terephthalate) (Dacron), have been successfully used in vascular surgery for several decades, no products are available in the case of small-diameter grafts (<6 mm) because of the high rate of restenosis [1, 2].

Rational design of small-diameter vascular grafts is fundamental for long-term vascular patency and homeostasis. In addition to the physical structure, bioactivity and function of the graft are more important for the vascular regeneration [3, 4]. In this regard, the grafts provide a favorable environment for the recruitment of autogenous vascular cells. After a full degradation of the polymer scaffold, "neo-artery" could be generated [5].

Feng Zhao and Kam W. Leong (eds.), *Vascular Tissue Engineering: Methods and Protocols*, Methods in Molecular Biology, vol. 2375, https://doi.org/10.1007/978-1-0716-1708-3_12,

As a Food and Drug Administration (FDA)-approved biodegradable polymer, PCL has been widely utilized for the fabrication of small-diameter vascular grafts that have been extensively evaluated in various animal models. Electrospun PCL vascular grafts demonstrated satisfactory patency in a rat model of abdominal aorta replacement [6, 7]. On the other hand, vascular regeneration was not satisfactory, and some adverse remodeling (such as calcification) was observed after long-term implantation (12 and 18 months), indicating the grafts cannot fulfill the requirement of vascular regeneration sufficiently [8].

Co-electrospinning is a versatile processing technique that affords the opportunity for preparing vascular graft with binary fibrous components [9]. By taking advantage of this technology, we fabricated a hybrid graft consisting of both PCL and PDS fibers. The rapid degradation of PDS fibers can enlarge the pore space to promote vascular regeneration and reduce adverse pathological remodeling to some extent [10].

Furthermore, extracellular matrix (ECM) proteins, such as collagen, can also be introduced into PCL vascular grafts by co-electrospinning. Of the two components, PCL fibers provide sufficient mechanical support and structural maintenance, while collagen fibers favor the adhesion and growth of vascular cells. More importantly, collagen can act as a carrier for delivery of vasoactive molecules, including cytokines or bioactive peptides that play a pivotal role in vascular regeneration [11, 12]. Recently, we demonstrated that a 7-amino-acid peptide (7A, MHSPGAD) is critical in modulating the mobilization and orientated differentiation of vascular stem/progenitor cells, thus enhancing the regeneration of vascular grafts.

2 Materials

2.1 Electrospinning

1. PCL solution (25% w/v): Dissolve 1.5 g PCL pellets (M_n = 80,000, Sigma, USA) in 6 mL mixture of chloroform and methanol (5:1, V/V) by stirring overnight at room temperature (*see* **Note 1**).
2. PDS solution (20% w/v): Dissolve 1.2 g PDS pellets (Tianjin Dongnanhesheng Medical Technologies, China) in 6 mL 1,1,1,3,3,3-fluoro-2-propanol (HFIP) by stirring for 6 h at room temperature.
3. Collagen I (8% w/v) solution: Dissolve 0.16 g type-I collagen (Sannie Bioengineering Technology Company, China) in 2 mL HFIP by stirring overnight at room temperature.
4. Bovine serum albumin (BSA) solution (0.5% w/v): Dissolve 5 mg BSA in 1 mL HFIP.

5. 7A peptide solution (1 mg/mL): Dissolve 1 mg 7A peptide (custom-synthesized by GL Biochem, China) in 1 mL phosphate-buffered saline (PBS).
6. Collagen I-peptide (Col-7A) solution: Mix 2 mL collagen I solution, 0.5 mL BSA solution, and 0.5 mL 7A peptide solution homogeneously by stirring overnight at room temperature.
7. DiI stock solution (1 mg/mL): Dissolve 1 mg 1, 1′-dioctadecyl-3, 3, 3′, 3′-tetramethylindocarbocyanine perchlorate (DiI) in 1 mL anhydrous ethanol, and store in a bottle wrapped with aluminum foil.
8. DiO stock solution (1 mg/mL): Dissolve 1 mg 3, 3′-dioctadecyloxacarbocyanine per- chlorate (DiO) in 1 mL of anhydrous ethanol, and store in a bottle wrapped with aluminum foil.

2.2 Surgical Preparation

1. Chloral hydrate solution (10% w/v): Dissolve 1 g chloral hydrate in 10 mL physiological saline. Filter and sterilize the solution. Store at 4 °C in a bottle wrapped with aluminum foil.
2. Heparin sodium solution (50 UI/mL): Dissolve 12,500 units of heparin sodium in 250 mL sterile physiological saline and store at 4 °C.
3. Gentamicin solution (320 UI/mL): Dissolve 80,000 units of gentamicin in 250 mL sterile physiological saline and store at 4 °C.

2.3 Histological Analysis

1. Phosphate-buffered saline (PBS) (pH 7.2–7.4): Add 100 mL deionized water to a 1 L glass beaker. Weigh 8.5 g NaCl, 2.9 g $Na_2HPO_4 \cdot 12H_2O$, and 0.296 g NaH_2PO_4, and transfer to the beaker. Make up to 1 L with deionized water.
2. 0.1% Triton-PBS: Add 100 mL PBS to a 1 L graduated cylinder. Weigh 1 g Triton X-100, and transfer to the cylinder. Make up to 1 L with PBS.
3. Primary antibody solution: Mouse anti-CD31 (1:100, Abcam, USA), Mouse anti-alpha smooth muscle actin α-SMA (1:100, Abcam, USA).
4. The secondary antibody: Alexa Fluor 488 goat anti-mouse IgG (1:200, Invitrogen) (*see* **Note 2**).
5. Normal goat serum (5% v/v), 4,6-diamidino-2-phenylindole (DAPI) containing mounting solution (5μg/mL), hydrochloric acid ethanol differentiation solution (1% v/v), and hematoxylin (0.8% w/v) and eosin (0.5% w/v) solutions.

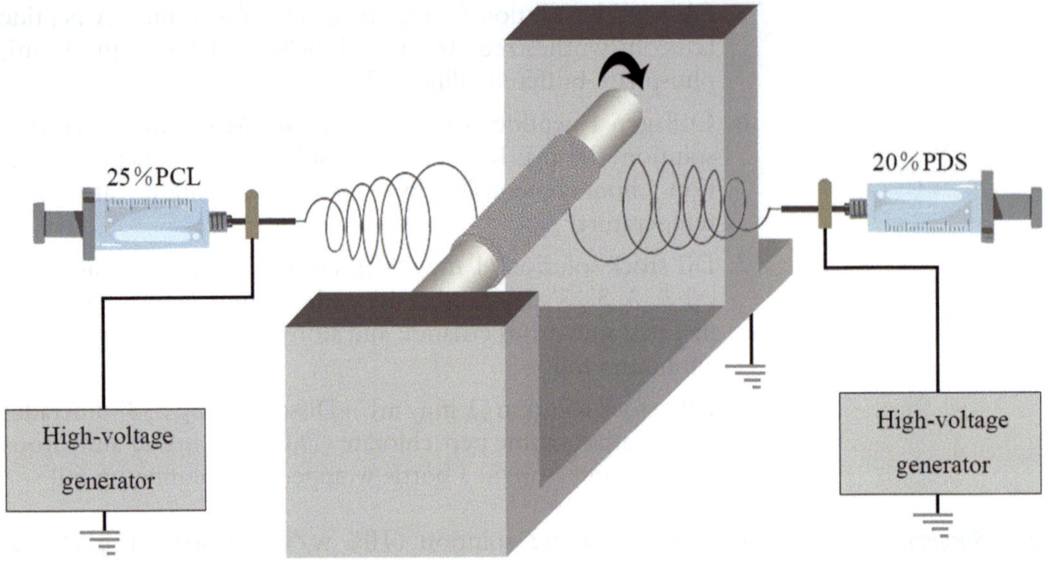

Fig. 1 Schematic illustration for the preparation of PCL/PDS vascular grafts. (Reproduced from ref. 10)

3 Methods

3.1 Co-electrospinning of Vascular Grafts

3.1.1 Fabrication of PCL/PDS Vascular Grafts

1. Load PCL and PDS solutions into 10 mL syringes, respectively.
2. Connect the syringe with a 21G blunt-ended needle that serves as the charged spinneret. Place the syringe on the syringe pump (*see* **Note 3**).
3. Put the stainless-steel mandrel (2 mm in diameter, 20 cm in length) on the grounded collection instrument (*see* **Note 4**).
4. Link the high-voltage generator with 21G blunt-ended needles.
5. Set the electrospinning parameters: flow rate = 8 mL/h (PCL) or 6 mL/h (PDS), voltages = 11 kV (PCL) or 15 kV (PDS), needle tip-collector distance = 25 cm (PCL) or 15 cm (PDS), and mandrel rotation speed = 50 rpm (*see* **Note 5**).
6. Perform the electrospinning (Fig. 1) (*see* **Note 6**).
7. Place the electrospun PCL/PDS vascular grafts in vacuum over 72 h to remove the residual solvent.
8. Prior to implantation, sterilize the graft by immersion in 75% ethanol for 30 min, and then expose the graft to UV light overnight.

3.1.2 Fabrication of PCL/COL-7A Vascular Grafts

1. Load the PCL solution into a 10 mL syringe, and load the Col-7A solution into a 2.5 mL syringe.
2. Connect the syringe with a 21G blunt-ended needle. Place the syringe on the syringe pump.

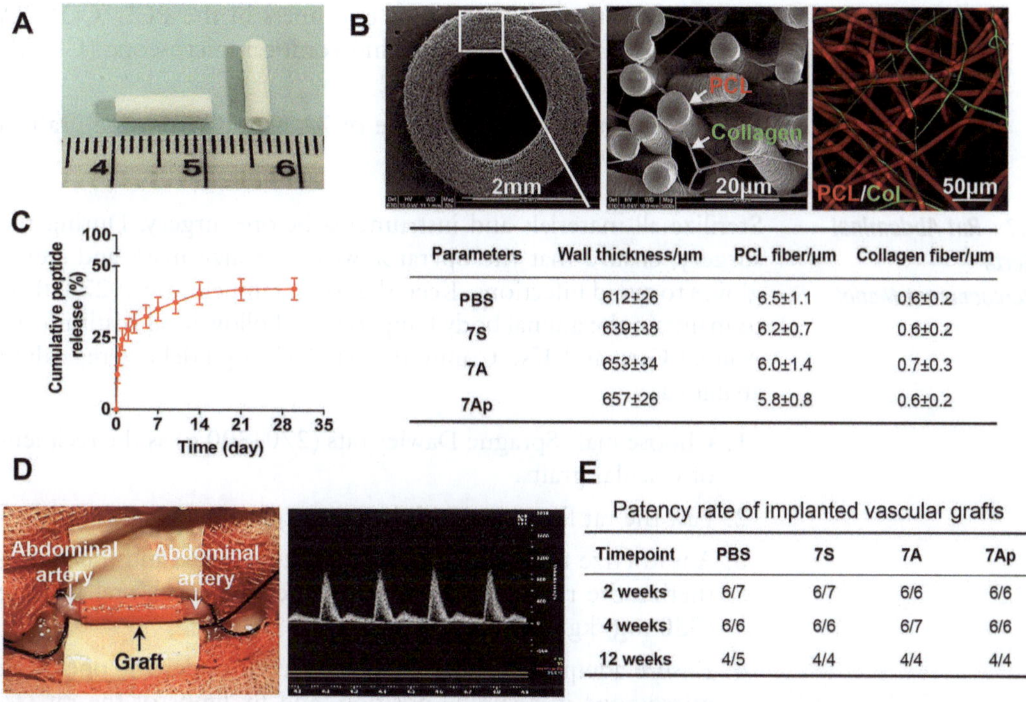

Parameters	Wall thickness/μm	PCL fiber/μm	Collagen fiber/μm
PBS	612±26	6.5±1.1	0.6±0.2
7S	639±38	6.2±0.7	0.6±0.2
7A	653±34	6.0±1.4	0.7±0.3
7Ap	657±26	5.8±0.8	0.6±0.2

Patency rate of implanted vascular grafts

Timepoint	PBS	7S	7A	7Ap
2 weeks	6/7	6/7	6/6	6/6
4 weeks	6/6	6/6	6/7	6/6
12 weeks	4/5	4/4	4/4	4/4

Fig. 2 The fabrication and evaluation of PCL/COL-7A vascular grafts.(**a**) The images show the morphology of the tubular vascular grafts, the microstructure as revealed by scanning electronic microscope, and (**b**) (upper panel) the distribution of the two types of fibers. (**b**) (lower panel) The graft parameters are summarized. (**c**) In vitro release of the peptide from the graft ($n = 3$). (**d**) Implantation of the graft in a rat to replace the abdominal artery (left) and the ultrasound scanning image of the implanted grafts (right). (**e**) The patency rate of implanted grafts at different time-points. (Reproduced from ref. 11)

3. Place the stainless-steel mandrel (2 mm in diameter, 20 cm in length) on the grounded collection instrument.
4. Link the high-voltage generator with 21G blunt-ended needles.
5. Set the electrospinning parameters: flow rate = 8 mL/h (PCL) or 0.6 mL/h (Col-7A), voltages = 11 kV (PCL) or 17 kV (Col-7A), needle tip-collector distance = 25 cm (PCL) or 15 cm (Col-7A), and mandrel rotation speed = 50 rpm.
6. Perform the electrospinning.
7. Place the electrospun PCL/COL-7A vascular grafts (Fig. 2a) in vacuum over 72 h to remove the residual solvent.
8. Prior to implantation, expose the grafts under the UV light overnight.
9. Characterize the morphology and structure of the electrospun vascular graft by scanning electron microscope (SEM) (Fig. 2b).

10. Observe the distribution of two fibers in the PCL/COL-7A vascular graft by laser scanning confocal microscope (CLSM) (Fig. 2b).
11. Evaluate the releasing profile of 7A peptide from the graft in PBS at 37 °C (Fig. 2c).

3.2 Rat Abdominal Aorta Replacement Model

Sterilize all materials and instruments before surgery. During the surgery, ensure that the operator wears a gauze mask and sterile gloves to avoid infections. Keep the room temperature at 27–30 °C to maintain the animal body temperature. Follow local Institutional Animal Care and Use Committee (IACUC) guidelines regarding analgesia.

1. Choose male Sprague Dawley rats (270–300 g) as the recipient of vascular grafts.
2. Fast the rat for about 24 h before surgery.
3. Anesthetize the rat quickly with isoflurane gas, and then anesthetize the rat by intraperitoneal injection of chloral hydrate (330 mg/kg) (*see* **Note 7**).
4. Ensure adequate anesthesia, place the rat under the operating microscope in a supine position, and fix limbs of the rat (*see* **Note 8**).
5. Perform anticoagulation treatment for the rat with heparin sodium solution (50 UI/mL) by tail vein injection (100 UI/kg).
6. Use a razor blade to shave off the fur in the anterior abdominal wall, and wipe the skin using iodine prep pad and medical alcohol solution.
7. Incise the skin and muscle layer along the midline of the abdomen. Ensure that the incision is about 4–5 cm long, and then expose the abdominal cavity.
8. Retract the intestines, and wrap them with moistened gauze.
9. Dissect the abdominal aorta carefully using a microtweezer (*see* **Note 9**).
10. Identify and ligate all small branches that encircled the aorta using 9-0 monofilament nylon sutures.
11. Clamp the isolated segment (up to 1 cm in length) of the aorta using two vascular clamps.
12. Transect the abdominal aorta between two clamps using a micro-scissor to create the anastomotic sites.
13. Flush the aortic segment between clamps immediately using heparin sodium solution (50 UI/mL) to remove the residual blood.
14. Peel off the adventitia using a micro-scissor meticulously.

15. Anastomose the graft (2 mm inner diameter and 1 cm in length) to the abdominal aorta in an end-to-end fashion with 8 interrupted stitches using 9-0 monofilament nylon sutures.
16. Perform four anastomoses in a sequence of 9, 3, 12, and 6 o'clock positions at the proximal side, then anastomose other four stitches between two sutures (*see* **Note 10**). Then suture the distal side in the same manner.
17. Remove the distal clamp to allow the blood to flow into the graft, identify the bleeding sites, and perform additional stitches immediately.
18. Ensure the complete hemostasis, remove the proximal clamp, and restore the blood flow.
19. Press the suture ends to stop the bleeding using a sterile cotton ball or a small gauze sponge. Press for about 3 min until hemostasis. Control the entire ischemic time of the surgery within 45 min (Fig. 2d).
20. Remove the gauze, and return the intestines back into the abdominal cavity gently.
21. Swash the abdominal cavity using warm physiological saline solution with gentamicin (320 UI/mL).
22. Sew up the abdominal wall using 3-0 nylon sutures in the muscle and skin layer, respectively (*see* **Note 11**).
23. Swab the wound with iodine to prevent infection.
24. Place the rat into a clean and dry cage on a heating pad to maintain animal body temperature, and then wait for the rat to recover from anesthesia.
25. After it regains consciousness, put the rat in a single cage with food and water.
26. Return the rat to the company of other animals until it fully recovers.
27. Monitor the patency of the vascular graft at various time-points by ultrasound (Fig. 2d).
28. Euthanize rats according to institutional guidelines at predetermined time-points. Sacrifice the rat with overdose chloral hydrate.
29. Incise the skin and muscle layer of the rat, and open the chest.
30. Cut the right ventricular apex, and then inject heparin sodium solution (25 UI/mL) into the left ventricle slowly for perfusion.
31. Explant the vascular graft, and peel off the extra connective tissue and fat layer outside the vessel wall.
32. Rinse the graft with saline, and bisect.

3.3 Histological Analysis

1. Embed the vascular graft in optimal cutting temperature compound, freeze in liquid nitrogen, and then cryo-section into 6μm in thickness.
2. Evaluate vascular regeneration by the standard histological analysis, including hematoxylin and eosin (H&E) staining and immunofluorescence staining with α-SMA and CD31 antibodies.

3.3.1 H&E Staining

1. Rinse the frozen section under tap water for 5 min.
2. Stain the nucleus in hematoxylin for 3 min, and rinse with tap water for 5 min.
3. Differentiate the frozen section in hydrochloric acid ethanol differentiation solution for 1–3 s, and rinse with tap water for 5 min.
4. Stain the extracellular matrix in eosin solution for 3 min, and rinse with tap water for 5 min (*see* **Note 12**).
5. Dip in 70%, 80%, 90%, 95% alcohol gradient for 1 s, followed by 100% ethanol dehydration twice, 1 min each time. Dry the ethanol in a ventilated cabinet.
6. Immerse the slice in xylene twice, 3 min each time.
7. Seal the slice with neutral gum and cover the slip.
8. Dry the xylene and neutral gum.
9. Photograph the slice with upright microscope.

3.3.2 Immunofluorescence Staining

1. Fix the frozen section in ice acetone for 10 min.
2. Keep the slice in the air for 10 min to ensure sufficient dry.
3. Rinse 2× with PBS, 5 min each time. Remove the residual PBS buffer (*see* **Note 13**).
4. For intracellular antigen staining, immerse the slice in 0.1% Triton-PBS at room temperature for 10 min. Wash 2× with PBS, 5 min each time (*see* **Note 14**).
5. Place the slice in a humidified plastic container and incubate the samples with 5% normal goat serum for 45 min at 4 °C.
6. Prepare the primary antibody working solution, and store at 4 °C (*see* **Note 15**).
7. Remove the blocked serum of the slice, and incubate the primary antibody on the surface of the slice. Place the slice in a humidified plastic container at 4 °C overnight.
8. After 12 h of incubation, re-warm the slice for 1 h at room temperature, and wash the slice 6× with PBS buffer, 5 min each time.
9. Remove the residual PBS buffer. Incubate the slice with secondary antibody, and keep it in the dark for 2 h. Wash the slice with 6× with PBS, 5 min each time.

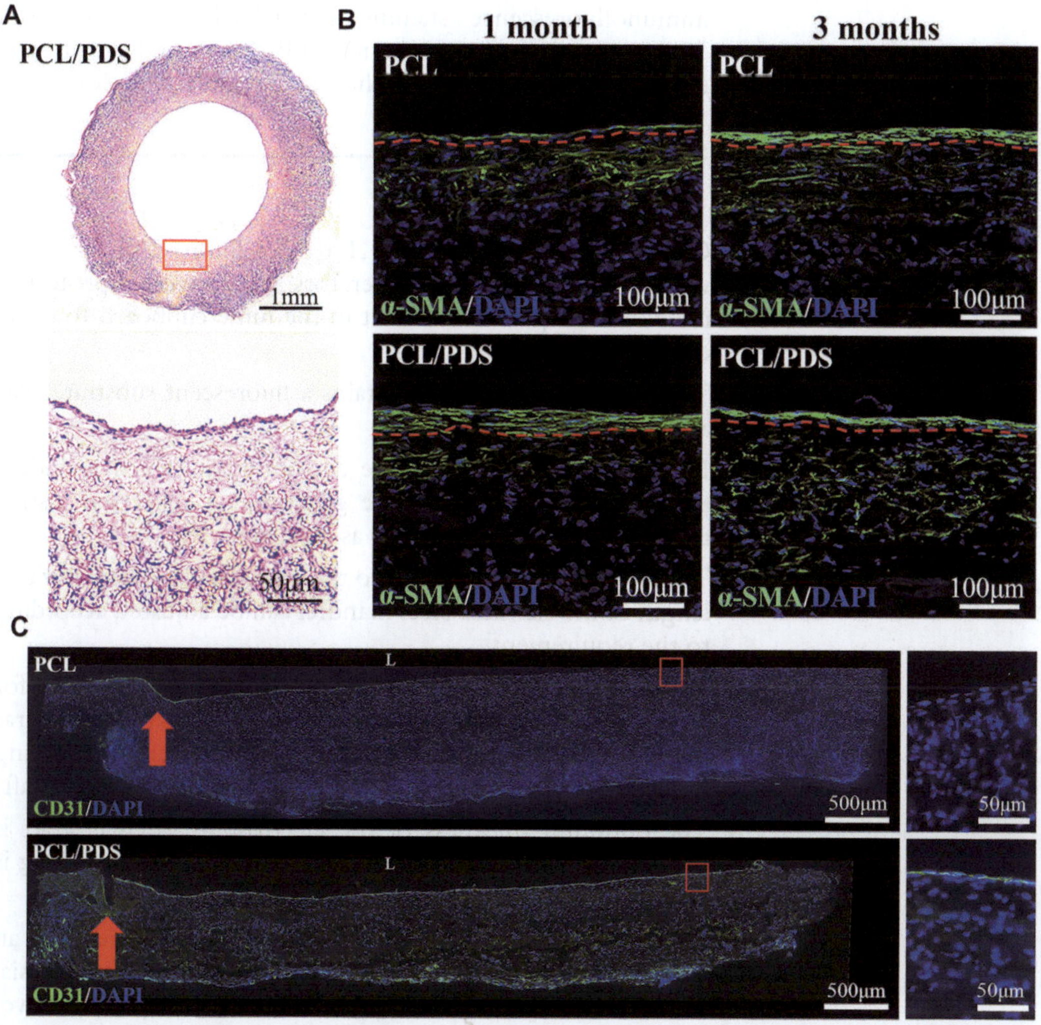

Fig. 3 Histological analysis of the explanted PCL/PDS vascular grafts. (**a**) H&E staining shows the tissue regeneration in the explanted graft. (**b**) Longitudinal sections were stained with anti-smooth muscle actin (α-SMA) antibody to identify the smooth muscle regeneration of grafts at 1 and 3 months. (**c**) Endothelialization was analyzed by immunofluorescence staining of longitudinal sections with CD31 antibody at 1 month. (L: lumen; red arrows: suture site)

10. Counterstain the nuclei with 4,6-diamidino-2-phenylindole (DAPI) containing mounting solution.
11. Observe the slice under a fluorescence microscope.
12. PCL/PDS vascular grafts showed good patency after implantation (Fig. 3a). The α-SMA$^+$ (α-anti-smooth muscle actin) cells were uniformly distributed in the area close to the lumen of the PCL/PDS graft, indicating the enhanced regeneration of smooth muscle because of the rapid degradation of PDS fibers compared to the PCL control (Fig. 3b). The endothelium regeneration was analyzed based on the

immunofluorescence staining for CD31, and the result demonstrated that incorporation of PDS fibers effectively promoted endothelialization in the vascular graft (Fig. 3c).

4 Notes

1. Chloroform, methanol, and 1,1,1,3,3,3-fluoro-2-propanol are volatile organic solvents. Operators should wear a protective mask, and prepare the solvent in the fume cupboard to avoid direct skin contact.
2. The secondary antibody contains a fluorescent substrate. Pay attention to avoid light.
3. Electrospinning apparatus are custom-made which include a syringe pump, a high-voltage generator, a collection instrument, and a rotating mandrel as collector.
4. The syringe is perpendicular to the mandrel. The diameter and length of the stainless-steel mandrel can be adjusted according to the requirement.
5. The environmental factors of electrospinning are as follows: for example, the humidity is preferably 40–50%, and the temperature is preferably about 20–25 °C. The time for electrospinning is determined on the flow rate and the dimension of the graft.
6. The optimum thickness of the electrospun vascular graft is between 350 and 500μm, and the time for electrospinning is about 6–8 min.
7. Grasp the back neck of the rat, and keep its head into an anesthesia valve with isoflurane gas; after the light anesthesia, insert the springe needle into the abdominal cavity of the lower abdomen, and induce the rat for deep anesthesia with chloral hydrate (330 mg/kg) by an intraperitoneal injection.
8. Ensure adequate anesthetization by observing that the rat has relaxed muscles and steady breathing. During the surgery, observe the breathing of the rat at any time. Increases in breathing rate or shallow breathing indicates that the rat is about to wake up. Anesthetize the rat with isoflurane until the end of the surgery.
9. Use a microtweezer to tear apart the connective tissue between the abdominal aorta and vein carefully.
10. Every stitch is required to ensure the native side is slightly embedded in the graft.
11. Suture the muscle layer continuously, and then suture the skin intermittently.
12. The soak time of hematoxylin and eosin solution can be adjusted according to dying effects of the slice.

13. Use the kerchief paper to erase the excess PBS buffer around the sample to avoid diluting the working solution.
14. For intracellular antigen staining, use 0.1% Triton-PBS to permeate the cell membrane before blocking with serum. For cell membrane antigen staining, this step can be omitted.
15. The dilution ratio of the primary antibody is determined according to the instruction. Dilute antibody solution is freshly prepared for each experiment.

Acknowledgments

This study is supported by the National Natural Science Foundation of China (Nos. 81925021, 81921004, and 81871500) and Science and Technology Project of Tianjin of China (No. 18JCJQJC46900).

References

1. Zheng W, Wang Z, Song L, Zhao Q, Zhang J, Li D, Wang S, Han J, Zheng XL, Yang Z, Kong D (2012) Endothelialization and patency of RGD-functionalized vascular grafts in a rabbit carotid artery model. Biomaterials 33 (10):2880–2891. https://doi.org/10.1016/j.biomaterials.2011.12.047
2. Zhang M, Wang K, Wang Z, Xing B, Zhao Q, Kong D (2012) Small-diameter tissue engineered vascular graft made of electrospun PCL/lecithin blend. J Mater Sci Mater Med 23(11):2639–2648. https://doi.org/10.1007/s10856-012-4721-4
3. Roh JD, Sawh-Martinez R, Brennan MP, Jay SM, Devine L, Rao DA, Yi T, Mirensky TL, Nalbandian A, Udelsman B, Hibino N, Shinoka T, Saltzman WM, Snyder E, Kyriakides TR, Pober JS, Breuer CK (2010) Tissue-engineered vascular grafts transform into mature blood vessels via an inflammation-mediated process of vascular remodeling. Proc Natl Acad Sci U S A 107 (10):4669–4674. https://doi.org/10.1073/pnas.0911465107
4. Yu J, Wang A, Tang Z, Henry J, Li-Ping Lee B, Zhu Y, Yuan F, Huang F, Li S (2012) The effect of stromal cell-derived factor-1alpha/heparin coating of biodegradable vascular grafts on the recruitment of both endothelial and smooth muscle progenitor cells for accelerated regeneration. Biomaterials 33 (32):8062–8074. https://doi.org/10.1016/j.biomaterials.2012.07.042
5. Wu W, Allen RA, Wang Y (2012) Fast-degrading elastomer enables rapid remodeling of a cell-free synthetic graft into a neoartery. Nat Med 18(7):1148–1153. https://doi.org/10.1038/nm.2821
6. Pektok E, Nottelet B, Tille JC, Gurny R, Kalangos A, Moeller M, Walpoth BH (2008) Degradation and healing characteristics of small-diameter poly(epsilon-caprolactone) vascular grafts in the rat systemic arterial circulation. Circulation 118(24):2563–2570. https://doi.org/10.1161/CIRCULATIONAHA.108.795732
7. Wang Z, Cui Y, Wang J, Yang X, Wu Y, Wang K, Gao X, Li D, Li Y, Zheng XL, Zhu Y, Kong D, Zhao Q (2014) The effect of thick fibers and large pores of electrospun poly (epsilon-caprolactone) vascular grafts on macrophage polarization and arterial regeneration. Biomaterials 35(22):5700–5710. https://doi.org/10.1016/j.biomaterials.2014.03.078
8. de Valence S, Tille JC, Mugnai D, Mrowczynski W, Gurny R, Moller M, Walpoth BH (2012) Long term performance of polycaprolactone vascular grafts in a rat abdominal aorta replacement model. Biomaterials 33 (1):38–47. https://doi.org/10.1016/j.biomaterials.2011.09.024
9. Yao Y, Wang J, Cui Y, Xu R, Wang Z, Zhang J, Wang K, Li Y, Zhao Q, Kong D (2014) Effect of sustained heparin release from PCL/chitosan hybrid small-diameter vascular grafts on anti-thrombogenic property and endothelialization. Acta Biomater 10(6):2739–2749.

https://doi.org/10.1016/j.actbio.2014.02.042

10. Pan Y, Zhou X, Wei Y, Zhang Q, Wang T, Zhu M, Li W, Huang R, Liu R, Chen J, Fan G, Wang K, Kong D, Zhao Q (2017) Small-diameter hybrid vascular grafts composed of polycaprolactone and polydioxanone fibers. Sci Rep 7(1):3615. https://doi.org/10.1038/s41598-017-03851-1
11. Pan Y, Yang J, Wei Y, Wang H, Jiao R, Moraga A, Zhang Z, Hu Y, Kong D, Xu Q, Zeng L, Zhao Q (2018) Histone deacetylase 7-derived peptides play a vital role in vascular repair and regeneration. Adv Sci 5 (8):1800006. https://doi.org/10.1002/advs.201800006
12. Issa Bhaloo S, Wu Y, Le Bras A, Yu B, Gu W, Xie Y, Deng J, Wang Z, Zhang Z, Kong D, Hu Y, Qu A, Zhao Q, Xu Q (2018) Binding of Dickkopf-3 to CXCR7 enhances vascular progenitor cell migration and degradable graft regeneration. Circ Res 123(4):451–466. https://doi.org/10.1161/CIRCRESAHA.118.312945

Chapter 13

Controlling Pore Size of Electrospun Vascular Grafts by Electrospraying of Poly(Ethylene Oxide) Microparticles

Muhammad Rafique, Adam C. Midgley, Tingting Wei, Lianyong Wang, Deling Kong, and Kai Wang

Abstract

Electrospinning has become a popular polymer processing technique for application in vascular tissue engineering due to its unique capability to fabricate porous vascular grafts with fibrous morphology closely mimicking the natural extracellular matrix (ECMs). However, the inherently small pore sizes of electrospun vascular grafts often inhibit cell infiltration and impede vascular regeneration. Here we describe an effective and controllable method to increase the pore size of electrospun poly(ε-caprolactone) (PCL) vascular graft. With this method, composite grafts are prepared by turning on or off electrospraying of poly(ethylene oxide) (PEO) microparticles during the process of electrospinning PCL fibers. The PEO microparticles are used as a porogen agent and can be subsequently selectively removed to create a porogenic layer within the electrospun PCL grafts. Three types of porogenic PCL grafts were constructed using this method. The porogenic layer was either the inner layer, the middle one, or the outer one.

Key words Electrospinning, Electrospraying, PCL, PEO, Vascular grafts, Pore size, Porogenic layer, Cellular infiltration

1 Introduction

Artificial vascular grafts with highly interconnected porous structure have been widely applied in vascular tissue engineering. Porous structure plays an important role in ensuring cell migration, nutrient and waste transport, and eventual integration of the implanted scaffold with the host tissues [1]. Many materials-fabrication techniques such as knitting [2], phase separation [3], 3D printing [4], wet-spinning [5], and electrospinning [6, 7] have been employed to fabricate porous scaffolds and artificial vascular grafts. Among these, electrospinning has become a popular technique for vascular graft fabrication because it has the unique features including simplicity, affordability, and wide-range selection of materials [8].

Feng Zhao and Kam W. Leong (eds.), *Vascular Tissue Engineering: Methods and Protocols*,
Methods in Molecular Biology, vol. 2375, https://doi.org/10.1007/978-1-0716-1708-3_13,

Electrospun vascular grafts possess appropriate characteristics for replacement of the damaged blood vessels; their high surface area-to-volume ratio and morphological similarity to the fibrillar ECM enable the enhancement of cell adhesion and proliferation [9]. Although pores in electrospun vascular grafts are highly interconnected, their inherently small size that arises from the dense deposition of fibers hinder cellular infiltration into scaffolds [10, 11], which is unfavorable for cellularization and tissue regeneration. Consequently, many efforts have been made to increase the pore size of electrospun scaffolds such as increasing the fiber diameter [12], using specific collector apparatus [13, 14], salt leaching [15], cryogenic electrospinning [16], and selective removal of sacrificial fibers [17]. Yet these approaches lack effectiveness and feasibility for regulating pore size and location in preparation of vascular graft.

Similar to electrospinning, electrospraying also involves the use of high voltage, which stretches the liquid to form a conical jet at the tip of the needle that further breaks into small droplets, as induced through an applied electrostatic field [18]. Nano/microparticles can be derived from various materials by using electrospraying techniques [19–21]. In our previous study, we successfully prepared PEO microparticles using this technique [22]. We fabricated silk/PEO composite scaffolds by co-electrospraying PEO microparticles during the process of electrospinning of silk fibers. The subsequent removal of PEO microparticles generated electrospun silk scaffolds with high porosity and large pores that significantly enhanced cellular infiltration [22].

PCL is a commonly used polymer for the fabrication of vascular grafts due to its excellent biocompatibility and suitable mechanical strength [23]. Here we describe a method of using PEO microparticles as a porogen agent to increase the pore size of electrospun PCL vascular grafts [24]. Firstly, composite grafts were prepared by turning on or off the electrospraying of PEO microparticles during the process of electrospinning PCL fibers. Then, PEO microparticles were selectively removed to create porogenic layers within the PCL grafts. By utilizing this PEO electrospraying method, we produced three types of porogenic PCL grafts, which include: three-layered grafts with a porogenic middle layer between dense inner and outer layers; two-layered grafts with a porogenic inner layer and a dense outer layer; and two-layered grafts with a dense inner layer and a porogenic outer layer. Our results indicated that electrospraying PEO microparticles is an effective and controllable method to regulate the pore structure of electrospun vascular grafts.

2 Materials

Prepare all solutions using analytical grade reagents. Prepare and store all reagents at room temperature (unless indicated otherwise). Carefully follow all waste disposal regulations when disposing the waste materials. The following sections list the materials required for fabrication of grafts and the morphological and mechanical analysis of vascular grafts.

2.1 Preparation of Polymer Solution

1. PCL solution: 10% w/v. Weigh 0.6 g of PCL, and transfer it to a 15 mL glass bottle. Add 5 mL of chloroform and 1 mL of methanol to the glass bottle. Add a magnetic stirrer bar to the vial, and place it on a magnetic stirrer. Keep the solution stirring for 12 h at room temperature (*see* **Note 1**).
2. PEO solution: 2.0 g/mL. Weigh 12 g of PEO, and transfer it to a 15 mL glass bottle. Add 6 mL of chloroform to the glass vial. Add a magnetic stirrer bar to the bottle, and place the bottle into a beaker containing water heated at 50 °C. Place the beaker on a magnetic stirrer with a temperature control function. Keep the solution stirring for 20 min, and maintain the PEO solution temperature at 50 °C. Transfer the clear PEO solution to a beaker containing ice-water mixture, and continue magnetic stirring PEO solution for 15 s until the solution becomes turbid (*see* **Note 2**).

2.2 Removal of PEO Microparticles

1. Gradient ethanol aqueous solutions: 50% v/v, 60% v/v, 70% v/v, 80% v/v, 90% v/v. Prepare distilled water containing 50, 60, 70, 80, and 90% v/v alcohol.

3 Methods

3.1 Fabrication of Electrospraying PEO Microparticles

1. Place a stainless-steel mandrel (2.0 mm in diameter, 25 cm in length) on the collection instrument.
2. Use an aluminum foil to wrap up the stainless-steel mandrel.
3. Load PEO solution into a separate 10 mL syringe. Bend a 7-G needle to a 120° angle using a pipe and tube bender. Attach the needle to the syringe head (*see* **Note 3**)
4. Electrospray PEO microparticles using the electrospraying setup as shown in Fig. 1. Electrospray PEO microparticles from oblique direction to a stainless-steel mandrel. Maintain the temperature and humidity of material preparation environment at around 25 °C and around 50%, respectively. Fix the syringe containing PEO solution (2.0 g/mL) at a working distance of 11 cm from needle tip to collector. Set the flow rate of PEO solution at 2 and 4 mL/h, respectively. Set the voltage at 13 kV.

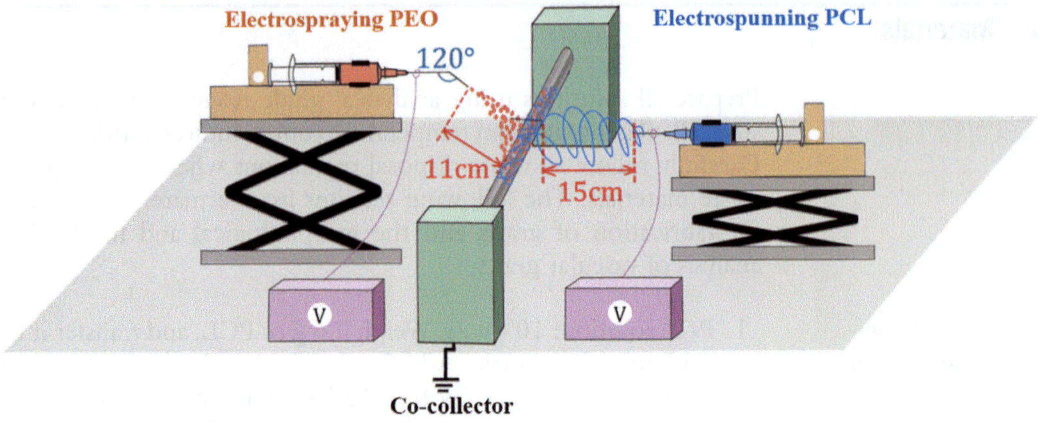

Fig. 1 Schematic of the setup for fabricating PCL/PEO composite vascular grafts

3.2 Characterization of PEO Microparticles

A scanning electron microscope (SEM) is required to observe the morphological properties of PEO microparticles. The Image-Pro Plus (IPP; Media Cybernetics, Rockville, MD, USA) software was used to measure the diameter of PEO microparticles based on SEM images.

1. Slightly take aluminum foil from stainless-steel mandrel. Cut the aluminum foil into small pieces. Mount pieces onto sample desk, and sputter coat with gold and palladium (*see* **Note 4**).
2. Observe the shape of PEO microparticles.
3. Calculate the diameter of PEO microparticles prepared at a flow rate of 4 mL/h and 2 mL/h based on the surface SEM images. Calculate the diameter of PEO microparticles by randomly selecting 20 PEO microparticles per image, three images per sample, and three samples per group, and analyze using IPP software.

3.3 Fabrication of PCL/PEO Composite Vascular Grafts

The average diameter of PEO microparticles was found to be $21.12 \pm 18.97\mu m$ at the flow rate of 4 mL/h and $14.10 \pm 6.2\mu m$ at 2 mL/h. Generally, pores greater than 10μm in size could allow adequate cellular infiltration [25]. The average diameter of PEO microparticles prepared at 4 mL/h and 2 mL/h is all larger than 10μm [24]. However, the average diameter of PEO microparticles at 4 mL/h is more inhomogeneous than that at 2 mL/h. So, the PEO microparticles prepared at 2 mL/h were selected to achieve the uniform distribution of pore size in porogenic layers.

1. Place a stainless-steel mandrel (2.0 mm in diameter, 25 cm in length) on the collection instrument.
2. Load the PCL solution into a 10 mL syringe. Attach a 21-G straight needle to the syringe head.

3. Load PEO solution into a separate 10 mL syringe. Bend a 7-G needle to a 120° angle using a pipe and tube bender. Attach the needle to the syringe head

4. Prepare the PCL/PEO composite vascular graft using the electrospinning/electrospraying setup as shown in Fig. 1. Electrospun PCL fibers from horizontal direction to a stainless-steel mandrel, and electrospray PEO microparticles from oblique direction to a stainless-steel mandrel. Set the rotation rate of stainless-steel mandrel at 150 rpm. Maintain the temperature and humidity of material preparation environment at around 25 °C and around 50%, respectively. The specific conditions of the preparation method are as follows: For electrospinning PCL fibers, fix the syringe containing the PCL solution (10% w/v) at a working distance of 15 cm from needle tip to the collector, with electrospinning parameters set to a flow rate of 2 mL/h and a voltage of 15 kV. For electrospraying of PEO microparticles, fix the syringe containing the PEO solution (2.0 g/mL) at a working distance of 11 cm from needle tip to collector, with electrospraying parameters set to a flow rate of 2 mL/h and voltage of 13 kV.

5. Fabricate different PCL-PEO composite vascular grafts by switching the electrospray on or off during the process of electrospinning PCL fibers. Examples of preparations are shown below (*see* **Note 5**):

   (a) Three-layered PCL/PEO composite grafts: Electrospinning PCL for 1.5 min to form the inner layer, electrospinning PCL with electrospraying PEO for 1.5 min to form the middle layer, and electrospinning PCL only for 30 min to form the outer layer.

   (b) Two-layered PCL/PEO composite grafts: Electrospinning PCL with electrospraying PEO for 1.5 min to form the inner layer and electrospinning PCL only for 31.5 min to form the outer layer.

   (c) Reversed two-layered PCL/PEO composite grafts: Electrospinning PCL only for 31.5 min to form the inner layer and electrospinning PCL with electrospraying PEO for 1.5 min to form the outer layer.

   (d) Regular PCL grafts: Electrospinning PCL only for 33 min.

6. Place the fabricated vascular grafts in a vacuum dryer for 2 days to remove residual solvents (*see* **Note 6**).

3.4 Removal of PEO Microparticles from PCL/PEO Composite Vascular Grafts

1. Cut off the thin parts of vascular grafts at both ends of the mandrel, and keep the middle of the graft that has uniform thickness.

2. Immerse the grafts in 100% alcohol for 1 min, and then remove the grafts from the steel mandrels by gentle pulling with the hand (*see* **Note 7**).
3. Cut off the graft end that was pinched for the removal from the mandrel.
4. Place grafts into glass tubes, and place the tubes onto a bench-top shaker. Set the speed of the shaker to 20 rpm. Wash the grafts in decreasing gradients of ethanol aqueous solutions for 5 min each to gradually remove PEO; start at 100% v/v ethanol, and then wash in 10% decrements until the final wash step in 50% v/v ethanol. Wash the grafts three times in distilled water for 5 min each, to further remove residual PEO and ethanol.
5. Place grafts in an open-topped stainless-steel box container, and add distilled water to completely immerse the grafts. Rapidly freeze the water and graft using liquid nitrogen (*see* **Note 8**).
6. Place a preservative parafilm cover across the steel box, and perforate the parafilm (*see* **Note 9**).
7. Place the steel box into a vacuum freeze-dryer, and freeze-dry the vascular grafts (*see* **Note 10**).
8. Leave some freeze-dried grafts at room temperature for current use.
9. Store the remaining grafts at 4 °C.

3.5 Characterization of Porogenic PCL Grafts

A SEM and a tensile-testing machine are required to test the morphological and mechanical properties of the porous PCL grafts.

3.5.1 SEM Analysis

1. Place the dried graft samples and a razor blade in liquid nitrogen for 1 min (*see* **Note 11**).
2. Fix the vascular graft sample with tweezers. Prepare a cross-section sample using the razor blade to cut the graft sample into lengths of 4–6 mm. Hold the small cross-sectioned sample with tweezers, and straighten the sample. Use a blade to cut the graft sample longitudinally to prepare a longitudinal sample (*see* **Note 12**).
3. Dry the samples at room temperature.
4. Mount samples onto aluminum foil, and sputter coat with gold and palladium (*see* **Note 13**).
5. Observe the thickness and location of the loose fiber layer within the porogenic PCL grafts, by the cross-sectional observation (Fig. 2).

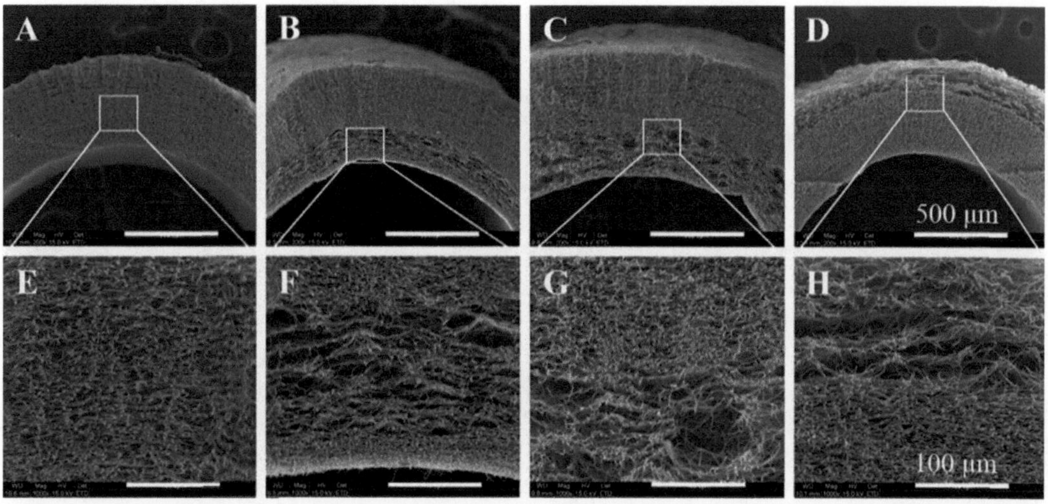

Fig. 2 Cross-sectional SEM images of regular and porogenic electrospun PCL grafts. (**a**, **e**) Regular grafts consisted of dense fibers without a porous layer. (**b**, **f**) Three-layered grafts showing a thin dense inner layer, a porogenic middle layer and a thick dense outer layer. (**c**, **g**) Grafts with a porogenic inner layer and a dense outer layer. (**d**, **h**) Grafts with a dense inner layer and a porogenic outer layer. **a–d**, original magnification ×200. **e–h**, original magnification ×1000. (reproduced from ref. 23 with permission from American Scientific Publishers)

6. Observe the fiber morphology and pore size of the outer surface and the inner surface, by longitudinal observation.
7. Calculate the fiber diameter and pore size based on the surface SEM images, according to the method described by Dong et al. [26]. Calculate the fiber diameter by randomly selecting ten fibers per image, three images per sample, and three samples per group, and measure fiber diameter using IPP software. Calculate the surface area of the pores using IPP software, and normalize it to a circular area (*see* **Note 14**). Regard circular diameter as the pore diameter. To calculate the pore diameter, include at least five pores per image, five images per sample, and three samples per group.

3.5.2 Mechanical Testing

1. Cut vascular grafts into 3-cm-length sections using a razor blade. Measure the thickness of grafts using a micrometer.
2. Hydrate the samples by immersing in gradient ethanol aqueous solutions, starting from 100% v/v ethanol, then subsequent 10% decrement steps until immersion in 50% v/v ethanol. Wash the samples three times in physiological saline for 5 min each, to further remove residual ethanol (*see* **Note 15**).
3. Incubate the samples in 37 °C physiological saline for 30 min prior to testing (*see* **Note 16**).
4. Place the vascular graft samples on a tensile-testing machine with a load capacity of 100 N.

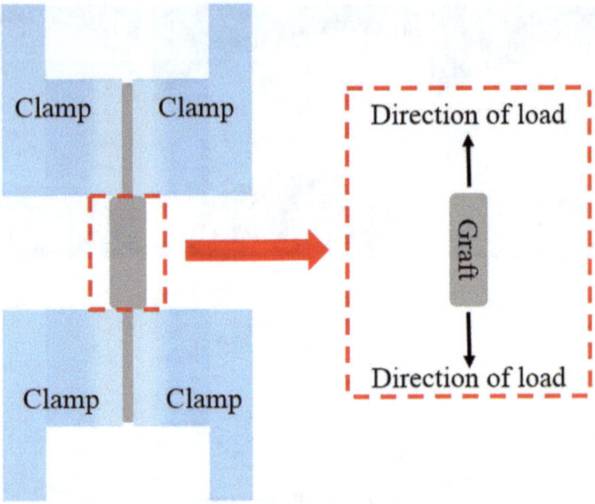

Fig. 3 Diagram of tensile-testing machine for longitudinal mechanical properties of grafts

5. Clamp the grafts with 1 cm inter-clamp distance, and pull longitudinally at a rate of 10 mm/min until rupture (Fig. 3) (*see* **Note 17**).
6. Measure the tensile strength and ultimate elongation at break. Calculate Young's modulus from the initial linear region (up to 5% strain) of the stress-strain curve (*see* **Note 18**).

4 Notes

1. PCL pellets are placed into the glass bottle prior to the addition of the organic solvent. Immediate agitation ensures full contact between PCL pellets and organic solvent. Ensure that the magnetic stirrer bar can rotate freely in the glass bottle as constant stirring is required to fully dissolve the PCL polymer. Fix the glass bottle in position with a flask clamp to prevent toppling during the stirring. Confirm that the cap is securely tightened and the parafilm is used to seal the cap to prevent evaporation of organic solvent, which may adversely affect the concentration of the PCL solution.
2. The complete dissolution of PEO can be ensured by loading PEO particles first and then adding organic solvent. Make sure that the cap is securely tightened and that the parafilm is used to seal the cap to prevent the organic solvent from evaporating and adversely affecting the concentration of the PEO solution. Ensure that the magnetic stirrer bar can freely rotate in the glass bottle, which will help to keep the overall solution temperature uniform. Immediately transfer the glass bottle into a beaker containing an ice-water mixture, after stirring for 20 min at

50 °C. The depth of the ice-water mixture should just be enough to immerse the PEO solution. Ensure that the magnetic stirrer bar in the glass vial can rotate with the magnetic stirrer in the ice-water mixture. This is conducive to the temperature uniformity of PEO polymer solution and ensures the homogeneity of PEO precipitation in the solution.

3. When bending the 7-G needle, make sure that the needle is unobstructed, avoid a dead fold (obstructed) that prevents PEO solution from flowing out. The thermal conductivity of the medical plastic syringe is poor. The use of a plastic syringe to contain the PEO solution can reduce heat conduction and help better maintain the partially precipitated state of PEO. The speed of PEO solution loading into plastic syringe should be fast so as to maintain the temperature of the PEO solution. After loading of the PEO solution into the plastic syringe, the graft fabrication process should started immediately.
4. The SEM preparation of PEO microparticles needs three steps: (1) Take aluminum foil from stainless-steel mandrel. (2) Cut the aluminum foil into small pieces. (3) Mount pieces onto sample desk, and sputter coat with gold and palladium. The process should be as far as slightly to avoid the shedding of PEO microparticles from aluminum foil. If the shedding occurs, the calculation for the average diameter of PEO microparticles will be inaccurate.
5. Pay special attention to the partially precipitated state of PEO solution in the plastic syringe during the process of fabricating PCL/PEO composite vascular grafts. If there is a marked change in the state of the PEO solution (such as too much dissolution or precipitation) during the process of fabricating composite vascular grafts, the PEO solution should be extruded from plastic syringe to a 15 mL glass bottle, and the materials preparation procedure 2.1 should be repeated (Subheading 2.1).
6. In the process of vacuum formation, ensure the composite vascular grafts wrapped around stainless-steel mandrels are kept upright, to avoid collision between the stainless-steel mandrels or the stainless-steel mandrels and other objects, which can lead to structural detriment of the vascular grafts. Excessive collision will cause depression of the vascular grafts wrapped around the stainless-steel mandrels and collapse the layered structure of vascular grafts.
7. During the process of pulling, keep the graft and the stainless-steel mandrel vertical, and do not bend the vascular graft, which is not beneficial to maintain the original structure of vascular grafts. The pulling process should not generate too

much resistance. If the resistance is too large, it may result from applying too much force when pinching the end of the graft.

8. The length of the stainless-steel box should be slightly longer than the length of the vascular graft to avoid the bending of grafts. Liquid nitrogen freezing can produce smaller ice crystals than conventional freezing and is more beneficial for maintaining the original structure of the vascular graft.
9. Parafilm sealing and perforation should be completed rapidly to avoid sample thawing.
10. Freeze-drying should be long enough to ensure the complete removal of ice crystals. After lyophilization, the freeze-dryer is returned to normal atmospheric pressure by releasing the gas valve, so that the container seal can be opened in order to access the sample. It is important to note that gas release should be slow, to prevent sudden pressure changes resulting in the sample breaking through the parafilm and becoming contaminated.
11. Rapid freezing with liquid nitrogen can make the PCL grafts stiff, which is easier to cut and maintain the original pore structure of vascular grafts.
12. All cutting should be completed in one procedure, and cuts should be made as cleanly as possible to keep the cross-section surface favorable to observe by SEM. For cross-sectioned samples, do not touch the cross section when moving the sample. For longitudinal samples, do not touch the inner or outer surfaces when moving the sample.
13. When fixing the sample to the aluminum foil, do not touch the surface that is to be observed. In order to avoid PCL grafts melting due to high temperature during gold spraying, low-temperature gold spraying equipment should be used for spraying gold onto the surface of regular and porogenic PCL grafts. Ensure that the sample is firmly adhered to the conductive adhesive tape to prevent the sample from falling off under the vacuum state during SEM observation, which may result in damage to the SEM instrument.
14. Electrospun grafts possessed three-dimensional structure consisting of lots of fibers. When measuring the pore size based on SEM images, the pore area should be formed by topmost fibers.
15. Ultimately, vascular grafts are required for in vivo implantation, so mechanical tests should be performed at temperature close to the body temperature. PCL is a hydrophobic material; thus, alcohol gradient rehydration is a necessary step. Rehydration promotes the influx of 37 °C physiological saline into the graft, which can allow temperature uniformity throughout the graft.

16. Ensure sufficient incubation time to allow the temperature of the grafts to reach a uniform temperature of 37 °C. However, pay special attention to the temperature of the physiological saline. Do not place the vascular grafts in excessively high-temperature physiological saline prior to cooling to 37 °C, as the melting point of PCL is 52 °C.
17. In the process of tensile testing, it is necessary to pay attention to whether the sample breaks or slips out from the clamp. The broken sample should be recorded and included in the statistical data. Rubber tape can be used to affix grafts to the clamp surface to increase the friction between the clamp and the vascular grafts and reduce risk of slippage.
18. In general, up to 5% (strain) of the stress strain curve will be selected for the determination of Young's modulus of materials. You can also choose other regions, such as 5–8%. In principle, generally do not exceed 10% of the stress-strain for calculation. Note which boundaries were chosen for the calculation of Young's modulus.

Acknowledgments

This work was supported by the National Natural Science Foundation of China (NSFC) projects (81772000, 81530059, 91939112), NSFC Research Fund for International Young Scientists (81850410552), and Tianjin Natural Science Foundation (18JCZDJC37600).

References

1. Ekaputra AK, Prestwich GD, Cool SM et al (2008) Combining electrospun scaffolds with electrosprayed hydrogels leads to three-dimensional cellularization of hybrid constructs. Biomacromolecules 9(8):2097–2103
2. Xie Y, Guan Y, Kim SH et al (2016) The mechanical performance of weft-knitted/electrospun bilayer small diameter vascular prostheses. J Mech Behav Biomed 61:410–418
3. Wang W, Nie W, Zhou X et al (2018) Fabrication of heterogeneous porous bilayered nanofibrous vascular grafts by two-step phase separation technique. Acta Biomater 79:168–181
4. Guo Z, Grijpma DW, Poot AA et al (2017) Preparation and characterization of flexible and elastic porous tubular PTMC scaffolds for vascular tissue engineering. Polym Advan Technol 28(10):1239–1244
5. Zhu M, Wu Y, Li W et al (2018) Biodegradable and elastomeric vascular grafts enable vascular remodeling. Biomaterials 183:306–318
6. Muylaert DE, van Almen GC, Talacua H et al (2016) Early in-situ cellularization of a supramolecular vascular graft is modified by synthetic stromal cell-derived factor-1α derived peptides. Biomaterials 76:187–195
7. Wang K, Zheng W, Pan Y et al (2016) Three-layered PCL grafts promoted vascular regeneration in a rabbit carotid artery model. Macromol Biosci 16(4):608–618
8. Zhong S, Zhang Y, Lim CT et al (2012) Fabrication of large pores in electrospun nanofibrous scaffolds for cellular infiltration: a review. Tissue Eng Part B Rev 18(2):77–87
9. Zhao L, Ma S, Pan Y et al (2016) Functional modification of fibrous PCL scaffolds with fusion protein VEGF-HGFI enhanced

cellularization and vascularization. Adv Healthc Mater 5(18):2376–2385

10. Leong MF, Rasheed MZ, Lim TC et al (2009) In vitro cell infiltration and in vivo cell infiltration and vascularization in a fibrous, highly porous poly(D,L-lactide) scaffold fabricated by cryogenic electrospinning technique. J Biomed Mater Res A 91A(1):232–240
11. Rnjak-Kovacina J, Weiss AS (2011) Increasing the pore size of electrospun scaffolds. Tissue Eng Part B Rev 17(5):365–372
12. Lowery JL, Datta N, Rutledge GC et al (2010) Effect of fiber diameter, pore size and seeding method on growth of human dermal fibroblasts in electrospun poly(ε-caprolactone) fibrous mats. Biomaterials 31(3):491–504
13. Blakeney BA, Tambralli A, Anderson JM et al (2011) Cell infiltration and growth in a low density, uncompressed three-dimensional electrospun nanofibrous scaffold. Biomaterials 32 (6):1583–1590
14. McClure MJ, Wolfe PS, Simpson DG et al (2012) The use of air-flow impedance to control fiber deposition patterns during electrospinning. Biomaterials 33(3):771–779
15. Wright LD, Andric T, Freeman JW et al (2011) Utilizing NaCl to increase the porosity of electrospun materials. Mater Sci Eng C 31 (1):30–36
16. Leong MF, Rasheed MZ, Lim TC et al (2009) In vitro cell infiltration and in vivo cell infiltration and vascularization in a fibrous, highly porous poly (D,L-lactide) scaffold fabricated by cryogenic electrospinning technique. J Biomed Mater Res Part A 91A(1):231–240
17. Baker BM, Gee AO, Metter RB et al (2008) The potential to improve cell infiltration in composite fiber-aligned electrospun scaffolds by the selective removal of sacrificial fibers. Biomaterials 29(15):2348–2358
18. Salata OV (2005) Tools of nanotechnology: electrospray. Curr Nanosci 1(1):25–33
19. Zhang S, Kawakami K (2010) One-step preparation of chitosan solid nanoparticles by electrospray deposition. Int J Pharm 397 (1):211–217
20. Lee YH, Mei F, Bai MY et al (2010) Release profile characteristics of biodegradable-polymer-coated drug particles fabricated by dual-capillary electrospray. J Control Release 145(1):58–65
21. Almería B, Deng W, Fahmy TM et al (2010) Controlling the morphology of electrospray-generated PLGA microparticles for drug delivery. J Colloid Interf Sci 343(1):125–133
22. Wang K, Xu M, Zhu M et al (2013) Creation of macropores in electrospun silk fibroin scaffolds using sacrificial PEO-microparticles to enhance cellular infiltration. J Biomed Mater Res Part A 101(12):3474–3481
23. Zhu L, Wang K, Ma T et al (2017) Noncovalent bonding of RGD and YIGSR to an electrospun poly(ε-caprolactone) conduit through peptide self-assembly to synergistically promote sciatic nerve regeneration in rats. Adv Healthc Mater 6(8):1600860
24. Wang K, Zhu M, Li T et al (2014) Improvement of cell infiltration in electrospun polycaprolactone scaffolds for the construction of vascular grafts. J Biomed Nanotechnol 10 (8):1588–1598
25. Zhong S, Zhang Y, Lim TC et al (2012) Fabrication of large pores in electrospun nanofibrous scaffolds for cellular infiltration: a review. Tissue Eng Part B Rev 18(2):77–87
26. Wang S, Zhang Y, Wang H et al (2009) Fabrication and properties of the electrospun polylactide/silk fibroin-gelatin composite tubular scaffold. Biomacromolecules 10 (8):2240–2244

Chapter 14

Injectable Hydrogels for Vascular Tissue Engineering

Fengqiao Li, William Ho, and Xiaoyang Xu

Abstract

Injectable scaffolds made of biodegradable biomaterials can stabilize a myocardial infarct and promote cardiac repair. Here, we describe an injectable, citrate-containing polyester hydrogel which can release citrate as a cell regulator via hydrogel degradation and simultaneously show sustained release of an encapsulated myeloid-derived growth factor (Mydgf). Xu et al. described the synthesis of hydrogel with biocompatible starting chemicals including citric acid and poly(ethylene glycol) diol. The characterization of materials demonstrated that the developed hydrogels possess tunable degradation and mechanical properties and exhibit sustained drug release. The authors also observed improved postmyocardial infarction (MI) heart repair in a rat MI model through coupling the therapeutic effect of the hydrogel degradation product (citrate) with encapsulated Mydgf. In their study, hematoxylin and eosin (H&E) staining and Masson's trichrome staining were performed on heart samples to evaluate the change in heart structure. Furthermore, immunohistochemistry was used to study neovascularization. Their results showed that the intramyocardial injection of Mydgf-loaded citrate-containing hydrogel significantly reduced scar formation and infarct size, increased wall thickness and neovascularization, and improved heart function.

Key words Myocardial infarction (MI), Injectable hydrogel, Cardiac remodeling, Mydgf

1 Introduction

Patients who have myocardial infarction are at a major risk for further cardiovascular events, but limited information exists on the effects of specific chemotherapy regimens used to treat these patients on their risk of death from myocardial infarction [1, 2]. Furthermore, the lack of blood supply to the cardiac tissues leads to significant cardiomyocyte death. The rapid wound healing process in the infarcted area forms fibrotic instead of functional cardiac tissue; ventricular remodeling and a loss of contractility follow. These processes cause the loss of heart function, which can lead to end-stage heart failure [3]. For the treatment of an infarction, delivering growth factors has been widely used for the regeneration of hearts.

Feng Zhao and Kam W. Leong (eds.), *Vascular Tissue Engineering: Methods and Protocols*, Methods in Molecular Biology, vol. 2375, https://doi.org/10.1007/978-1-0716-1708-3_14,

Hydrogels are promising for the treatment of myocardial infarction as they can function as delivery vehicles for cells and growth factors. Hydrogels, when injected into the body, exist in close proximity with biological systems and thus need to interface with a complex, dynamically changing environment. Cells surrounding the implant are able to sense the materials' mechanical stiffness, surface topography, and chemistry and translate these cues into signaling events that govern cell fate [4–7]. Therefore, injectable hydrogels have received much attention as 3D scaffolds due to their widespread applications in the fields of drug delivery, tissue engineering, and regenerative medicine [5].

Mydgf is an evolutionarily conserved, 142-amino acid secreted protein with previously unknown biological functions. Mydgf does not belong to any known growth factor or cytokine family. Recently, Mydgf was identified in a human bone marrow cell secretome screened based on its ability to promote cardiomyocyte survival and angiogenesis [8, 9]. In mice, bone marrow-derived monocytes and macrophages were shown to secrete Mydgf as part of an endogenous adaptive response to acute MI, and treatment with recombinant Mydgf was found to improve heart function after MI [9].

In this chapter, we describe the process of developing a robust citrate-presenting injectable hydrogel preparation method and an approach for local sustained delivery of Mydgf for post-MI heart regeneration. We will also introduce different methods to characterize these hydrogel formulations. For example, a discovery hybrid rheometer (DHR) is used to evaluate the mechanical properties of hydrogels. Scanning electron microscopy (SEM) is performed to study the morphology of hydrogels. The release kinetics are investigated by high-performance liquid chromatography (HPLC). Furthermore, to evaluate biocompatibility in vitro, the XTT assay is utilized to evaluate the cytotoxicity of poly(propylene carbonate)-ET (PPC-ET)/poly(ethylene glycol) (PEG) hydrogels. The endothelial tube formation assay in HUVECs provides a tool for the assessment of angiogenesis in vitro. Moreover, a rat acute myocardial infarction (AMI) model is established by polypropylene induction. Echocardiographic images, H&E and Masson's trichrome staining, and immunofluorescence assessment are applied to analyze how the developed hydrogel can improve cardiac repair after MI in vivo.

2 Materials

2.1 Reagents

1. Citric acid.
2. PEG (Mw ≈ 200 kDa).
3. Ethyl thioglycolate.

4. Candida antarctica lipase B (CALB).
5. Bovine serum albumin (Mw: 68 kDa).
6. Cell Proliferation Kits (XTT).
7. 8-Arm PEG with terminal maleimide groups (Mw = 10 kDa).
8. Ellman's reagent.

2.2 Instruments

1. Discovery hybrid rheometer (DHR).
2. SEM.
3. HPLC.
4. Confocal microscope.
5. Two-dimensional echocardiography.
6. Ultrasound imaging system.
7. Small-animal positron emission tomography (PET) system.

2.3 Cell Lines

1. NIH 3T3 cell: NIH 3T3 cells are cultured in Dulbecco's Modified Eagle's Medium (DMEM) containing 10% FBS, 100 units/mL penicillin, and 100 mg/mL streptomycin. NIH 3T3 cells are grown in a humidified 5% CO_2 atmosphere at 37 °C in an incubator (*see* **Note 1**).
2. Human umbilical vein endothelial cells (HUVECs): HUVECs cells are cultured in DMEM-F12 supplemented with 10% heat-inactivated FBS and 1% pen/strep and incubated at 37 °C in a humidified incubator with 5% CO_2 (*see* **Note 2**).

2.4 Animals

Obtain male Sprague-Dawley (SD) rats (220–250 g, 6–8 weeks) from a commercial company (*see* **Note 3**).

3 Methods

Carry out all procedures at room temperature unless otherwise specified.

3.1 Synthesis of Biomaterials

3.1.1 Synthesis of PPC

React citric acid with an equimolar amount of PEG under mild stirring at 145 °C for 75 min.

Cool mixture solution to room temperature.

Dialyze the solution for 3 days to purify the PPC.

Freeze-dry sample overnight.

3.1.2 Synthesis of PPC-ET

Dissolve the PPC in acetonitrile to obtain 100 mg/mL PPC solution.

After PPC is completely dissolved in acetonitrile, add ethyl thioglycolate and 1 g CALB (5:1 molar ratio of ethyl thioglycolate to citric acid) to 10 mL PPC solution.

Stir the reaction at 60 °C for 16 h under nitrogen protection.

Stop the reaction by filtering CALB out.

Dialyze the solution for 3 days to purify the resultant PPC-ET oligomer.

Freeze-dry sample overnight.

3.1.3 Synthesis of CRGDS-Functionalized 8-Arm PEG Derivatives

Dissolve CRGDS (1 mg) and 8-arm PEG-maleimide (molar ratio 1:1) in 1 mL PBS (pH = 7.4) with 400 mM triethanolamine for 60 min.

Dialyze the solution for 3 days to purify the product by a dialysis bag (3500 kDa).

Freeze-dry sample for 3 days.

3.1.4 Hydrogel Fabrication

Dissolve oligomer and multiarm PEG derivatives individually in PBS to obtain 50 mg/mL oligomer solution and 1 mg/mL PEG derivative solution.

Form hydrogels in situ within seconds to hours by mixing the two solutions (molar ratio of oligomer to PEG derivative is 2:1), depending on formulation compositions and gelation conditions.

3.2 Characterization of Biomaterials

3.2.1 Dynamic Rheology Measurement

Mix hydrogel precursors sufficiently.

Apply a volume of 280μL to a temperature-controlled stage at preset temperature (25 and 37 °C). Lower the 25 mm parallel stainless steel plate to a predetermined gap to initiate the measurement (*see* **Note 4**).

Record the storage modulus (G′) and loss modulus (G″) as a function of time.

Use the time point where tan (δ) = 1 (i.e., G′/G″ crossover) to define the gelation time.

Add the operation time of 15 s to the data.

3.2.2 Swelling and Porosity Measurement

Incubate cured hydrogel disk samples (≈50μL, diameter = 5.6 mm, height = 2.9 mm) in PBS at 37 °C for 24 h to leach out the unreacted sol fraction.

Measure the initial dry hydrogel weight (W_0) after freeze-drying.

Immerse samples in PBS (pH = 7.4) solution at 37 °C/5% CO_2 for at least 6 h to achieve equilibrium status.

Weigh the swollen hydrogels (W_t) after gently removing the excess water by the filter paper. Calculate the swelling ratio (Q_m) by comparing the wet weight with the dry weight of the hydrogel sample:

$$Q_m = \left(\frac{W_t}{W_0}\right) \times 100\%$$

Immerse the freeze-dried hydrogel samples in DI water (volume known, V_1) in the beaker for 1 h.

Record the total volume (hydrogels + DI water) as V_2, and record the volume of DI water remaining in the beaker as V_3 after removing the hydrated hydrogel samples.

Calculate the porosity (P) of the scaffold:

$$P(\%) = \left[\frac{V1 - V3}{V2 - V3}\right] \times 100\%$$

3.2.3 Morphology of Hydrogels

Use SEM to observe the morphology of the hydrogel samples.

Lyophilize and coat the hydrogel samples with a 20 nm layer of gold using a sputter coater.

Blow test samples clean using compressed air before SEM characterization.

3.2.4 In Vitro Degradation of Hydrogels

Lyophilize hydrogels (~120μL), and record their weights as W_0.

Incubate at 37 °C/5% CO_2 in 48-well cell culture plates containing PBS (pH = 7.4) solution (PBS solution changed daily).

At predetermined time points, remove the hydrogel samples from the incubation, and record their weights after lyophilization as W_t.

Evaluate the degradation of the hydrogel by comparing the initial weight (W_0) and the weight measured at certain time points:

$$\text{weight loss } (\%) = \left[\frac{W_0 - W_t}{W_0}\right] \times 100\%$$

3.3 Release Kinetics Study

Add BSA to PPC-ET precursor solution.

Mix with PEG solution to form protein-loaded hydrogels in situ as described above.

Incubate the yielded BSA-loaded hydrogel disk samples in 2 mL of PBS (changed daily) at 37 °C ($n = 5$ per group).

Analyze a 200μL aliquot of the incubation media for each sample at defined time points using HPLC.

3.4 In Vitro Studies

3.4.1 Cytotoxicity Study

Before the study, sterile filter precursor solutions and sterilize hydrogels by 70% ethanol and UV light. Immerse each sample in 1 mL DMEM at 37 °C/5% CO_2 for 72 h.

Add fetal bovine serum (FBS) to the hydrogel extracts to the final concentration of 10%.

Seed the NIH 3T3 cells at a density of 5×10^4 in cell culture plates (96-well), and incubate at 37 °C/5% CO_2 in high-glucose DMEM containing 10% FBS, 100μg/mL streptomycin, and 100 units/mL penicillin for 24 h.

Next, remove the cell culture medium from the cell culture plates, and replace them with the hydrogel extract.

Perform XTT assay according to the manufacturer's instructions after an additional 48 h culture, and use cell culture plates as control.

3.4.2 Cell Adhesion Study

Sterilize the PC-ET/CRGDS-PEG hydrogel samples by 70% ethanol and UV light before use.

Seed 10,000 cells/cm^2 of NIH 3T3 cells onto each hydrogel sample, and allow them to incubate at 37 °C/5% CO_2 to facilitate cell attachment (cell culture media changed daily).

At defined time points, observe and image cell morphology using a microscope.

3.4.3 Effects of Mydgf Composite Hydrogels on the Vascularization of HUVECs

Use the growth factor-reduced basement membrane extract Matrigel as a culture substratum.

Sterilize hydrogel solutions by filtration before use, and lay a layer of PPC-ET/PEG hydrogels (300μL) and Mydgf-loaded hydrogels (50 ng/mL) over the prechilled 96-well culture plates, while the control group is coated with growth factor-reduced Matrigel matrix until completely solidified.

Culture HUVECs in Dulbecco's Modified Eagle's Medium/F12.

Resuspend trypsinized HUVECs in DMEM medium with only PBS, and add Mydgf (50 ng/mL) to the preconditioned culture plates at a density of 2×10^4 cells/well.

After incubation at 37 °C with 5% CO_2 for 8 h, wash nonadherent HUVECs in the culture plates with PBS.

Image HUVEC cells, and count the tubular numbers to evaluate the tube formation ability (Fig. 1).

Use Image-Pro Plus software to evaluate the tubular length and tubular intersecting nodes in five random fields (200×) based on Mirshahi's method.

Perform each experiment three times.

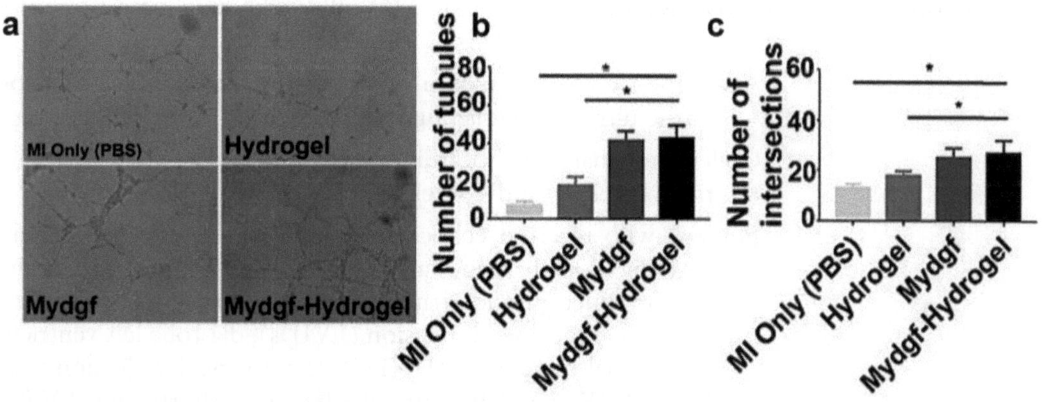

Fig. 1 In vitro cytocompatibility studies of the hydrogel. (**a**) Richly formed tubular structure was observed in the Mydgf and Mydgf-hydrogel groups compared with the PBS and hydrogel groups (scale bar 100μm). Bar charts show numbers of (**b**) tubules and (**c**) intersecting nodes between different groups. (Reproduced from ref. 4)

3.5 Rat Acute Myocardial Infarction (AMI) Model and Evaluation of Heart Functions After Hydrogel Injection

3.5.1 AMI Model

Anesthetize rats with sodium pentobarbital (30 mg/kg, i.p.), orally tracheal intubate, and ventilate with air at a respiratory rate of 80 times/min.

After finishing a left thoracotomy in the fourth ribs intercostal, expose the left anterior descending artery (LAD) of the heart.

Ligate permanently with polypropylene for induction of acute myocardial infarction (AMI) model.

Confirm successful AMI by left ventricle (LV) ligation regional wall pale and ST-segment elevation on an electrocardiogram (ECG) device.

3.5.2 Hydrogel Injection

Prepare samples: syringes with PBS (MI-only), control hydrogel (no Mydgf), Mydgf, and Mydgf-hydrogel (10μg Mydgf is encapsulated in 120μL sterile well-mixed hydrogel precursor solutions). 10 wt% gel concentration is used here.

After LAD ligation, the heart is to receive one of the above four treatments randomly: utilize a 29-gauge insulin needle to insert the injection into the peri-infarct LV wall and direct toward the MI zone.

Inject a total volume of 120μL directly into three locations after LAD ligation, and withdraw the needle slowly to avoid material leakage.

Close the thorax.

Place the rat on a warm plate.

3.5.3 Cardiac Echocardiography and Micro-PET Measurements

Two days after experiencing MI, collect the baseline echocardiographic measurements.

Assess the LV cardiac function and structure through the operation of the ultrasound imaging system.

Collect and analyze echocardiographic data every 2 days after ligation (baseline) and 4 weeks after injection.

For rats who received echocardiography 2 days after permanent LAD ligation (baseline) and 4 weeks post-treatment, measure cardiac function and structure indicators such as systolic left ventricular internal dimension (LVIDs), diastolic left ventricular internal dimension (LVIDd), left ventricular ejection fraction (LVEF), and left ventricular fractional shortening (LVFS).

Measure all parameters based on the mean of three consecutive cardiac cycles.

Determine these high-quality images from short-axis two-dimensional imaging at the midaxillary level.

Measure myocardial viability with a dedicated small-animal PET system (Fig. 2).

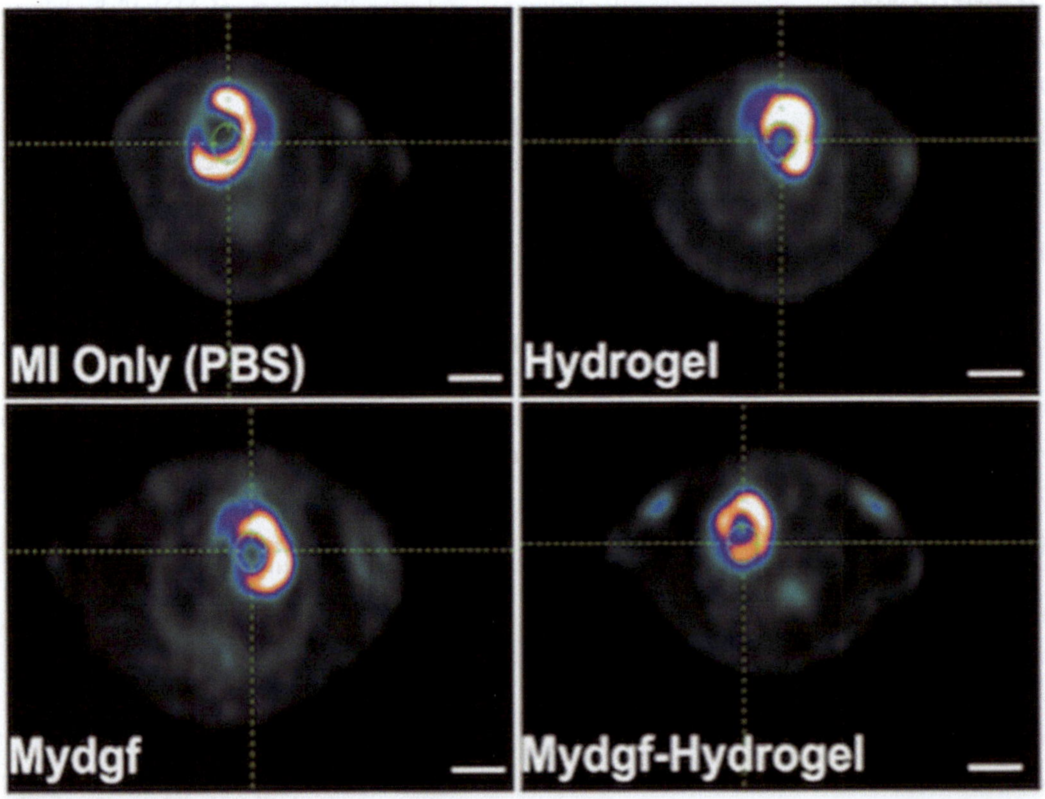

Fig. 2 Evaluation of myocardial function. Short-axis images of myocardial 18F-fluorodeoxyglucose (FDG) uptake acquired with a small-animal PET system. Scale bar: 10 mm. (Reproduced from ref. 4)

After being anesthetized, the rats are given an intravenous injection of 18F-FDG (37 MBq) through the tail vein.

Then, the rats are subjected to intraperitoneal administration of glucose (1 mg/g body weight) and insulin (8 mUI/g body weight) to improve the myocardium uptake tracer in 30 min before the 18F-FDG injection.

After 18F-FDG administration and continued for 30 min, acquire the 18F-FDG images.

Operate the small-animal micro-PET system.

Define LV defect size by the fraction of polar map elements with decreased tracer uptake in the total polar map derived from the 18F-FDG uptake.

3.6 Morphometric and Histological Studies for Heart Structure Evaluation

At 4 weeks following the injection, euthanize the rats ($n = 8$ in each group), and collect the heart tissue.

Rinse the hearts with 4% formaldehyde for 20 min, and then rinse with PBS for about 10 min.

Then, divide the hearts into five pieces in the transverse orientation to the apex at 2 mm intervals.

Fix the heart samples, embed in paraffin, and then section into 5μm slices for subsequent histopathological and morphometric analyses (Fig. 3).

Perform Masson's trichrome-stained slides from the closest section plane.

Measure the following data using ImageJ software: the ventricular wall thickness, scar thickness (mm), septum thickness (mm), LV cavity area (mm^2), and whole LV area (mm^2).

Perform each measurement three times.

Calculate relative scar thickness as the ratio of the average thickness of the scar wall to the average thickness of the uninfarcted wall.

Calculate the infarct expansion index as follows: (LV cavity area/whole LV area)/relative scar thickness.

3.7 Immunohisto chemistry in the Evaluation of Neovascularization

Stain blood vessels with a primary antibody against CD31 and an antibody against alpha-smooth muscle actin (α-SMA) (*see* **Note 5**).

Utilize a confocal microscope for confocal images (Fig. 4).

Calculate the quantification of the arterioles and capillaries separately.

Calculate the capillary density values as the mean within 1 mm^2 area in five random fields, and calculate the maturation index based

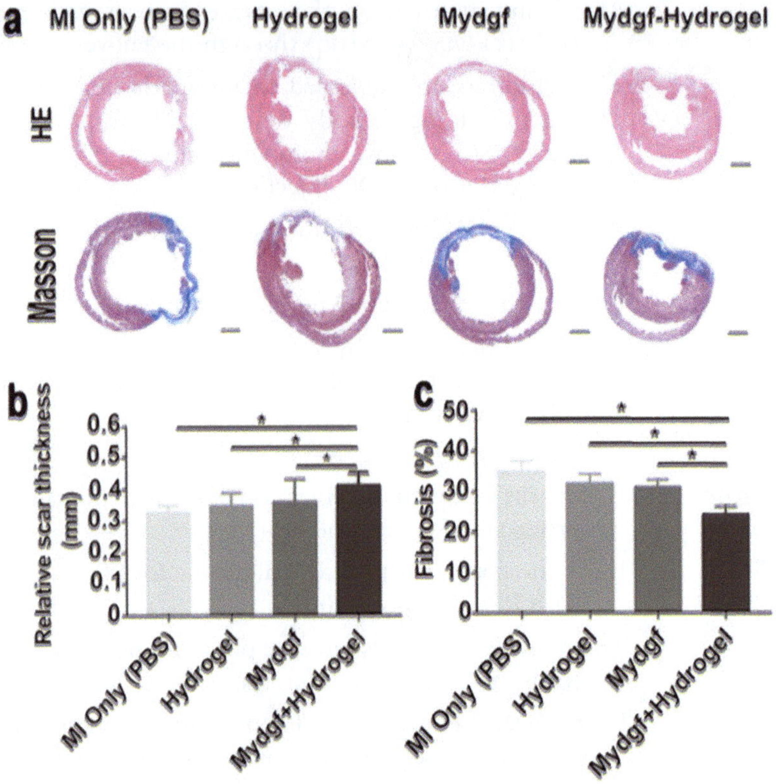

Fig. 3 Morphologies and histology of the LV. (**a**) H&E and Masson's trichrome staining. Scale bars, 2 mm. (**b**) Quantification analysis of the relative scar thickness. (**c**) Percent fibrosis of the severity of interstitial fibrosis. Data were represented as mean ± SEM; (*)$p < 0.05$. (Reproduced from ref. 4)

on the number of α-SMA-positive vessels relative to the total number of vessels.

3.8 Terminal Deoxynucleotidyl Transferase-Mediated dUTP Nick End Labeling (TUNEL) Staining to Measure Apoptosis

Perform TUNEL staining of tissue cross sections from LV papillary muscle regions of all groups.

For each slide, select five separate fields randomly.

Define cells with clear nuclear labeling as TUNEL-positive cells.

Calculate the numbers of TUNEL-positive nuclei or total nuclei with ImageJ.

Define the apoptotic index as the ratio of the number of TUNEL-positive nuclei versus the total number of nuclei.

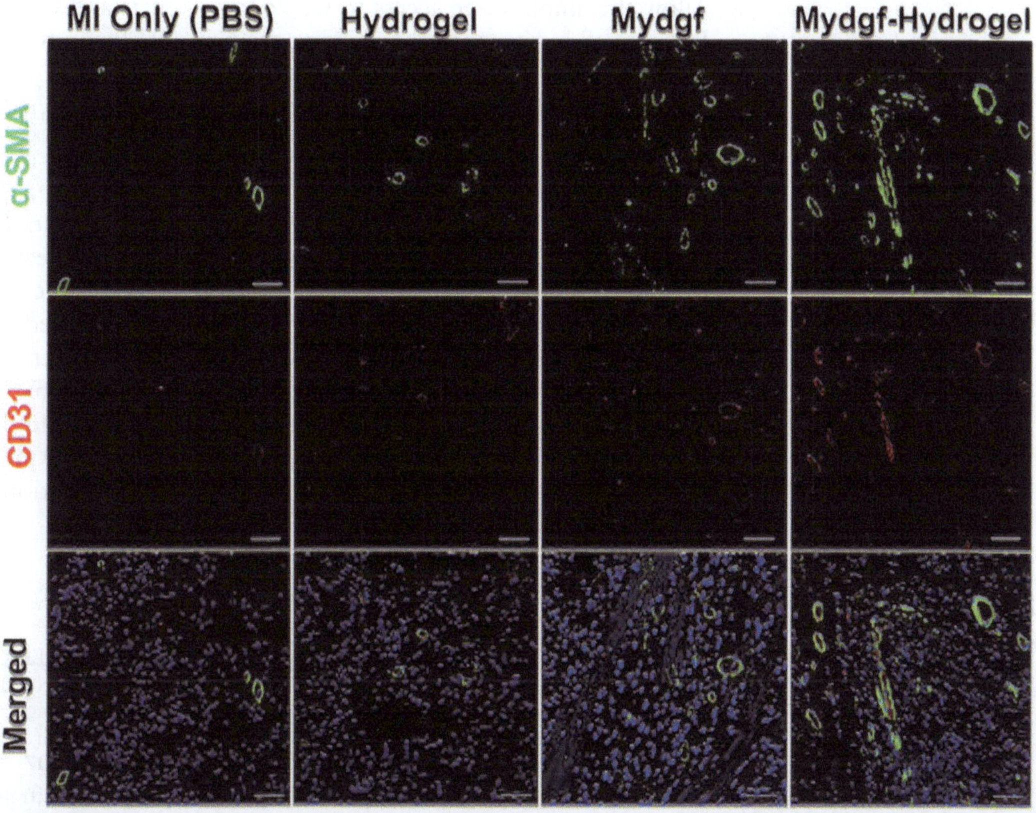

Fig. 4 Immunofluorescence assessment of the vascularization 4 weeks after treatment. CD31 (red), α-SMA (green) and cell nucleus (blue) staining for the blood vessels. Scale bar: 50μm. (Reproduced from ref. 4)

4 Notes

1. NIH 3T3 cells are most suitable for culture in a 75 cm^2 flask. To avoid clumping do not agitate the cells by hitting or shaking the flask while waiting for the cells to detach. Cells that are difficult to detach may be placed at 37 °C to facilitate dispersal. Considering the nature of these cells, it is suggested that freezer stocks be made as soon as a new vial of low-passage numbered cells is cultured.

2. When HUVECs cultures reach 80% confluence, most of the cells should have an epithelioid morphology and should be associated with each other in colonies. Some irregularly sized and shaped cells may be observed. The time required to reach 80% confluence will vary with the starting cell density and the methods employed by individual investigators. Damage to cultured HUVECs can occur during trypsinization. This damage can result from exposure of the cells to the trypsin/EDTA solution for excessive lengths of time, trypsinization at temperatures exceeding room temperature, and/or excessive mechanical agitation.

3. Allow the animals free access to sterile food pellets and water. Perform all in vivo studies following the National Institutes of Health Animal Care guidelines.
4. Make sure dynamic time sweep measurements are within the linear viscoelastic region (strain = 5%, angular frequency = 1 rad/s).
5. During the immunohistochemistry experiment, it is commonly accepted that the volume of fixative should be 50 times greater than the size of the immersed tissue. Avoid fixing the tissue for greater than 24 h since tissue antigens may either be masked or destroyed. Cut 5 to 15-μm-thick tissue sections using a cryostat. The suggested cryostat temperature is between −15 °C and −23 °C. The section will curl if the specimen is too cold. If it is too warm, it will stick to the knife. Dry the slides for 30 min on a slide warmer at 37 °C. Slides containing cryostat sections can be stored at −20 °C to −70 °C for up to 12 months.

References

1. Swerdlow AJ, Higgins CD, Smith P, Cunningham D, Hancock BW, Horwich A, Hoskin PJ, Lister A, Radford JA, Rohatiner AZ, Linch DC (2007) Myocardial infarction mortality risk after treatment for Hodgkin disease: a collaborative British cohort study. J Natl Cancer Inst 99(3):206–214. https://doi.org/10.1093/jnci/djk029
2. Kitakaze M, Asakura M, Kim J, Shintani Y, Asanuma H, Hamasaki T, Seguchi O, Myoishi M, Minamino T, Ohara T, Nagai Y, Nanto S, Watanabe K, Fukuzawa S, Hirayama A, Nakamura N, Kimura K, Fujii K, Ishihara M, Saito Y, Tomoike H, Kitamura S (2007) Human atrial natriuretic peptide and nicorandil as adjuncts to reperfusion treatment for acute myocardial infarction (J-WIND): two randomised trials. Lancet 370 (9597):1483–1493. https://doi.org/10.1016/s0140-6736(07)61634-1
3. Yoon SJ, Fang YH, Lim CH, Kim BS, Son HS, Park Y, Sun K (2009) Regeneration of ischemic heart using hyaluronic acid-based injectable hydrogel. J Biomed Mater Res B Appl Biomater 91(1):163–171. https://doi.org/10.1002/jbm.b.31386
4. Yuan Z, Tsou YH, Zhang XQ, Huang S, Yang Y, Gao M, Ho W, Zhao Q, Ye X, Xu X (2019) Injectable citrate-based hydrogel as an angiogenic biomaterial improves cardiac repair after myocardial infarction. ACS Appl Mater Interfaces 11(42):38429–38439. https://doi.org/10.1021/acsami.9b12043
5. Tsou Y-H, Zhang X-Q, Bai X, Zhu H, Li Z, Liu Y, Shi J, Xu X (2018) Dopant-free hydrogels with intrinsic photoluminescence and biodegradable properties. Adv Funct Mater 28(34). https://doi.org/10.1002/adfm.201802607
6. Ma C, Gerhard E, Lu D, Yang J (2018) Citrate chemistry and biology for biomaterials design. Biomaterials 178:383–400. https://doi.org/10.1016/j.biomaterials.2018.05.003
7. Tsou YH, Khoneisser J, Huang PC, Xu X (2016) Hydrogel as a bioactive material to regulate stem cell fate. Bioact Mater 1(1):39–55. https://doi.org/10.1016/j.bioactmat.2016.05.001
8. Polten F, Reboll MR, Widera C, Kempf T, Bethmann K, Gupta P, Miglietta J, Pekcec A, Tillmanns J, Bauersachs J, Giannitsis E, Pich A, Wollert KC (2019) Plasma concentrations of myeloid-derived growth factor in healthy individuals and patients with acute myocardial infarction as assessed by multiple reaction monitoring-mass spectrometry. Anal Chem 91 (2):1302–1308. https://doi.org/10.1021/acs.analchem.8b03041
9. Korf-Klingebiel M, Reboll MR, Klede S, Brod T, Pich A, Polten F, Napp LC, Bauersachs J, Ganser A, Brinkmann E, Reimann I, Kempf T, Niessen HW, Mizrahi J, Schonfeld HJ, Iglesias A, Bobadilla M, Wang Y, Wollert KC (2015) Myeloid-derived growth factor (C19orf10) mediates cardiac repair following myocardial infarction. Nat Med 21 (2):140–149. https://doi.org/10.1038/nm.3778

Chapter 15

Rabbit Surgery Protocol for End-to-End and End-to-Side Vascular Graft Anastomosis

Yuan Yao, YeJin Jeong, Aung Moe Zaw, Marek Kukumberg, and Evelyn K. F. Yim

Abstract

Preclinical testing in animal model is a required stage of vascular device development. Among small animal models, rabbits provide vasculature with relative larger caliber for anastomotic implantation of vascular grafts as preclinical testing before conducting large animal studies. Rabbits have similar hemostatic mechanism with human and can accommodate vascular grafts with various diameters at different locations, and thus provide a valid model to assess small-diameter vascular grafts. This chapter will describe the procedures and materials required to conduct survival surgery in rabbit carotid artery models for implantation of small-diameter tubular grafts with an end-to-side and end-to-end anastomotic technique.

Key words Rabbit carotid artery, End-to-side anastomosis, End-to-end anastomosis, Small-diameter vascular graft, Surgical technique

1 Introduction

Cardiovascular diseases (CVDs) still remain as one of the leading natural causes of death worldwide [1]. To combat various CVDs, large-diameter synthetic vascular grafts—vascular grafts with inner diameter larger than 6 mm—have been developed using expanded polytetrafluoroethylene (ePTFE) or Dacron [2, 3]. While ePTFE and Dacron showed good clinical success, small-diameter vascular grafts (SDVG)—vascular grafts with inner diameter less than 6 mm—still remain to have low success rate [4, 5]. Different materials are being developed for the SDVG application by many labs to overcome the current lack of clinical success. To verify the performance of newly developed SDVGs, it is important to choose the correct animal model.

While testing in large animal models remains a required preclinical testing, small animal models offer advantages in terms of

Feng Zhao and Kam W. Leong (eds.), *Vascular Tissue Engineering: Methods and Protocols*, Methods in Molecular Biology, vol. 2375, https://doi.org/10.1007/978-1-0716-1708-3_15,

low cost to purchase and maintain, ease in handling, and ability to be tested in a larger number of animals as preliminary in vivo testing before large animal studies. Rabbit common carotid graft implantation model, with end-to-end anastomosis, is a well-established model that has been used for the assessment of vascular graft patency in small animal model [6]. A study using expanded polytetrafluoroethylene (ePTFE) grafts implanted in the rabbit carotid artery showed that the results from the rabbit model closely matched the clinical responses in humans. Rabbit models are especially ideal for small-diameter vascular graft application as this species allows for a wide range of vascular graft implantations ranging from 1 mm to 5 mm in diameter [6–12]. Recent studies also demonstrated the feasibility of end-to-side anastomosis, which is a more clinically relevant in vivo model, in rabbits [13, 14].

In this chapter, we will discuss the materials required prior to and during the surgery, pre-surgical preparation protocol, surgical procedures for both end-to-end and end-to-side vascular graft implantation, and post-operational animal care protocol.

2 Materials

Ensure that all of the supplies are well stocked and prepared appropriately. The materials and steps mentioned in this protocol were applied to male New Zealand white rabbits with body weight of 3.5–4 kg.

2.1 Supplies for Pre-surgical Procedures

1. Gauze sponges and eye spears.
2. Surgical tie-downs.
3. Clippers (*see* **Note 1**).
4. 22- and 24-gauge catheter.
5. 25G hypodermic needles.
6. 1 cc and 3 cc syringes.
7. Disinfectants and antiseptics: povidone-iodine scrub and chlorhexidine scrub.
8. Eye ointment: ingredients include sterile water, sorbitol, carbomer, sodium hydroxide, disodium EDTA, and cetrimide.
9. Antibiotics: enrofloxacin (10 mg/kg, Baytril).
10. Anesthetics: ketamine (20 mg/kg), xylazine (2 mg/kg), and isoflurane (1–3%) with 400–800 mL/min oxygen flow through an inhalation mask.
11. Anticholinergics: glycopyrrolate (0.05 mg/kg).
12. Anesthetic reversal: atipamezole (0.5 mg/kg), apply subcutaneously only if the animal has xylazine-induced bradycardia.

13. Anaphylaxis reversal: epinephrine (0.1 mg/kg), apply subcutaneously only if the animal has a severe allergic reaction.
14. Analeptic: doxapram (2 mg/kg), apply subcutaneously only if the animal has drug-induced ventilatory depression and apnea.
15. Analgesic: buprenorphine (0.05–0.1 mg/kg).
16. Crystalloid fluids: 0.9% sodium chloride.
17. Colloid fluids: Tetraspan or Voluven (60 mg/mL hydroxyethyl starch).
18. Surgical tools (scalpel, scissors, microsurgery instruments) (*see* **Note 2**).

2.2 Supplies for the Tubular Graft Implantation Surgery

1. 1 cc and 10 cc syringes.
2. Bupivacaine (0.25%), apply topically on carotid sinus only if the sinus is exposed.
3. Anticoagulant: heparin (200 IU/kg).
4. Anesthetics: isoflurane (1–3%) with 400–800 mL/min oxygen flow through an inhalation mask.
5. Antispasmodic: papaverine and lidocaine.
6. Crystalloid fluids: 0.9% sodium chloride.
7. Colloid fluids: Tetraspan or Voluven (60 mg/mL hydroxyethyl starch).
8. Nonsteroidal anti-inflammatory drugs: carprofen (5 mg/kg, Rimadyl).
9. Antibiotics: enrofloxacin (10 mg/kg, Baytril).
10. 7-0 to 9-0 nylon or other non-absorbable sutures.
11. 3-0/4-0 polypropylene or other non-absorbable sutures.
12. 3-0/4-0 polyglycolide or other absorbable sutures.
13. Tubular graft: PVA vascular graft with inner diameter of 1.7 mm and outer diameter of 2.1 mm.
14. Surgical tools: clamps (#3), scalpel (#10) and scalpel handle, scissors (4–6 in, shape and blunt end), retractors, cauterizer, forceps (3–5 in, curved tip and straight tip), mosquito forceps, towel forceps, tissue forceps, and micro-scissors (3 in.).
15. Personal protective equipment: caps, masks, surgical gloves, and surgical gowns.
16. Surgery instrument: electrocardiography (ECG) machine, electrical thermometer, surgical microscope, and heating pad.

2.3 Supplies for Post-operation Animal Care

1. Crystalloid fluids: 0.9% sodium chloride.
2. Colloid fluids: Tetraspan or Voluven (60 mg/mL hydroxyethyl starch).

3. Nonsteroidal anti-inflammatory drugs: carprofen (5 mg/kg, Rimadyl).
4. Antibiotics: enrofloxacin (10 mg/kg, Baytril).
5. Anticoagulant/antiplatelet (optional).

3 Methods

3.1 Pre-surgical Procedures

Unlike other animals, rabbits are more sensitive to stress and anesthesia. Thus, a pre-anesthetic evaluation is critically important. Pre-operative assessment should include regular weighing, minimal laboratory tests (renal function and hematocrit), and a good physical examination. Stress should be kept as minimal as possible. Obtaining baseline readings increases the chances of successful anesthesia in rabbits. Perform any pre-operative assessments, such as ultrasound blood flow measurement and blood collection for necessary tests, when appropriate.

3.1.1 Two to Seven Days Before Surgery

1. Evaluate the mechanical functionality of each surgical instrument necessary for surgical procedure.
2. Autoclave all surgical instruments, surgical gowns, towels, and drapes.
3. Ensure to have the correct sizes and quantities of surgical gloves, caps, and sterile gauzes ready.

3.1.2 One Day Before Surgery

1. Twenty four hours before the surgery, weigh the rabbit, and administer enrofloxacin (10 mg/kg, Baytril) intramuscularly to the animal (*see* **Note 3**).
2. Clean and disinfect the surgical room and table before operation.

3.1.3 Surgery Day Pre-surgery Preparation

1. A quick assessment of the stress level and physical examination is recommended prior to surgery. Fasting of rabbits for more than 1 h prior to surgery is unnecessary and dangerous and should be avoided.
2. Administer glycopyrrolate (0.05 mg/kg) subcutaneously to the animal 30 min before surgery.
3. Anesthetize the rabbit by administering ketamine (20 mg/kg) and xylazine (2 mg/kg) intramuscularly. Rabbit is kept in quiet and dark(er) place after administration of anesthetics for 5–10 min to allow for sedation to take effect (*see* **Note 4**).
4. Place the heating pad on the surgical table, and warm up the heating pad to 37 °C. Also, warm up the water bath to 37 °C, and place the necessary solutions (e.g., saline) into the water batch.

5. With careful handling (spine support), move the animal to a surgical table, and carefully intubate with a correct size endotracheal tube. As rabbit endotracheal intubation can be tricky, it is recommended to evaluate the breathing rate with a monitor. In the case of a non-present breathing curve, reposition the endotracheal tube immediately. Induce gas anesthesia with 400–800 mL/min oxygen and gas isoflurane (2–3%). Depending on the animal's response (heart and breathing rate, blood pressure), lower the gas amount "maintenance" amount of 1–1.5% isoflurane (*see* **Note 5**).
6. Monitor the animal by placing the ECG electrodes, blood pressure cuff, and rectal thermometer. Also monitor heart rate, respiration rate, blood oxygen level (SPO_2), blood pressure, and body temperature for successful rabbit anesthesia.
7. Apply eye ointment to both eyes of the animal to protect the cornea from drying out. In the case of prolonged surgery, re-apply the ointment.
8. Once the animal is stabilized in the anesthesia, carefully restrain the animal in a supine position on the operating table. Restraining ropes should be put above joints (carpal and tarsal) if possible (*see* **Note 6**).
9. As rabbits are sensitive to change of temperature, a warming mat positioned under the animal is recommended. A regulated water circulation warming mat is superior to other devices in order to avoid skin burns and further complications. Alternatively, a forced air-warmer provides safe perioperative temperature maintenance and control.
10. Using the electric clippers, shave the area larger than the expected incision over the ventral neck area (*see* **Note 7**). Remove excess clipped hair from the area using a vacuum.
11. Administer analgesic (buprenorphine, 0.05–0.1 mg/kg) subcutaneously, and move the animal to the surgical table.
12. Draw 2 mL of blood from the auricular artery of the animal for complete blood count and 3 mL blood for coagulation profile.
13. Insert a 22G catheter into the marginal ear vein, and tape the catheter on the ear with hypoallergenic plastic medical tapes. Give 3–6 mL/kg/h 0.9% sodium chloride intravenously to supply fluid through the catheter.
14. Using a gauze, wipe the animal's skin with alternating povidone-iodine (PI) scrub. Scrub the surgical area in a back-and-forth motion starting at the incision site and radiating out. Do not re-wipe the central area once the periphery has been wiped. Fold the gauze sponge over exposing a clean portion prior to wiping a new area. Discard the gauze sponge after each wipe. Repeat the scrubbing of surgical area two more times.

Before the last round of PI scrub, one round of chlorhexidine-alcohol (CHA) scrub is recommended.

15. Pinch the webbing between the toes of the animal, and make sure it is maintained in a surgical plane of anesthesia. Monitor the animal's vital signs, including respiration, skin or mucous membrane color, and lack of responsiveness. Monitor the rabbit's level of anesthesia and consciousness continuously after the intubation. If there should be any change in breathing rhythm and/or heartbeat, especially in the case of tachycardia and/or tachypnea, adjust the inhalation anesthetic drug.
16. Using aseptic technique, cover the animal with sterile drape so that only the surgical field is exposed.
17. Surgeons for the procedure should wear appropriate personal protective equipment, such as face masks and caps. Before surgery, surgeons should thoroughly scrub hands with a bactericidal scrub (povidone-iodine, chlorhexidine) and wear sterile gloves and gowns.
18. The microvascular surgery on the common carotid artery (CCA) will be performed either with surgical loupes (CCA with diameter > 2 mm) or with a stereomicroscope (CCA with diameter < 2 mm) with varying magnification (10–20×). If a microscope is being used, it has to be thoroughly but carefully prepared with two rounds of CHA disinfection. Additionally, cover the microscope with sterile drapes. Do not allow it to come into contact with any part of the animal.

3.2 End-to-Side Tubular Graft Implantation Surgical Procedure

1. Prepare the vascular graft according to the length and implantation angle based on experimental design.
2. Apply lidocaine subcutaneously as local anesthesia on the location of skin incision.
3. Make a ventral midline incision at the neck area; blunt dissect sternohyoid and sternomastoid muscles to expose the common carotid artery (CCA). Rabbit carotid artery structure and implantation location are shown in Fig. 1.
4. Isolate the CCA, and expose CCA by using 4–6 round wire handmade retractors. Secure the retractors with mosquito clamps at the end of the retractor, as shown in Fig. 2 (top). The length of isolated CCA should be at least 3 cm longer than required length for graft implantation.
5. Using fine forceps and a fine pointed cotton swab, gently pierce through the membranous sheath to expose the neurovascular bundle. Then, using a clean set of fine forceps and cotton swab, dissect and separate the common carotid artery (CCA) and the external carotid artery (ECA).

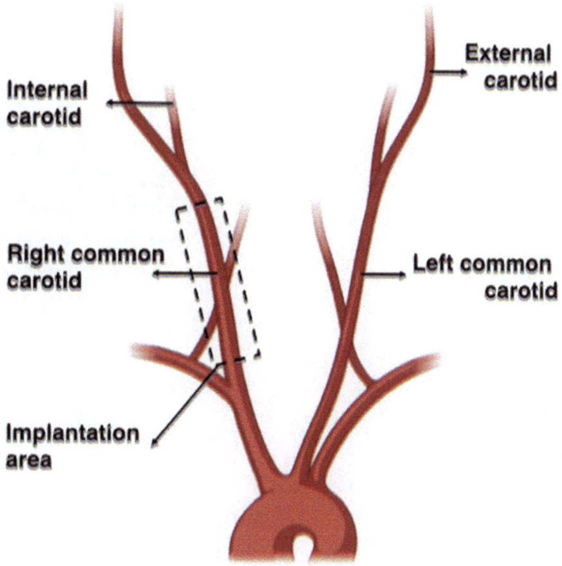

Fig. 1 Diagram of rabbit carotid artery and graft implantation (created in BioRender)

6. Administer heparin (200 IU/kg) intravenously prior to arterial clamping (*see* **Note 8**). As arteries react quickly to change of temperature and mechanical stimulation, it is recommended to handle arteries very carefully in order to avoid vasoconstriction. In the event of vasoconstriction, apply 5–10 mL of warm (37 °C), sterile saline solution topically on the constricted vessel. Wait for 1–2 min to observe if the vessel is dilated. If vasodilation is not observed, a few drops (total quantity of 0.1–0.3 mL) of antispasmodic drug such as papaverine can be applied topically on the vessel.
7. Clamp the CCA using two vascular clamps (size-3 clamps were used for 1.5–2 mm diameter CCA) to temporarily occlude the CCA. Put the first clamp on the distal side of the CCA and the second on the proximal side of the CCA. The distance between the two clamps should be 1.5–2 cm longer than the length of the vascular graft, as shown in Fig. 2 (bottom). Ligate all the thyroid artery branches in between the two clamping sites.
8. Angiostomy and anastomosis. Align the graft with the CCA, and identify the locations of the proximal and distal anastomosis. Starting from proximal side, clean the surface of the vessel by removing the excess adventitia (*see* **Note 9**). Make a stitch with the suture on the vessel wall, lift up the stitch, and excise the suture stitch using micro-scissors to make a hole (steps 1 and 2 in Fig. 3). It is recommended to make a smaller incision with a smaller hole first. Fine trim the hole to match the size of the graft end. Using 7/0–9/0 non-absorbable suture, anchor

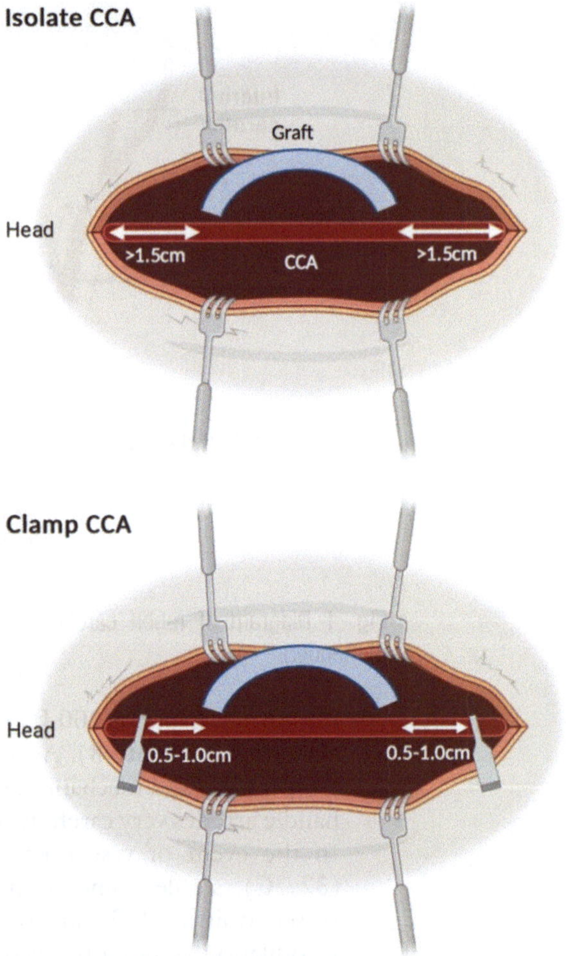

Fig. 2 Diagram illustration of carotid artery isolation and clamping (created in BioRender)

one side of the graft onto the vessel with a simple suture followed by a second anchoring suture on the opposite side (steps 3 and 4 in Fig. 3). To stitch the graft on the vessel, continue with simple interrupted suture evenly spread between the anchoring suture (in average 4–5 sutures), as shown in steps 5 and 6 in Fig. 3.

9. After the first anastomosis is completed, make a second hole in the distal side. Stitch the other side of the graft onto the distal side of the CCA with the same technique in **step 8**. Before closing all the stitches on one side, leave two stitches open, and insert a 24G catheter. Remove any air bubbles by injecting sterile saline into the graft lumen, and close the anastomosis, as shown in Fig. 4. Tie off the native artery between the two

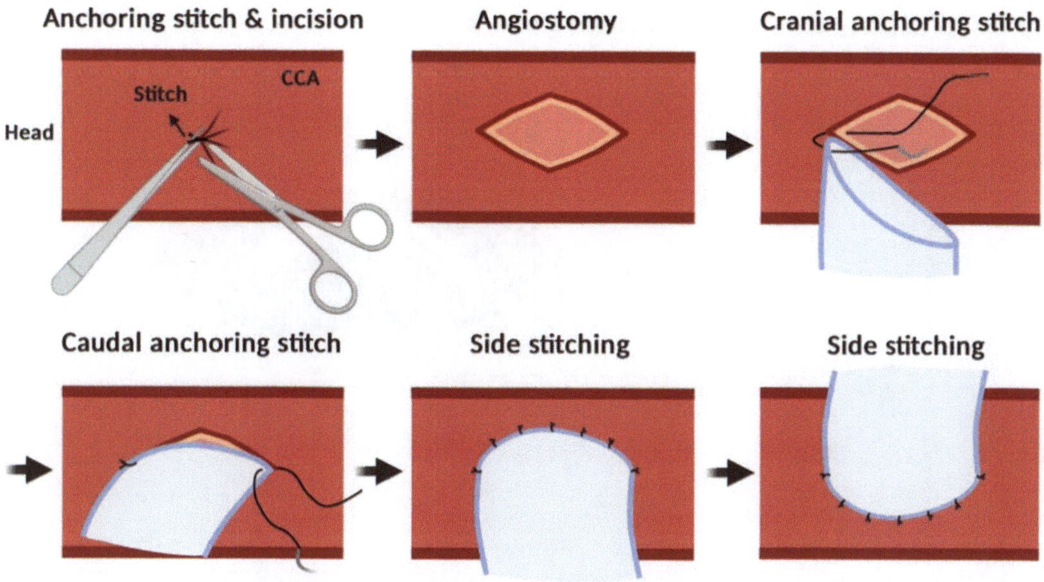

Fig. 3 Steps to stitch graft onto carotid artery (CCA) (created in BioRender)

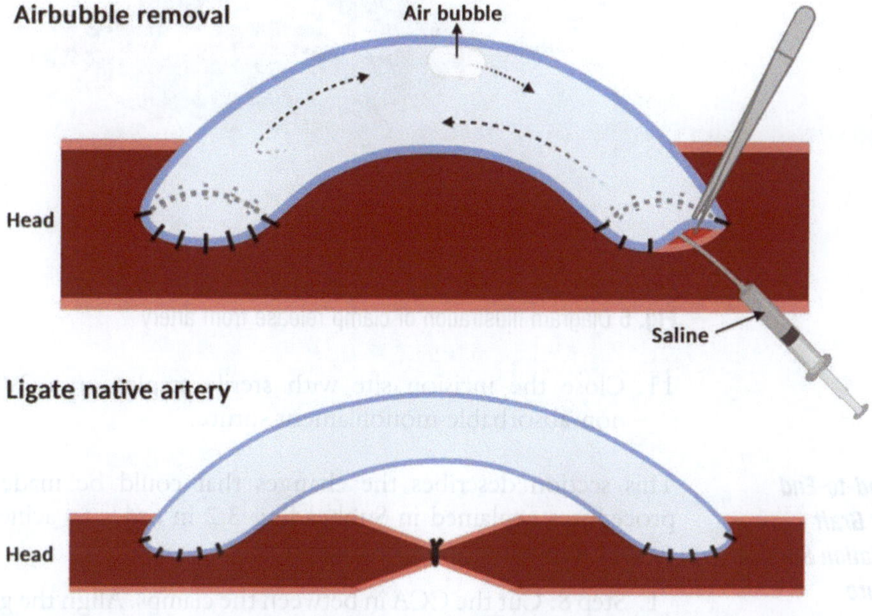

Fig. 4 Diagram illustration of air bubble removal and native artery ligation (created in BioRender)

anastomoses to prevent blood from going through the native artery, as shown in Fig. 4.

10. Release the clamps to allow blood flow (Fig. 5). Release the clamp at the distal side to allow blood backflow to fill the graft. When no severe leaking is observed, release the proximal clamp to allow blood flow going through the graft.

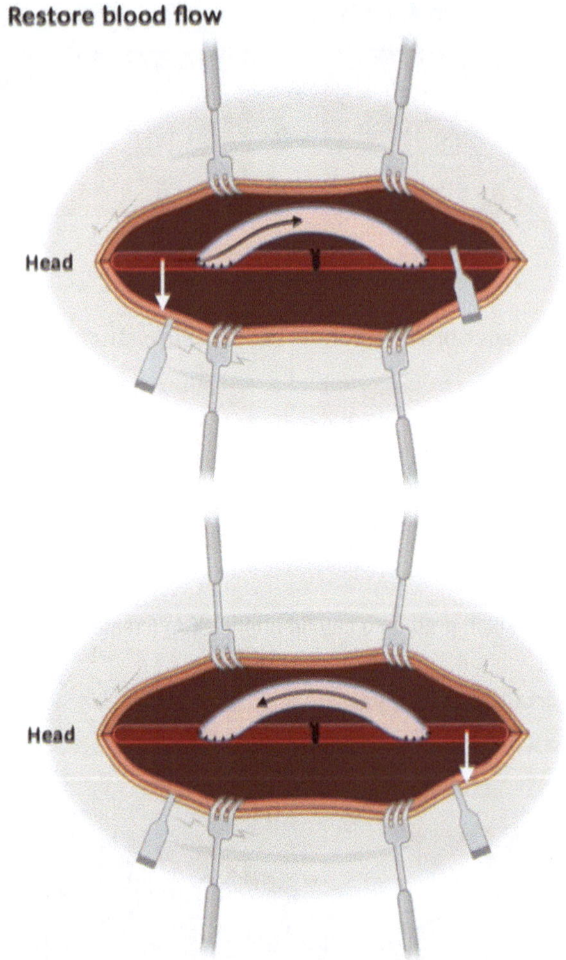

Fig. 5 Diagram illustration of clamp release from artery

11. Close the incision site with sterile staples or a 3-0/4-0 - non-absorbable monofilament suture.

3.3 End-to-End Tubular Graft Implantation Surgical Procedure

This section describes the changes that could be made to the procedures explained in Subheading 3.2 in order to achieve end-to-end implantation of vascular grafts:

1. Step 8: Cut the CCA in between the clamps. Align the graft end with the one of the open vessel ends. Using an 8-0 polyamide suture, stitch the graft onto the vessel with two anchoring simple interrupted sutures followed by 8–10 simple interrupted sutures between the anchoring sutures as described in Subheading 3.2, **step 10**. After the first anastomosis is completed, align the other side of the graft with the other open vessel end, and suture the graft onto the vessel with 8–10

simple interrupted sutures for CCA with a diameter around 2 mm.

The above-described method is an invasive surgical procedure. As such, it requires a team consisting of a trained vascular surgeon, a surgical assistant, and an anesthesiologist preferably with veterinary anesthesiology experience. To emphasize, this procedure should not be performed by personnel without any prior vascular surgical training.

3.4 Postoperative Animal Care

1. Set a heating pad in the postoperative care area, and warm up the pad to 37 °C.
2. Move the rabbit to the postoperative care area, and monitor during recovery. Constantly monitor the rabbits for vital signs during this critical time period.
3. In case of a significant blood loss, in order to avoid a hypovolemic shock, treat the animal with colloid solutions by supplementing the saline solution with plasma colloid fluids (Gelofusine, Tetraspan, or Voluven) in a bolus dose of 10–15 mL/kg in a slow pace within 5–10 min through the ear vein. When blood pressure is stabilized, supplement the animal with crystalloids only. All fluids should be warmed to 37 °C prior to administration, and the animal should be regularly checked on blood pressure and body temperature until stabilized.
4. After the animal has fully recovered from anesthesia, return the animal to the routine housing cage (*see* **Note 10**).
5. Treat the animal postoperatively with nonsteroidal anti-inflammatory drug carprofen (4–5 mg/kg SC SID) and buprenorphine (0.1 mg/kg, SC SID) for 3 days to provide analgesia. Additionally, provide the animal with antibiotics—enrofloxacin (10 mg/kg, Baytril, SC SID)—for 7 days.
6. Optional: antiplatelet/anticoagulant can be given according to the experimental design.
7. Monitor the incision site daily from the day of the surgery until the endpoint, and observe for dehiscence of sutures (opening of the incision line), any discharge, swelling/edema, redness, or pain.

4 Notes

1. Different sized clippers are available for rabbits. Choose the size of the clipper that best fits the type of rabbit used and for the individual that will be shaving the animal.
2. Surgical tools must be sterilized and placed in a sterile container prior to surgery.

3. It is performed to minimize the possibility of infection due to the surgery.
4. Xylazine overdose could be fatal. The dose of xylazine is not recommended to be higher than 2 mg/kg. In the case of emergency, atipamezole should be used to reverse xylazine adverse effect. If the animal showed abnormal slow heart rate and breathing rate after the administration of xylazine, administer atipamezole (0.5 mg/kg) subcutaneously to ensure safety of the animal.
5. Isoflurane gas and oxygen mixture can also be applied using inhalation mask.
6. Place the animal in a supine position only after the animal is completely anesthetized.
7. Adequate shaving area should be done to reduce the possibility of unwanted hair entering the incision. If a considerable amount of hair remains, a depilatory cream could be applied on the area for 1–2 min then washed thoroughly with lukewarm water.
8. Heparin has a half-life about 1–2 h after infusion and should be given every 2 h intravenously during the surgery.
9. In the case of animal that has fragile or extra thin vessel wall, minimize this step to maintain the blood vessel's integrity.
10. A full recovery of animal from anesthesia can be determined by: (1) if the animal is able to sit on its own; (2) if the animal has immediate reflection to touching; (3) it is important to note that rabbit must show appetite almost immediately post-surgery. Any prolonged time without food leads to dramatic changes in gastrointestinal tract and subsequently worsens recovery prognosis. Animal with longer surgery time and excess intravenous fluid infusion will take longer to wake up.

Acknowledgments

This work was supported by the National Institutes of Health (NIH R01 HL130274-01A1) and NSERC-CREATE, Training in Global Biomedical Technology Research and Innovation at the University of Waterloo [CREATE 509950-2018]. We acknowledge Dr. Filip Konecny, Dr. Kathleen Delaney, and Ms. Jean Flanagan for discussion of surgery methods and Ms. Grace Pohan, Ms. Sabrina Mattiassi, Ms. Rebecca Mac, and the University of Waterloo Central Animal Facility for their technical assistance and discussion.

References

1. Willis M, Homeister JW, Stone JR (2013) Cellular and molecular pathobiology of cardiovascular disease. Academic Press, Cambridge, Massachusetts
2. Deutsch M, Meinhart J, Fischlein T, Preiss P, Zilla P (1999) Clinical autologous in vitro endothelialization of infrainguinal ePTFE grafts in 100 patients: a 9-year experience. Surgery 126(5):847–855
3. Lieberman JR, Lozman J, Czajka J, Dougherty J (1988) Repair of Achilles tendon ruptures with Dacron vascular graft. Clin Orthop Relat Res 234:204–208
4. Enomoto S, Sumi M, Kajimoto K, Nakazawa Y, Takahashi R, Takabayashi C, Asakura T, Sata M (2010) Long-term patency of small-diameter vascular graft made from fibroin, a silk-based biodegradable material. J Vasc Surg 51(1):155–164
5. Mousa AY, Patterson W, Abu-Halimah S, Hass SM, Alhalbouni S, Sadek BT, Nanjundappa A, Modak A, Stone PA, Emmett M (2013) Patency in arteriovenous grafts in hemodialysis patients. Vasc Endovasc Surg 47(6):438–443
6. Byrom MJ, Bannon PG, White GH, Ng MK (2010) Animal models for the assessment of novel vascular conduits. J Vasc Surg 52 (1):176–195
7. Cassel WS, Mason RA, Campbell R, Newton GB, Hui JC, Giron F (1989) An animal model for small-diameter arterial grafts. J Investig Surg 2(2):181–186
8. Gherardini G, Haegerstrand A, Matarasso A, Gurlek A, Evans G, Lundeberg T (1997) Cell adhesion and short-term patency in human endothelium preseeded 1.5-mm polytetrafluoroethylene vascular grafts: an experimental study. Plast Reconstr Surg 99(2):472–478
9. Sparks SR, Tripathy U, Broudy A, Bergan JJ, Kumins NH, Owens EL (2002) Small-caliber mesothelial cell-layered polytetraflouroethylene vascular grafts in New Zealand white rabbits. Ann Vasc Surg 16(1):73–76
10. Contreras MA, Quist WC, Logerfo FW (2000) Effect of porosity on small-diameter vascular graft healing. Microsurgery 20(1):15–21
11. Huynh T, Abraham G, Murray J, Brockbank K, Hagen P-O, Sullivan S (1999) Remodeling of an acellular collagen graft into a physiologically responsive neovessel. Nat Biotechnol 17 (11):1083–1086
12. Babatasi G, Massetti M, Bara L, Mazmanian M, Samama MM, Khayat A (1997) Graft thrombosis in small diameter vascular prosthesis: a laboratory model. Int J Angiol 6(2):118–123
13. Anderson DE, Pohan G, Raman J, Konecny F, Yim EK, Hinds MT (2018) Improving surgical methods for studying vascular grafts in animal models. Tissue Eng Part C Methods 24 (8):457–464
14. Yao Y, Zaw AM, Anderson DE, Hinds MT, Yim EK (2020) Fucoidan functionalization on poly (vinyl alcohol) hydrogels for improved endothelialization and hemocompatibility. Biomaterials 249:120011

References

Chapter 16

Vascular Imaging in Small Animals Using Clinical Ultrasound Scanners

Aung Moe Zaw, Richard Shangguan, Yuan Yao, YeJin Jeong, Billy Y. S. Yiu, Adrian J. Y. Chee, Alfred C. H. Yu, and Evelyn K. F. Yim

Abstract

Conventional ultrasound with frequency (2–15 MHz) has been a global diagnostic and therapeutic tool in clinical medicine, and high-frequency ultrasound (>30 MHz) has been a powerful investigative device for preclinical studies such as cardiovascular research. In this chapter, we describe the use of conventional ultrasound with a 2.5–10 MHz transducer as an investigative device for the measurement/detection of blood flow in rabbit model. The chapter will describe the procedures for the preparation of sonographer, imaging locations, and the details of the rabbits used as well as detailed imaging steps for the preoperative, immediately after operation, and postoperative follow-up ultrasound for vascular surgery, using a vascular graft implantation as an example. We also provide useful notes to avoid pitfalls for successful imaging. The overall goal of this chapter is to deliver the steps in using low-cost, non-invasive, and highly versatile clinical ultrasound imaging in preclinical small animal testing.

Key words Preclinical ultrasound, Perioperative ultrasound, Small-animal ultrasound, Rabbit vascular graft model, Conventional ultrasound in animal model

1 Introduction

Conventional ultrasound system in the clinical setting can be used to perform easy, non-invasive, and affordable imaging to assess vascular blood flow in preclinical animal models. The ultrasound had been widely used for diagnostic imaging as well as a therapeutic tool in clinical settings for many decades [1]. It enables the clinicians and healthcare providers a reliable and non-invasive method for diagnostic and therapeutic purposes [2]. As the preclinical studies and translational research in medicine flourished, various animal models had been developed [3]. Consequently, the application of ultrasound extended beyond clinical applications and established in various in vivo animal models [4]. High-frequency ultrasound (ultrasound imaging system with >30 MHz) imaging

Feng Zhao and Kam W. Leong (eds.), *Vascular Tissue Engineering: Methods and Protocols*, Methods in Molecular Biology, vol. 2375, https://doi.org/10.1007/978-1-0716-1708-3_16,

devices have been specifically designed for animal models for high resolution and various modality measurements [5]. These systems usually come with onsite training and technical support tailored for animal imaging. Additionally, the high-frequency ultrasound system needs expensive equipment setup (e.g., inclusive ECG, body temperature and anesthesia platforms, laser and photoacoustic measurement modalities, etc.), sonographer trained for high-frequency system, and designated location/room to use. Thus, a less expensive and more versatile imaging platform is preferred oftentimes if resources are limited. For successful ultrasound imaging in animal models, the appropriate handling of the instrument, imaging procedure, and experimental animal will be critical.

The advantages of using the conventional ultrasound system are low cost, easy operation, and portability that allows for onsite non-invasive intraoperative ultrasound quickly and efficiently. For example, in vascular graft implantation, the vascular graft patency can be checked immediately after implantation. However, the conventional clinical ultrasound system does not have an adjustable stand and a transducer mount as used in the sophisticated high-frequency animal ultrasound systems. To get the clear images or clips with appropriate length for analysis, the skillful sonographer is important and should operate the imaging with stable hand (or stable handling). In this chapter, we describe the use of a conventional clinical ultrasound system and transducer in the rabbit model to assess carotid vascular blood flow. Using vascular graft implantation as an example vascular procedure, the step-by-step procedures for imaging baseline vascular dimensions and flow parameters before the surgery, confirmation of successful graft implantation immediately after the surgery, and follow-up monitoring of the vascular dimensions and blood flow parameter changes in the rabbit model will be described. We also provided the relevant notes for every critical step so that pitfalls can be avoided for successful imaging.

2 Materials

2.1 Supplies for Imaging

1. 3.5–4 kg male New Zealand white rabbit for vascular surgery and ultrasound imaging.
2. Scale to measure rabbit body weight.
3. Gauze and 70% ethanol to clean the imaging area/table and the animal.
4. Surgical tie-downs to position the animal during imaging.
5. Syringes and needles for subcutaneous and intravenous injections.
6. Acepromazine (1 mg/kg) to sedate the animal.

7. Oxygen (400–800 mL/min) and gas isoflurane (1–3%) to anesthetize the animal.
8. Eye ointment (artificial tears) to cover the eyes.
9. Large towel for restraining the animal.
10. Clipper to shave the hair.
11. Ultrasound gel.
12. A contrast agent (e.g., microbubbles).

2.2 Equipment for Imaging

1. Ultrasound instrument: VF 8–3+ (2.5–10 MHz) linear transducer, Siemens Acuson X300 ultrasound system.
2. Animal monitor equipment: core temperature monitor with a probe for animal body temperature monitoring and pulse oximeter with a heart rate monitor to monitor the blood oxygen level and heart rates.
3. Temperature-controlled pad.
4. Temperature-controlled water bath.

3 Methods

3.1 Pre-imaging Preparation

1. Clean the area of imaging (e.g., surgical table).
2. Don appropriate personal protective equipment (PPE) (*see* **Note 1**).
3. Measure the animal body weight, and sedate the animal by administering 1 mg/kg of acepromazine subcutaneously.
4. Switch on the temperature-controlled pad, and adjust the temperature to 37 °C to prevent hypothermia or hyperthermia in the animal (*see* **Note 2**).
5. Switch on the temperature-controlled water bath, and set the bath temperature to be 37 °C; then, warm the ultrasound gel in the 37 °C water bath, or warm up the ultrasound gel using the gel warming chamber in the ultrasound machine.
6. Turn on the ultrasound machine, and input the animal, study information, and date and time of experiment using the "New Patient" key.
7. Choose the desired transducer VF 8–3+ (8–3 MHz) using the "Transducer" key, and select the image acquisition parameters such as select imaging frequency to 10 MHz with the toggle key.

3.2 Preoperative Ultrasound Imaging

Carry out the preoperative ultrasound measurement >7 days prior to the surgery.

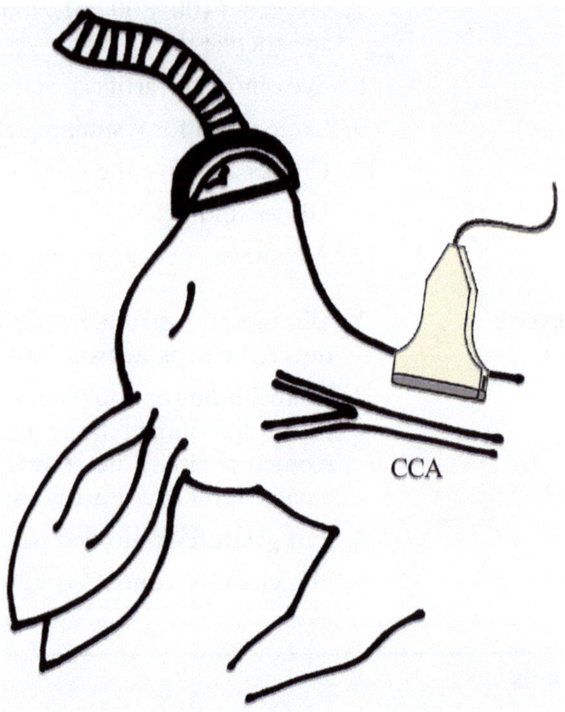

Fig. 1 A schematic diagram showing the positioning of the probe and the common carotid artery in the anesthetized rabbit neck

1. Once the animal is fully sedated, place the animal on the warmed-up temperature-controlled pad.
2. Administer isoflurane (1–3%) with 400–800 mL/min oxygen flow through an inhalation mask. Confirm the unconsciousness with negative eye reflex. Place the animal in a supine position exposing the ventral neck area (Fig. 1).
3. Lubricate the rectal temperature probe, and put it into the anus to monitor the core body temperature throughout the imaging procedure.
4. Put the pulse oximeter with heart rate monitor on the bare area of the skin (e.g., pinna) to monitor the oxygen saturation in the peripheral blood (SpO_2) and heart rate throughout the imaging procedure.
5. Apply eye ointment (artificial tears) on the eyes (*see* **Note 3**).
6. Fix the limbs with tie-downs to maintain the animal position throughout the imaging procedure.
7. Shave the area over the ventral neck area using clippers (*see* **Note 4**).
8. Apply prewarmed ultrasound gel onto the ultrasound transducer (probe) and on the shaved area (*see* **Note 5**).

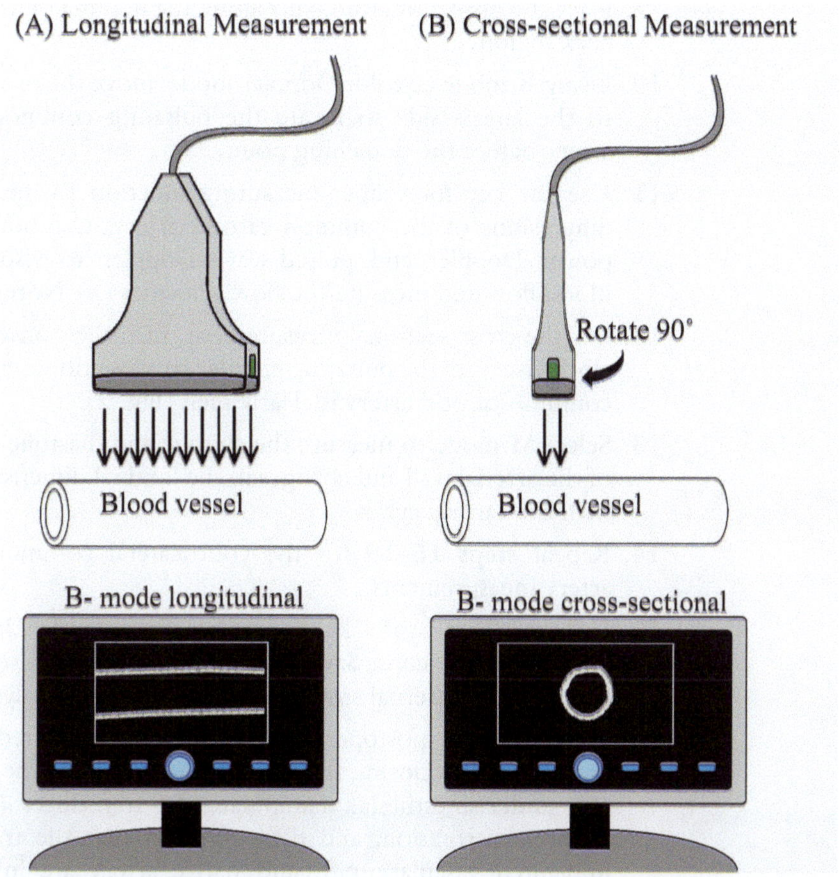

Fig. 2 Longitudinal and cross-sectional measurement of the blood vessel. (**a**) The probe orientation is aligned with the vessel, and the longitudinal section of the vessel is clearly seen in the B-mode sketch. (**b**) The probe is rotated 90° clockwise at the same position, and the cross-sectional section of the vessel can be seen in B-mode sketch

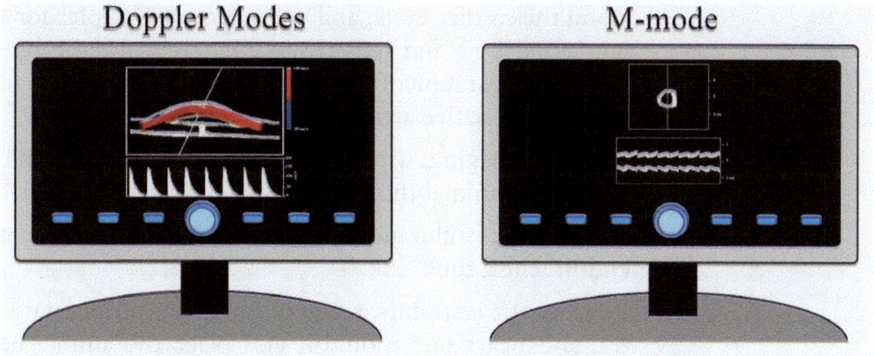

Fig. 3 Schematic diagrams of Doppler and M-modes showing wall thickness, blood flow, and velocity. (**a**) The color Doppler mode showing the blood flow (upper panel of screen) and pulsed wave Doppler mode showing blood flow velocity (lower panel of screen). (**b**) M-mode showing the wall thickness

9. Place the probe longitudinally along the midline on the ventral neck region.
10. Using B-mode or color Doppler mode, move the probe slowly to the lateral side to locate the pulsating common carotid artery before the branching point.
11. Use the key for caliper measuring function to measure the dimensions of the common carotid artery. Use both color/power Doppler and pulsed-wave Doppler to visualize the blood flow and measure the flow velocities (*see* **Note 6**).
12. For the cross-sectional measurement, turn the transducer 90° clockwise until the pulsating circular cross-sectional view of the common carotid artery is clearly seen (Fig. 2).
13. Select M-mode to measure the systolic and diastolic thickness of the arterial wall and changes in the luminal dimension along with the cardiac cycles.
14. Repeat **steps 15–19** for the contralateral common carotid artery measurements.
15. Use the "Print/store" key to save the images and "Clip/store" key to save the clips. Save the data in the internal storage, or export to an external storage device for further analysis.
16. For immediate postoperative imaging for implanted vascular grafts just after closing the surgical wound, while the animal is still under anesthesia, manipulate the transducer (pressure, alignment, rotation, and tilt) gently so that the transducer, implanted vascular graft, and native artery are in a single plane to get the clear view of the graft, anastomosis locations, and native artery [6] (*see* **Notes 7–10**).
17. Use B-mode to scan the longitudinal and cross-sectional graft dimensions, anastomosis locations, and native artery, M-mode for cross-sectional lumen and wall thickness measurements along the cardiac cycle, and color/power Doppler for visualizing blood flow and pulsed-wave Doppler for the blood flow velocity measurement in the graft and the proximal and distal parts of the native artery (Fig. 3).
18. After the imaging, wipe off the ultrasound gel around the neck area of the animal thoroughly with a gauze.
19. Turn off the isoflurane and oxygen, and remove the gas mask/endotracheal tube.
20. Remove the restraints, rectal temperature probe, pulse oximeter, and heart rate monitor, and place the animal back into the cage.
21. Monitor until the animal fully recovers from anesthesia.

3.3 Postoperative Ultrasound Imaging Without Anesthesia

3.3.1 Checklist for Suitability of Imaging Without Anesthesia

If ultrasound imaging is required within a short time (within ~7 days) after surgery, anesthetizing the animal is not recommended. The following checklist can be used for suitability determination:

1. The animal's overall health condition is good; e.g., the animal is alert and active.
2. The animal is eating and drinking properly and achieving pre-operative body weight.
3. There is no hair growth in the previously shaved area for surgery and imaging.

3.3.2 Imaging Steps for Suitable Animals

Steps to perform ultrasound imaging without anesthesia are described as below:

1. Lightly restrain the rabbit with a large towel by covering the face, body, and limbs (*see* **Note 11**).
2. Place the rabbit in a prone position exposing the hairless ventral and lateral neck areas which had been shaved during the surgery.
3. Using a gauze, disinfect the animal's skin with 70% ethanol.
4. Apply prewarmed ultrasound gel on the imaging area and ultrasound probe.
5. Slightly lift the chin of the animal, place the probe longitudinally on the lateral side of the neck, and move slowly and gently to the medial until the pulsating common carotid artery is seen in B-mode or blood flow is observed in color Doppler mode. Adjust the pressure, and tilt the probe as necessary.
6. Perform the detailed ultrasound imaging **steps 8–17** as described in Subheading 3.2 (*see* **Notes 12** and **13**).
7. After imaging, clean the neck area of the animal thoroughly.
8. Unwrap the animal, and place it back into the cage.
9. Monitor the animal until fully recovered from sedation.

3.4 Ultrasound Imaging at Endpoint with Contrast Agent

1. Ensure that the Mechanical Index (MI) displayed on the scanner remains below 0.8 [7], by adjusting ultrasound power prior to injection of contrast agents and at all points of contrast imaging (*see* **Notes 14** and **15**).
2. Prepare microbubble [8] contrast agents following manufacturer instructions.
3. As a baseline contrast concentration, introduce the recommended clinical bolus dosage (e.g., Definity™, 10μL/kg [9]) for the product into the intravenous at a rate of 1 mL/s. Smaller aliquots can be diluted to 1 mL of solution with 0.9% saline to aid in lowering injection rate (*see* **Notes 16** and **17**).

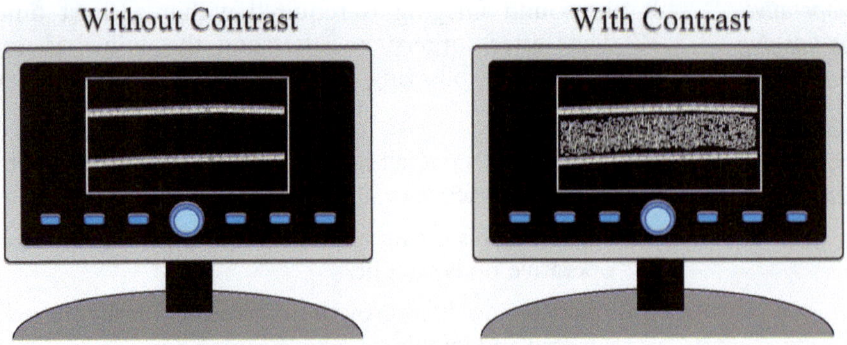

Fig. 4 Illustration of the expected visualization of contrast agents in an ultrasound image. (A) Without contrast, a B-mode clinical ultrasound image with longitudinal view of a blood vessel would render the vessel lumen as an echolucent (dark) region. (B) After microbubble contrast agents are administered, the vessel lumen is visualized as an echogenic (bright) region

4. Note the time on the ultrasound scanner.
5. Follow the contrast injection with a 5 mL 0.9% saline flush introduced at a rate of 1 mL/s. Alternatively, a sterile three-way stop cock can be used to deliver the contrast solution and the saline flush.
6. Identify contrast on the image. Contrast signal should appear 30–45 s after injection and should persist for up to 5 min (Fig. 4).
7. The contrast imaging mode (e.g., Cadence™ CPS on Siemens systems) can be used if supported by the scanner.
8. Repeat the imaging steps for the graft, native artery, and control artery (*see* **Note 18**).
9. Repeat the injection if necessary.
10. After completion of endpoint imaging, proceed to euthanasia according to the experimental protocol.

4 Notes

1. These include but are not limited to lab coats, surgical gowns, gloves, and masks.
2. Hypothermia or hyperthermia can affect respiratory and heart rate.
3. This is to prevent the cornea dryness which may lead to cornea injury.
4. Try to shave as clean as possible since the hair can be the potential transmission barrier for the ultrasound. If a considerable amount of hair remains, a depilatory cream could be

applied to the area for 1–2 min then washed thoroughly with lukewarm water to avoid chemical burn from the cream [10].

5. If there are air bubbles in the gel between the probe and the skin, this can create shadows in images. Try to remove the bubbles by gently rubbing the probe against the animal skin.
6. Adjust the acquisition parameters such as zoom, focus, gain, gate/sample volume, and ultrasound angle accordingly to get optimal image quality.
7. Perform pre-imaging preparation prior so that ultrasound imaging can be ready and performed immediately.
8. Before closing the subcutaneous tissue and skin, gently flush the operation site with sterile saline, and make sure there is no air bubble and fluid trapping which can hinder the ultrasound image quality [11].
9. Avoid suturing directly on top of the graft implant but as close as possible to the midline when closing the wound since the sutures can be the barrier for ultrasound transmission.
10. The proper probe manipulation (pressure, alignment, rotation, and tilt) can considerably improve the image quality [6], and this could take a few to several minutes depending on the skill of the sonographer.
11. This is to prevent the kicking out of the rabbit.
12. Ultrasound imaging should be done as quickly as possible to avoid unnecessary distress since the rabbit is not under anesthesia. The purpose of week-1 imaging is to check the patency and flow in the graft and control artery.
13. Remove the suture from the animal skin using appropriate forceps and scissors according to surgical protocol prior to ultrasound imaging.
14. Contrast imaging step should be performed at the endpoint to avoid adverse consequences of contrast imaging such as potential vascular damage caused by microbubbles [12] or infection from intravenous injection if an unsterile contrast agent is used.
15. Reducing the mechanical index will decrease microbubble burst phenomena and reduce incidences of vascular damage. This will also increase the half-life of the agents within the subject.
16. In-house microbubbles should be diluted to the prepared concentration of their closest clinical approximate (e.g., phospholipid-perfluorocarbon formulations approximate to Definity™) in #bubbles/mL and delivered following the dosage recommendation.
17. To evenly suspend the microbubbles within the syringe prior to intravenous injection, contrast agents should be mixed by

rolling the syringe between two hands. Optionally after mixing, macroscale foam in a diluted solution can be reduced by drawing the solution into the barrel of the syringe, sharply striking the barrel at the air-solution interface, and discarding the small layer of froth that immediately forms.

18. Note the appearance of any shadowing artifacts or noise. Halve the dosage of the next injection if encountered. Dosage should be increased if using a high-frequency transducer. If contrast enhancement is weak, the operator should check that the MI is low, ensure that the IV tubing is not clamped, ensure that the contrast agents are activated, ensure that no filter is present in the IV tubing, and ensure that contrast mode is on if applicable.

Acknowledgments

This work is supported by the National Institutes of Health (NIH R01 HL130274-01A1) and the Natural Sciences and Engineering Council of Canada (CREATE-509950-2018 and CREATE-528202-2019). Staff at the Central Animal Facility of the University of Waterloo for animal care, Dr. Filip Konecny for surgery, Sabrina Mattiassi, Sarah Wen-Hui Chan, Laura Edmonds, and John Tse for the transportation.

References

1. Leighton TG (2007) What is ultrasound? Prog Biophys Mol Biol 93(1–3):3–83. https://doi.org/10.1016/j.pbiomolbio.2006.07.026
2. Miller DL, Smith NB, Bailey MR, Czarnota GJ, Hynynen K, Makin IR, Bioeffects Committee of the American Institute of Ultrasound in M (2012) Overview of therapeutic ultrasound applications and safety considerations. J Ultrasound Med 31(4):623–634. https://doi.org/10.7863/jum.2012.31.4.623
3. Denayer T, Stöhr T, Van Roy M (2014) Animal models in translational medicine: validation and prediction. New Horiz Transl Med 2 (1):5–11. https://doi.org/10.1016/j.nhtm.2014.08.001
4. Kagadis GC, Loudos G, Katsanos K, Langer SG, Nikiforidis GC (2010) In vivo small animal imaging: current status and future prospects. Med Phys 37(12):6421–6442. https://doi.org/10.1118/1.3515456
5. Shung KK (2009) High frequency ultrasonic Imaging. J Med Ultrasound 17(1):25–30. https://doi.org/10.1016/S0929-6441(09)60012-6
6. Ihnatsenka B, Boezaart AP (2010) Ultrasound: basic understanding and learning the language. Int J Shoulder Surg 4(3):55–62. https://doi.org/10.4103/0973-6042.76960
7. Frinking P, Segers T, Luan Y, Tranquart F (2020) Three decades of ultrasound contrast agents: a review of the past present and future improvements. Ultrasound Med Biol 46 (4):892–908. https://doi.org/10.1016/j.ultrasmedbio.2019.12.008
8. Klibanov AL (2006) Microbubble contrast agents: targeted ultrasound imaging and ultrasound-assisted drug-delivery applications. Investig Radiol 41(3):354–362. https://doi.org/10.1097/01.rli.0000199292.88189.0f
9. Lantheus Medical Imaging (2015) Definity/Luminity Product Label. https://www.accessdata.fda.gov/drugsatfda_docs/label/2015/021064s017lbl.pdf
10. Karegoudar JS, Prabhakar PJ, Vijayanath V, Anitha MR, Surpur RR, Patil VM (2012) Shaving versus depilation cream for pre-operative skin preparation. Indian J Surg

74(4):294–297. https://doi.org/10.1007/s12262-011-0368-5

11. Feldman MK, Katyal S, Blackwood MS (2009) US artifacts. Radiographics 29(4):1179–1189. https://doi.org/10.1148/rg.294085199
12. Hosseinkhah N, Chen H, Matula TJ, Burns PN, Hynynen K (2013) Mechanisms of microbubble-vessel interactions and induced stresses: a numerical study. J Acoust Soc Am 134(3):1875–1885. https://doi.org/10.1121/1.4817843

Chapter 17

Vascular Graft Implantation Using a Bilateral End-to-Side Aortoiliac Preclinical Model

Khanh P. Nguyen, Monica T. Hinds, and Deirdre E. J. Anderson

Abstract

Arterial bypass grafts are a standard preclinical model for evaluating physiology and pathophysiology at graft-material interfaces. Implantations of vascular grafts are commonly done as end-to-end grafts in small animal models. Here we detail bilateral end-to-side aortoiliac graft implantation, which requires open surgery and the creation of vascular anastomoses between the graft material and the infrarenal aorta and iliac artery in a nonhuman primate model. In this model, the aortoiliac graft configuration is created using two 4 mm inner diameter vascular grafts (e.g., ePTFE). After exposure and control of the infrarenal aorta and bilateral common iliac arteries and heparinization, the proximal aortic-graft anastomosis is sewn on the lateral wall of the aorta, and subsequently the distal graft-common iliac anastomosis is sewn on the anterior wall of the common iliac artery with one tube graft. Another tube graft is sewn on the contralateral side in the same manner.

Key words Vascular graft, Anastomosis, End-to-side, Vascular surgery, ePTFE, Biomaterial

1 Introduction

Vascular grafts are widely used in the clinical setting to treat coronary artery and peripheral vascular disease and in the creation of arteriovenous grafts for dialysis access for patients with end-stage renal disease. The materials used for these grafts have remained largely unchanged for decades, despite numerous preclinical biomaterial studies. The current standard of preclinical testing consists of implantation in small and large animal models with interpositional grafts using end-to-end anastomoses [1]. The unidirectional blood flow within the interpositional vascular grafts makes attaining patency more likely using this model of preclinical testing. Yet, this is likely to hamper the translation to the clinical setting. The large changes in blood flow-induced wall shear stresses on the end-to-side anastomosed vascular grafts and arteries are a more accurate representation of the human circulatory system and are more likely

Feng Zhao and Kam W. Leong (eds.), *Vascular Tissue Engineering: Methods and Protocols*,
Methods in Molecular Biology, vol. 2375, https://doi.org/10.1007/978-1-0716-1708-3_17,

to induce the neointimal formation, which is a known cause of graft failure [2, 3]. Thus, the aortoiliac arterial bypass procedure described herein better replicates human physiology and pathophysiology as well as the clinical challenges and limitations of current surgical management in caring for patients who require vascular grafts.

The nonhuman primate model for aortoiliac arterial bypass is a useful in vivo cardiovascular model with many applications comprising biomaterial science including testing novel vascular conduits, altered hemodynamic flow conditions, and drug testing. While other animal models exist such as rabbit, sheep, pig, and dog, the nonhuman primate model most closely resembles human coagulation and pathophysiology, including thrombosis and intimal hyperplasia formation [4–6]. Thus, we describe the aortoiliac graft preclinical model using two 4 mm vascular grafts (e.g., ePTFE) in the *Papio anubis* baboon. After exposure of the infrarenal aorta and bilateral common iliac arteries and administration of heparin, the vessels are controlled with vascular clamps or vessel loops, an arteriotomy is made in the lateral wall of the proximal aorta, and the proximal aortic-graft anastomosis is sewn to the arteriotomy (Fig. 1). Subsequently, the distal graft-common iliac end-to-side anastomosis is created connecting the graft to the anterior wall of the common iliac artery, completing the aortoiliac bypass on one side (Fig. 2). A second vascular graft is the sewn on the contralateral side (Fig. 3). Serial imaging with ultrasound, CTSCAN, or MRI allows for noninvasive evaluation of vascular graft patency. Explantation and analysis of the grafts enable measurement of the extent of endothelialization and neointimal formation (Figs. 4 and 5). This model is useful for the evaluation of biomaterials [7–13] and the impact of fluid flow preconditioning of endothelial cells on neointimal formation [14]. We have demonstrated the ability to translate this model to smaller animals, including rabbits [15, 16]. Importantly, the nonhuman primate model is the closest to human and therefore vital in the translation of new vascular biomaterials to the clinic.

2 Materials

2.1 Graft Implant

1. Animals: *Papio anubis* baboon model with juvenile baboons, aged 3–5 years, either male or female, weighing between 12 and 25 kg.
2. Induction of anesthesia and setup: Anesthetics including ketamine (10–20 mg/kg) or Telazol (3–5 mg/kg) with isoflurane (1–3%), operating table with heating pad at 37 °C, size 6 endotracheal tube (28 mm), Foley catheter (5 French, 1.7 mm × 16″ (41 cm)), sterile towels and drapes, antibiotics (e.g., cefazolin

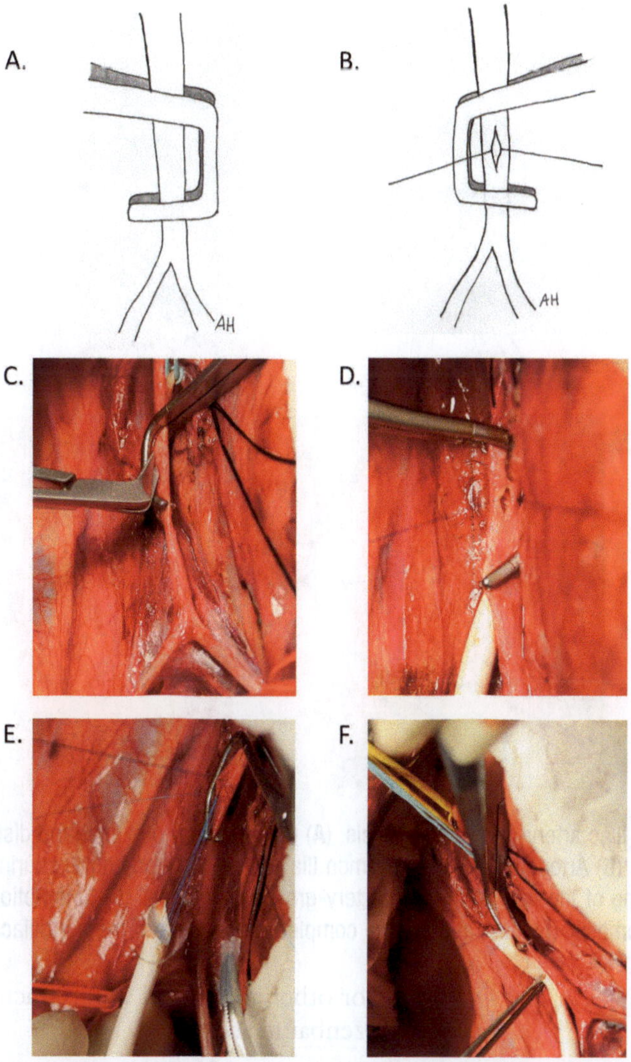

Fig. 1 Aortic-graft anastomosis. (**A**) Placement of proximal clamp in lateral position. (**B**) Placement of proximal clamp after anterior rotation. (**C**) Arteriotomy of proximal aorta performed with ophthalmic blade and extended with Potts scissors. (**D**) Stay sutures opening arteriotomy of proximal aorta. (**E**) Parachute suturing technique for heel of the graft. (**F**) Suturing of anastomosis continues after heel has been sutured first

25 mg/kg or enrofloxacin 10 mg/kg), and analgesics including buprenorphine (0.01–0.1 mg/kg) and/or hydromorphone (0.05–0.4 mg/kg).

3. Surgical exposure of the aorta and iliac arteries and closure of abdominal wall: Anesthetics including 0.5% bupivacaine or 1–3% lidocaine, 10 blade scalpel, electrocautery, Adson forceps, malleable retractor, lap sponges, warm saline, Balfour retractor

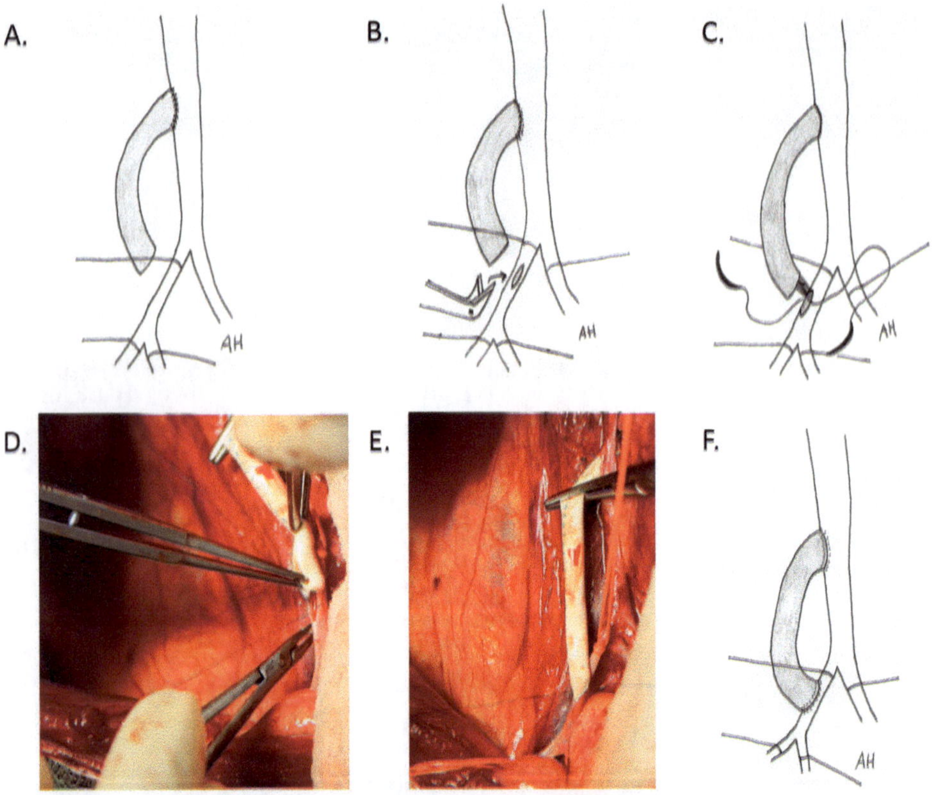

Fig. 2 Iliac artery-graft anastomosis. (**A**) Placement of proximal and distal vessel loops on the common iliac artery. (**B**) Arteriotomy of the common iliac artery. (**C**) Parachute suturing technique for heel of the graft. (**D**) Suturing of the heel of the iliac artery-graft anastomosis. (**E**) Completion of suturing of the iliac artery-graft anastomosis. (**F**) Illustration of the completed right side of the aortoiliac graft bypass

or other abdominal wall retractor, DeBakey forceps, and Metzenbaum scissors.

4. Dissection of arteries of interest: DeBakey vascular tissue forceps, Metzenbaum scissors, right-angled forceps, and small-vessel loops (silicone, 18″, 1.3 mm × 0.9 mm) (Devon Surgical Vessel Loops, Mini, Cardinal Health, Dublin, OH).
5. Creation of the proximal graft to aorta anastomosis: Heparin initial bolus of 300 units/kg and maintenance dose of 100 units/kg every 1 h, heparinized saline (2000 units/L), two grafts with inner diameters of 4 mm (experimental biomaterial or a clinical control such as ePTFE (W. L. Gore & Associates, Inc., VT04010, Flagstaff, AZ, USA)) 5–7 cm in length, Metzenbaum scissors, DeBakey forceps, a side-biting clamp (e.g., DeBakey side-biting U clamp or Cooley Satinsky anastomosis clamp), vascular clamps (angled DeBakey clamp, DeBakey atraumatic tangential clamp, or Reynolds vascular clamp), ophthalmic scalpel (straight, round, double-bevel blade with

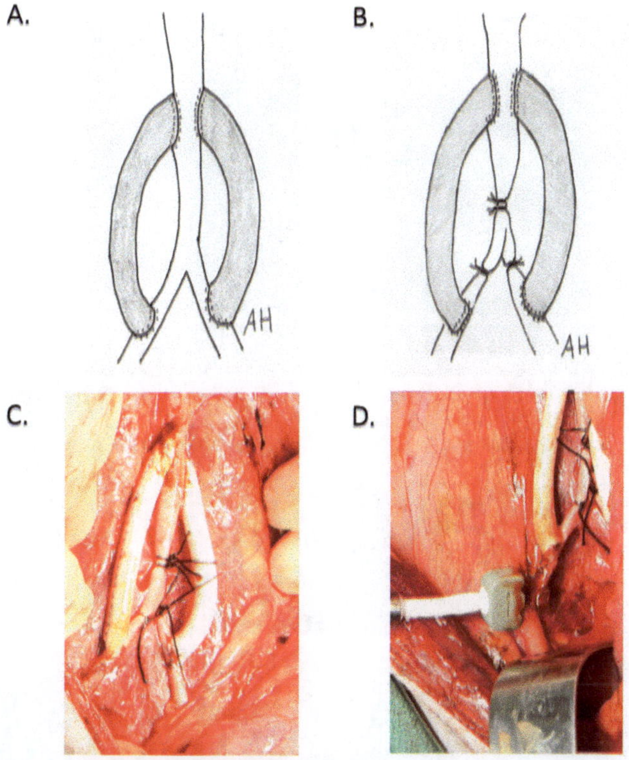

Fig. 3 Bilateral aortoiliac graft configuration. (**A**) Illustrated depiction of the final bilateral aortoiliac graft configuration. (**B**) Aortoiliac graft configuration with ligation of the aorta and common iliac arteries. (**C**) Example of an in situ implanted aortoiliac graft. (**D**) Blood flow meter assessment of iliac artery to assess graft patency

180-degree cutting surface), angled Potts scissors, hemostats with vascular booties, double-armed (two needles) 7-0 monofilament, polypropylene sutures (45–60 cm length), Gerald forceps, Castroviejo needle holder, and small-vessel loops (silicone, 18″, 1.3 mm × 0.9 mm).

6. Creation of the distal graft to iliac artery anastomosis: Metzenbaum scissors, DeBakey forceps, ophthalmic scalpel (straight, round, double-bevel blade with 180-degree cutting surface), angled Potts scissors, heparinized saline (2000 units/L), vascular clamps (DeBakey clamp or side-biting clamp), Gerald forceps, Castroviejo needle holder, double-armed (two needles) 7-0 monofilament, polypropylene sutures (15 cm length), small-vessel loops (silicone, 18″, 1.3 mm × 0.9 mm), and 2-0 silk ties.

7. Assessment of graft patency: Blood flow meters, including a transonic flow probe (Transonic®, Ithaca, NY, USA) or Doppler ultrasound.

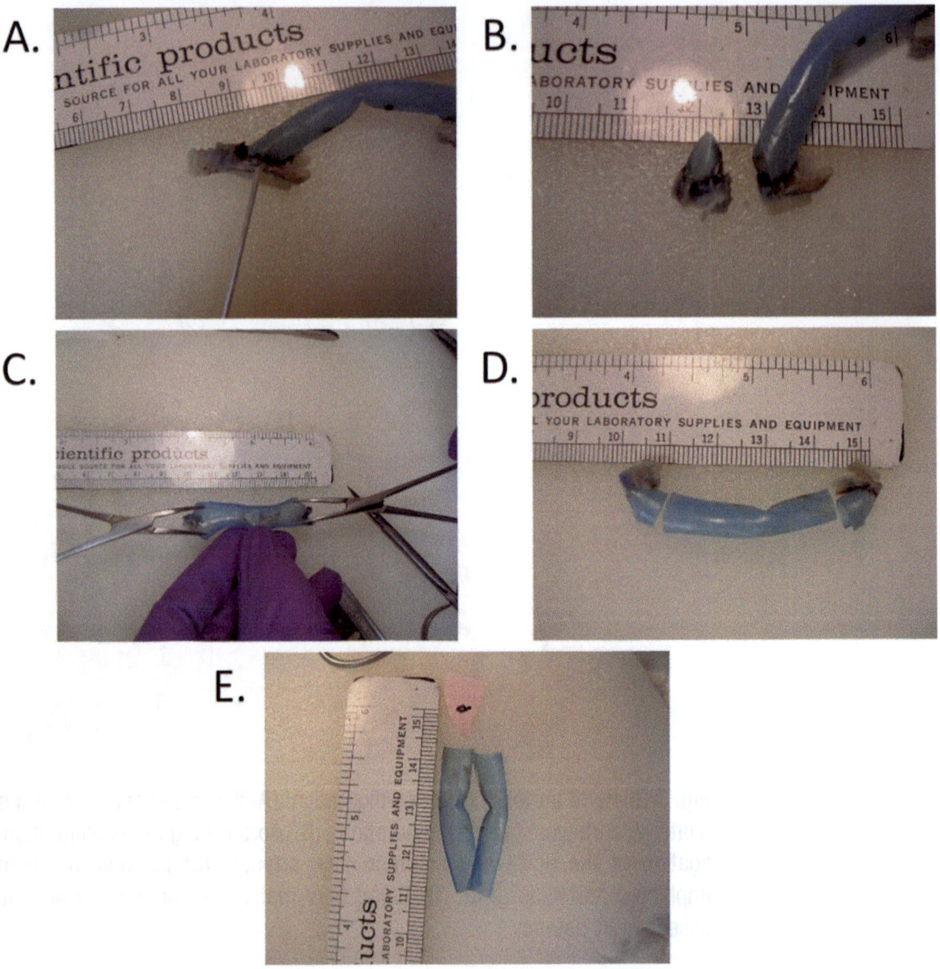

Fig. 4 Method for explant dissection. (**A**) Each anastomosis is cut perpendicular to the native artery with a razor blade. (**B**) This example shows the proximal anastomosis, which contains the aorta that has been cut to separate left- and right-side grafts. (**C**) Set up for longitudinal transection of the graft to expose the interior. (**D**) Cutting for the distal anastomosis block ensures there are two blocks for each anastomosis. (**E**) Completed longitudinal transection of the graft

8. Abdominal wound closure: 3-0 Vicryl sutures with needles, 2-0 absorbable monofilament suture, 4-0 monocryl sutures with needles, liposomal bupivacaine 0.4 ml/kg, and surgical skin glue.
9. Postoperative care and follow-up: hydromorphone (0.1–0.3 mg/kg), buprenorphine (0.03–0.1 mg/kg), antibiotics (cefazolin 25 mg/kg), optional oral antiplatelets such as aspirin (10 mg/kg) or clopidogrel (2 mg/kg), and optional duplex ultrasound machine.

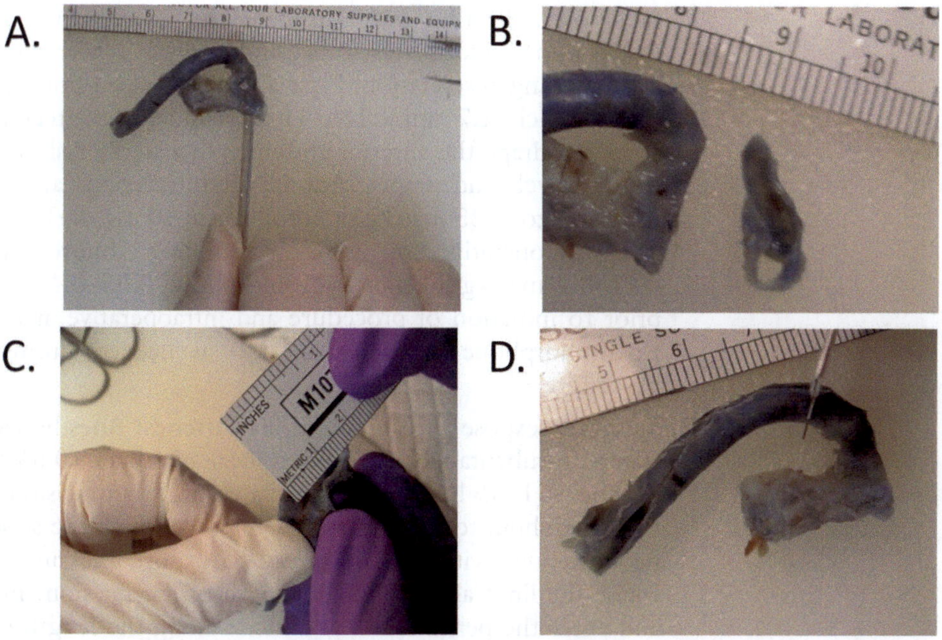

Fig. 5 Method for measuring endothelialization. (**A**) Each anastomosis is cut perpendicular to the native artery with a razor blade. (**B**) The distal anastomosis example shown here contains the entire cross section of the iliac artery. (**C**) Endothelialization is measured using a ruler or calipers to quantify the white region not stained with Evans blue. (**D**) After measurement, the remaining anastomosis block is removed for sectioning

2.2 Graft Explant

1. Medications used for graft explantation include ketamine (20 mg/kg), phenobarbital (25 mg/kg), heparin (300 units/kg), Evans blue (Sigma-Aldrich, 50 mg/kg body weight, 50 mg/ml in 0.9% sodium chloride USP, for injection), and lactated Ringer's solution.
2. Isolation of grafts: Metzenbaum scissors (7 in.) and DeBakey forceps for dissection, vascular clamps, and silk suture.
3. Fixation: Normal 0.9% saline solution, 10% formalin in phosphate-buffered saline (PBS), and cannula.
4. Histological processing: Standard sectioning and staining materials (e.g., histological cassettes, paraffin embedding medium, microtome, slides, staining reagents, microscope with camera) for histological processing based on the desired stains.

3 Methods

3.1 Implantation of Bilateral Aortoiliac Arterial Bypass Grafts

1. Induce anesthesia and set up (*see* **Note 1**): Induce general anesthesia with an intramuscular injection of either ketamine (10–20 mg/kg) or telazol (3–5 mg/kg) and isoflurane (1–3%). Place the animal in the supine position on the operating room

table on a heating pad. Intubate the animal with a size 6 (28 mm) endotracheal tube, and confirm placement with bilateral lung auscultation. Place a pediatric-sized Foley catheter (5 French (1.7 mm × 16″ (41 cm)) under sterile technique. Prep and drape the anterior abdomen in a sterile fashion with sterile towels and drapes. Administer intravenous antibiotics (e.g., cefazolin 25 mg/kg or enrofloxacin 10 mg/kg). Administer preoperative analgesics (including buprenorphine 0.01–0.1 mg/kg and/or hydromorphone 0.05–0.4 mg/kg) prior to initiation of procedure and intraoperative analgesics (hydromorphone 0.05–0.4 mg/kg) as needed during the procedure.

2. Surgically expose the aorta and iliac arteries: Anesthetize the skin with subcutaneously injected local anesthetic (0.5% bupivacaine or 1–3% lidocaine). Make a vertical skin incision from the subxiphoid to the suprapubic region. Divide the subcutaneous tissues with electrocautery to expose the linea alba. Incise the linea alba along the length of the incision. Elevate and enter the peritoneum, and incise along the length of the incision. Perform an abdominal cavity exploration. Check the bladder for adequate decompression with the Foley catheter. Lyse intra-abdominal adhesions with sharp Metzenbaum scissors, and investigate any abnormalities, such as fluid collections, masses, enlarged lymph nodes, aberrant vasculature, or abnormal organs. Elevate the transverse colon superiorly, mobilize the small bowel to the right, and gently retract with lap pad sponges moistened with warm normal saline using the Balfour retractor. Retract the splenic flexure of the colon to the left with a lap pad moistened with warm saline. Position the Balfour retractors to optimally expose the retroperitoneum. Incise the retroperitoneum overlying the aorta with Metzenbaum scissors and electrocautery. Expose the aorta using sharp dissection with Metzenbaum scissors from the level of the renal vessels to the aortic bifurcation. Place a small silastic vessel loop circumferentially to encircle the proximal aorta below the renal arteries. Circumferentially dissect the bilateral common iliac arteries with Metzenbaum scissors, and encircle the proximal and distal iliac arteries with small silastic vessels loops. After silastic loops are in place for vascular control, administer heparin at least 3–5 min prior to occlusion or clamping of the arteries. Administer heparin (300 units/kg) once intravenously, and then administer 100 units/kg every 1 h to maintain adequate anticoagulation during surgical time when vessel occlusion is occurring.

3. Create proximal graft to aorta anastomosis: Prepare the 4 mm diameter grafts with variable lengths (5–7 cm) by soaking in heparinized saline. Bevel one end of the graft with

Metzenbaum scissors so that the toe of the anastomosis is positioned on the outer curvature of the graft and the heel of the anastomosis is positioned on the inner curvature of the graft (*see* graft in Fig. 1e). Apply the proximal DeBakey side-biting U clamp on the aorta below the renal arteries on the right lateral wall of the aorta at a lateral angle (*see* Fig. 1a). Rotate the clamp to the left side of the animal to better expose the right lateral wall of the aorta, and bring the lateral wall more anterior (*see* Fig. 1b). Create a longitudinal arteriotomy in the right lateral wall of the aorta using an ophthalmic scalpel or 11 blade scalpel, and enlarge it with angled Potts scissors to an approximate length of 1.5 cm (*see* Fig. 1c, d). Flush the aortic lumen with heparinized saline. With Gerald forceps and Castroviejo needle holder, perform an end-to-side anastomosis using a continuous 7-0 monofilament, polypropylene nonabsorbable suture by starting at the inferior heel of the graft and using the parachute technique (*see* Fig. 1e). Carry the running suture up each side of the arteriotomy (*see* Fig. 1f). Hemostats with vascular booties are used to atraumatically clamp and apply gentle traction on the polypropylene sutures to promote exposure and maintain tension at the suture. Prior to completing the anastomosis, flush the graft with heparinized saline, and forward and back bleed the aorta. Clamp the distal end of the graft (*see* **Note 2**), and flush the aorta and graft with heparinized saline. Complete the anastomosis, and tie the suture with at least 7 knots. After the anastomosis is completed, check the suture line for hemostasis (*see* **Note 3**).

4. Create distal graft to iliac anastomosis: Direct attention to the right common iliac artery. Dissect the common iliac artery circumferentially both proximally and distally. Control the proximal and distal common iliac artery with vessel loops or vascular clamps (angled DeBakey clamp, DeBakey atraumatic tangential clamp, or Reynolds vascular clamp (*see* Fig. 2a)). Create a longitudinal arteriotomy in the right common iliac artery using an ophthalmic scalpel or 11 blade scalpel, and enlarge with Potts scissors to an approximate length of 1.5 cm (*see* Fig. 2b). Flush the arterial lumen with heparinized saline. Reclamp the proximal aorta. Flush the graft with heparinized saline. Size the length of the graft, and bevel the distal end of the graft to reach the mid-to-distal common iliac artery using Metzenbaum scissors and Potts scissors. With Gerald forceps and Castroviejo needle holder, perform an end-to-side anastomosis using a continuous 7-0 monofilament, nonabsorbable suture. Start the anastomosis at the inferior heel of the graft using the parachute technique. Carry the running suture up each side of the arteriotomy (*see* Fig. 2c, d). Prior to completing the anastomosis, flush the graft with heparinized

saline, and forward and back bleed the common iliac artery. Forward bleed the common iliac artery by releasing the aortic clamp and proximal common iliac silastic vessel loop, and then reapply the clamp. Back bleed the common iliac artery by temporarily releasing the silastic loop around the distal common iliac artery. Flush the common iliac artery and graft with heparinized saline. Release the proximal clamp and vessel loop from the aorta. After the anastomosis is completed, check the suture line for hemostasis (*see* Fig. 2f).

5. Implant second aortoiliac bypass graft: Repeat the entire procedure on the left side, beginning with the proximal anastomosis to the aorta and followed by the distal anastomosis to the left common iliac artery (*see* Fig. 3a). Tie off the aorta and iliac arteries to reroute blood into grafts using 2-0 silk ties (*see* Fig. 3b, c).
6. Assess graft patency: Assess the patency of grafts subjectively and objectively after completion of graft anastomoses. Examine the grafts by palpation for a pulse in the graft and bilateral femoral arteries (*see* **Note 4**). Directly assess blood flow in the grafts and/or proximal and distal arteries intraoperative using blood flow meters, including a transonic flow probe (Transonic®) or Doppler ultrasound (*see* Fig. 3d) (*see* **Note 5**).
7. Close the abdominal wound: After assuring hemostasis of the operative field, re-approximate the retroperitoneum with a running 3-0 absorbable suture. Release the retractors, and inspect the bowel for viability. Close the abdominal fascia and wall with 2-0 absorbable monofilament suture, and close the skin with 4-0 absorbable monofilament suture. Anesthetize the skin with a long-acting local anesthetic (liposomal bupivacaine 0.4 ml/kg) along the incision, if preoperative local anesthetic was not used at the beginning of the procedure. Dress the surgical incision with surgical skin glue.
8. Postoperative care and follow-up: Monitor the animal in the operating room, and extubate once it is able to protect its airway with a swallowing reflex (*see* **Note 6**). Take the animal to the post-anesthesia care area for close monitoring. Following extubation, watch the animal continually until sitting up. Thereafter, observe animal at hourly intervals until alert and awake. Continuously monitor postsurgical animals during the recovery period in conjunction with veterinary staff using assessment of vital signs, food intake, activity level, and assessment of surgical wounds according to local and international guidelines. Give postoperative medications under guidance of the local veterinary care team. Postoperative pain management consists of hydromorphone (0.1–0.3 mg/kg every 1–4 h as needed) and buprenorphine (0.03–0.1 mg/kg every 1 h as

needed). Administer postoperative antibiotics (cefazolin 25 mg/kg twice daily intramuscular injection) for up to 5 days after surgery. Under sedation or anesthesia, perform duplex ultrasound interrogation of the graft to check for patency at serial time points.

3.2 Humane Euthanasia and Explantation of Arterial Bypass Grafts

1. Animals are euthanized according to the American Veterinary Medical Association (AVMA) Guidelines for the Euthanasia of Animals depending on the time point of interest and nature of scientific inquiry. Typically, animals sacrificed between 4 weeks and 4 months after the procedure will have a significant neointimal layer on the graft. Sedate the animal 30 min prior to explant with ketamine (20 mg/kg). Inject heparin (300 units/kg) and Evans blue into the cephalic vein (*see* **Note** 7). After 30 min, euthanize animal with an anesthetic overdose (25 mg/kg phenobarbital).
2. Dissect the surgical area to reveal the grafts using Metzenbaum scissors and DeBakey forceps. Irrigate to clear the vascular tree with saline solution for approximately 2 min by injection via cannula into the abdominal aorta approximately 5 cm proximal to the proximal anastomosis. Fix arteries with a perfusion of 10% formalin in PBS (*see* **Note 8**) for 5 min via the inserted cannula (volume ~ 400 ml) (*see* **Note 9**).
3. Retrieve the grafts and surrounding vessels dividing the aorta above the renal arteries and the distal iliac arteries using Metzenbaum scissors, and immerse the specimen in additional fixative.
4. After at least 48 h of immersion fixation, trim the specimens of excess tissue. Collect blocks from the proximal and distal section for histology: Cut down the center of the aorta (to separate left and right grafts) using a razor blade (*see* Figs. 4a and 5a). Using a razor blade, cut through the center of the anastomosis (*see* Figs. 4b and 5b). Cut perpendicular to the native vessel. Make a second cut parallel to the initial cut to allow for the best orientation and lay flat. Measure the distance between the toe of the proximal or distal anastomosis and the adjacent blue-white interface on grafts opened longitudinally to determine the extent of endothelialization (*see* Fig. 5c). Maintain orientation of the proximal and distal sections, and collect longitudinal sections (*see* Figs. 4d, 4e, and 5d) (*see* **Note 10**). Submerge histology cassettes in 70% EtOH, and process with standard histological procedures, as desired staining dictates.

4 Notes

1. An experienced surgeon is recommended to perform the surgery. This surgeon should preferably have greater than 2 years of experience in performing microvascular anastomoses and have performed a minimum of 30 procedures.
2. Repeated reclamping of the aorta and iliac may be necessary if the graft material cannot be clamped. Clamping and reclamping may injure the intima of the aorta; therefore, carefully apply and remove the clamp, and minimize the number of times the arterial clamps are applied and reapplied to avoid graft thrombosis and vessel injury.
3. Small amounts of bleeding can be controlled with direct pressure, or the application of hemostatic agents such as Surgicel (Ref 1953, Ethicon, Johnson & Johnson) or Surgifoam (Ref 1973, Ethicon, Johnson & Johnson). Additional sutures may be required when hemostasis cannot be maintained.
4. Variations to graft implant: Aortoiliac bypass can be performed unilaterally or bilaterally. However, unilateral and bilateral configurations have significant differences in flow dynamics and shear stresses.
5. Heparinization during the procedure is necessary to avoid intraoperative thrombosis of native arteries and grafts. In addition, perioperative antiplatelet agents such as aspirin and/or clopidogrel can be used to assist with patency of grafts.
6. The potential complications of the surgery include bleeding, acute or chronic blood loss, graft kinking, graft thrombosis, graft infection, pseudoaneurysm at graft anastomoses, embolism and/or ischemia, wound infection, wound dehiscence, ureteral injury, and anesthesia-related complications including cardiorespiratory compromise, thermo-dysregulation, and stroke or neurological event.
7. Evans blue may have side effects in the animal, and thus monitoring should be performed after administration.
8. Alternative fixatives may be used, e.g., Karnovsky's fixative, if examination with scanning electron microscopy is of interest.
9. Arterial pressure may be monitored via an axillary artery cannula to ensure 100 mmHg of pressure during fixation.
10. Cross-sectional sections may be taken instead of, or in addition to, longitudinal sections.

Acknowledgments

We gratefully acknowledge the critical technical assistance of Ms. Jennifer Johnson and the entire staff of the Oregon National Primate Research Center, funded by NIH award P51OD011092. This work was funded by NIH grants R01HL130274, R01HL144113, and R01DE026170. We recognize the artistic contributions of Mr. John Alexander Hinds for creating the figures of this chapter.

References

1. Zilla P, Bezuidenhout D, Human P (2007) Prosthetic vascular grafts: wrong models, wrong questions and no healing. Biomaterials 28:5009–5027
2. Ballyk PD, Walsh C, Butany J, Ojha M (1998) Compliance mismatch may promote graft-artery intimal hyperplasia by altering suture-line stresses. J Biomech 31:229–237
3. Loth F, Fischer PF, Bassiouny HS (2008) Blood flow in end-to-side anastomoses. Annu Rev Fluid Mech 40:367–393
4. Hampton J, Matthews C (1966) Similarities between baboon and human blood clotting. J Appl Physiol 21:1713–1716
5. Harker LA, Kelly AB, Hanson SR (1991) Experimental arterial thrombosis in nonhuman primates. Circulation 83:IV41–IV55
6. Todd ME, McDevitt E, Goldsmith EI (1972) Blood-clotting mechanisms of nonhuman primates. Choice of the baboon model to simulate man. J Med Primatol 1:132–141
7. Chen C, Hanson SR, Lumsden AB (2000) Boundary layer infusion of heparin prevents thrombosis and reduces neointimal hyperplasia in venous polytetrafluoroethylene grafts without systemic anticoagulation. J Vasc Surg 31:237–245; discussion 246–247
8. Lin PH, Chen C, Bush RL et al (2004) Small-caliber heparin-coated ePTFE grafts reduce platelet deposition and neointimal hyperplasia in a baboon model. J Vasc Surg 39:1322–1328
9. Golden MA, Hanson SR, Kirkman TR et al (1990) Healing of polytetrafluoroethylene arterial grafts is influenced by graft porosity. J Vasc Surg 11:838–844; discussion 845
10. Hanson SR, Powell JS, Dodson T et al (1991) Effects of angiotensin converting enzyme inhibition with cilazapril on intimal hyperplasia in injured arteries and vascular grafts in the baboon. Hypertension 18:II70
11. Chen C, Lumsden AB, Hanson SR (2000) Local infusion of heparin reduces anastomotic neointimal hyperplasia in aortoiliac expanded polytetrafluoroethylene bypass grafts in baboons. J Vasc Surg 31:354–363
12. Clowes AW, Gown AM, Hanson SR, Reidy MA (1985) Mechanisms of arterial graft failure. 1. Role of cellular proliferation in early healing of PTFE prostheses. Am J Pathol 118:43–54
13. Clowes AW, Kirkman TR, Clowes MM (1986) Mechanisms of arterial graft failure. II. Chronic endothelial and smooth muscle cell proliferation in healing polytetrafluoroethylene prostheses. J Vasc Surg 3:877–884
14. Anderson DEJ, Glynn JJ, Song HK, Hinds MT (2014) Engineering an endothelialized vascular graft: a rational approach to study design in a non-human primate model. PLoS One 9: e115163
15. Anderson DEJ, Pohan G, Raman J et al (2018) Improving surgical methods for studying vascular grafts in animal models. Tissue Eng Part C Methods 24:457–464
16. Yao Y, Zaw AM, Anderson DEJ et al (2020) Fucoidan functionalization on poly(vinyl alcohol) hydrogels for improved endothelialization and hemocompatibility. Biomaterials 249:120011

Chapter 18

Quantification of In Vitro Blood-Brain Barrier Permeability

Bingmei M. Fu

Abstract

In vitro blood-brain barrier (BBB) models have been widely used to simulate in vivo models due to their low cost, feasibility, and repeatability. To serve as a valid substitute, the in vitro BBB should have the similar barrier function as that in vivo. This chapter summarizes the detailed methods for quantifying the barrier function, e.g., the permeability of the BBB to water, ions, and solutes for an in vitro BBB generated on the Transwell filter.

Key words In vitro blood-brain barrier, Transwell filter, Hydraulic conductivity, Diffusive solute permeability, Trans-BBB electric resistance (TER)

1 Introduction

The blood-brain barrier (BBB) is the name for the wall of the cerebral microvessels in the brain parenchyma. It is a unique dynamic regulatory interface between the cerebral circulation and the brain tissue and is essential for maintaining the micro-environment within the brain. The BBB is composed of an endothelial layer with embedment of pericytes at irregular intervals with the complete wrapping of the basement membrane, which is ensheathed by astrocyte foot processes [1–3]. Due to its special anatomical structure, the BBB strictly regulates the exchange of substances between blood and the cerebral parenchyma. The permeability of substances across the BBB is thus maintained to a very low level. This is good for protecting the central nervous system (CNS) from the potentially blood-borne harmful toxins. However, they also impede the passage of pharmacological substances across the BBB that otherwise could be useful for the treatment of CNS diseases.

To investigate the mechanisms by which the BBB maintains and regulates its permeability in health and disease, various in vivo and in vitro models have been used to study the transport across the

Feng Zhao and Kam W. Leong (eds.), *Vascular Tissue Engineering: Methods and Protocols*, Methods in Molecular Biology, vol. 2375, https://doi.org/10.1007/978-1-0716-1708-3_18,

BBB. The development of in vitro models for the BBB has enabled the study of transport phenomena at the molecular and cellular levels. The aim of such in vitro BBB models is to functionally resemble as many as possible the unique characteristics of the BBB. Compared with in vivo animal models, the in vitro models are relatively accessible, flexible, reproducible, and abundantly available. So far, two major types of in vitro BBB models have been developed: endothelial cell monolayer and coculture of endothelial cells with glial cells (the non-nerve cells in the brain) [4]. The cells for these models are basically obtained from primary/sub-passaged or immortalized cell cultures. The origins of the cells are also very diverse: human, primate, bovine, porcine, rodent, and murine species [5]. Primary brain microvascular endothelial cells have the closest resemblance to the BBB phenotype in vivo and exhibit excellent characteristics of the BBB at early passages [3]. They, however, have inherent disadvantages such as being extremely time-consuming and costly to generate, being easily contaminated by other neurovascular unit cells, losing their BBB characteristics over passages, and requiring high technical skills for extraction from brain tissue [6, 7]. Recently, an immortalized mouse brain endothelial cell line, bEnd3, and an immortalized human cerebral microvascular endothelial cell line, HCMEC/D3, have been used for in vitro BBB models because of their numerous advantages over primary cell culture: the ability to maintain BBB characteristics over many passages, easy growth and low cost, formation of functional barriers, and amenability to numerous molecular interventions [4, 7–12]. Most recently, human induced pluripotent stem cells (hiPSCs) have been differentiated into the brain microvascular endothelial cells, astrocytes, and pericytes, which offer a new approach to generate patient specific BBB as well as healthy ones [13].

There are various indicators for the permeability of the BBB. The BBB can be viewed as a permeable membrane, whose transport properties are described by Kedem-Katchalsky equations derived from the theory of irreversible thermodynamics [14]

$$J_s = PRT\Delta C + (1 - \sigma_f)CJ_v$$

$$J_v = L_p(\Delta p - \sigma_d RT\Delta C)$$

where J_s and J_v are the solute and volumetric fluxes and ΔC and Δp are the concentration and pressure differences across the membrane. L_p, the hydraulic conductivity, describes the membrane permeability to water. Thus, the definition of L_p is the water flux over the driving force (hydrostatic pressure difference minus the osmotic pressure difference across the membrane). P, the diffusive permeability, describes the permeability to solutes. The definition of P is thus the solute flux over driving force (concentration difference across the membrane when no water flux). σ_f is the solvent drag or ultrafiltration coefficient that describes the retardation of solutes due to membrane restriction, and σ_d, the reflection coefficient, describes the selectivity of a membrane to solutes. In many transport processes, σ_f is equal to σ_d, and thus we often use σ, the reflection coefficient, to represent both of them. These three coefficients can be determined experimentally and theoretically. The transmembrane electric resistance (TER) describes the resistance of a membrane to ions. The reverse of the TER reflects the membrane permeability to ions. In addition to these quantitative coefficients, there are other less quantitative permeability indicators for the BBB, e.g., brain uptake index (BUI) and brain efflux index (BEI) [15].

The most commonly used cell culture substrate consists of a porous membrane support submerged in the culture medium (Transwell apparatus or filter). The Transwell filter system is characterized by a horizontal side-by-side or vertical diffusion system. During the experiment, the flux of tracers into the abluminal compartment of the Transwell system is recorded as a function of the time, and the solute permeability is calculated from the slope of the flux. The tracers used in the transport experiments are labeled by a fluorescent dye or isotope whose intensity can be measured quantitatively [4]. Another index, trans-endothelial (or trans-BBB) electrical resistance (TER), or the ionic conductance of the monolayer/multilayer (BBB), is also a measurement of the "tightness" of the in vitro BBB models [4, 16]. The following presents the detailed methods for quantifying the solute permeability (P), hydraulic conductivity (L_p), and TER.

2 Materials

Either mammalian Ringer's solution or the cell culture medium (purchased from the company for your specific cells) can be used as the solvent for making the solution with the fluorescently labeled solutes. All the solutions additionally contain 10 mg/mL bovine serum albumin (BSA) or 1% BSA to simulate the normal oncotic pressure in the blood [17]. The solutions are made at room temperature 25 °C and warmed up to 37 °C when added to the Transwell filter with the cultured cell monolayer/multilayer (in vitro BBB). The following quantity of the solution is made for the in vitro BBB cultured on 10 Transwell filters.

2.1 Mammalian Ringer's Solution

Dissolve 7.714 g NaCl, 0.3432 KCl, 0.294 $CaCl_2 \cdot 2H_2O$, 0.2957 g $MgSO_4 \cdot 7H_2O$, 0.99 g glucose, 0.42 g $NaHCO_3$, 2.37 g Hepes (salt), and 2.6 g Hepes (acid) in 1 L double-distilled water [12]. Store at 4 °C for a week.

2.2 1% BSA Solution

Dissolve 500 mg BSA in 50 mL mammalian Ringer to make 50 mL 10 mg/mL BSA (1% BSA) solution [12].

2.3 Fluorescently Labeled Solute Solutions

FITC or TRITC or other fluorescently labeled solutes, e.g., dextrans with various molecular weights, sodium fluorescein, TAMRA, etc. are often used to represent solutes with a range of sizes. We use sodium fluorescein (NaFl, molecular weight 376 Da) and FITC-dextran 70k (molecular weight 70 kDa) as a typical small- and a large-sized solute to demonstrate the permeability measurement method. The excitation/emission wavelengths are set to 485/535 nm for these two solutes in the plate reader.

1. Make 30 mL of 10μM sodium fluorescein (NaFl): Dissolve 3.76 mg NaFl in 1 mL 1% BSA solution to make 1 mL 10^4μM, take 30μL of 10^4μM solution, put it into 30 mL 1% BSA solution to make 30 mL NaFL at 10μM, and filter the solution through a 0.2μm syringe filter.
2. Make 30 mL of 10μM FITC-dextran 70k: Dissolve 21 mg dextran 70k into 30 mL 1% BSA solution to make 30 mL FITC-dextran 70k solution at 10μM, and then filter the solution through a 0.2μm syringe filter.
3. Make diluted solutions for the above 10μM NaFl/dextran-70k following the procedure listed in Table 1.

2.4 Cells Needed to Generate the In Vitro BBB

Immortalized mouse brain microvascular endothelial cells (bEnd3) and rat astrocytes [4].

3 Methods

The method for quantification of the solute permeability is from Yuan et al. [12], and those for quantification of the hydraulic conductivity and TER are from Li et al. [4]. Carry out all the procedures at room temperature unless otherwise specified.

3.1 Generation of In Vitro BBB

Two types of in vitro BBB are generated. One is monoculture of bEnd3 cells, and another is coculture bEnd3 and astrocytes. For the monoculture, the luminal side of the Transwell filter (0.4μm pore size, 1.2 cm^2 insert bottom area, Corning) is coated with fibronectin (30μg/mL). The bEnd3 cells are seeded at 66k cells per filter and grown for 3–4 days. For the coculture, the luminal side of the filter is coated with fibronectin (30μg/mL) and the abluminal side with collagen type I and IV mixture or Matrigel. The Transwell filter is first inverted, and astrocytes are seeded onto the abluminal side of the filter at a density of 27.5k cells per filter and allowed to adhere for 2 h. After that, the filter is flipped back, and the astrocytes are cultured for 2 days. At the end of 2 days, bEnd3 cells are seeded onto the luminal side of the filter at 66k cells per filter and coculture with astrocytes for an additional 3–4 days [4].

Table 1
Making diluted solutions for calibration

C_A	C_B	1% BSA	C
C_0	0.1 mL*C_A	0.9 mL	1 mL 0.1C_0
0.1C_0	0.5 mL*C_A	0.5 mL	1 mL 0.05C_0
0.05C_0	0.5 mL*C_A	0.5 mL	1 mL 0.025C_0
0.025C_0	0.5 mL*C_A	0.5 mL	1 mL 0.0125C_0
0.0125C_0	0.8 mL*C_A	0.2 mL	1 mL 0.01C_0
0.01C_0	0.75 mL*C_A	0.25 mL	1 mL 0.0075C_0
0.01C_0	0.5 mL*C_A	0.5 mL	1 mL 0.005C_0
0.005C_0	0.5 mL*C_A	0.5 mL	1 mL 0.0025C_0
0.0025C_0	0.5 mL*C_A	0.5 mL	1 mL 0.00125C_0
0.00125C_0	0.8 mL*C_A	0.2 mL	1 mL 0.001C_0

$C = C_B$ + 1% BSA
List of concentrations for the diluted fluorescence solution used to calibrate the linear range of concentration vs. intensity curve for the fluorescence plate reader

3.2 Calibration Curve for Solute Permeability Measurement

The premise for using the fluorescence intensity as an indicator of the solute concentration is that the intensity changes linearly with the concentration measured by the instrument (e.g., fluorescence intensity plate reader) for the range of concentrations in the experiment. We use here the BioTek Synergy HT plate reader (Winooski, VT) in our measurement.

1. Take 50μL each of the diluted solutions, put it into 96-well plate reader, select excitation 485 nm and emission 535 nm, and get the readings.
2. Plot the Intensity vs. Concentration curve as shown in Fig. 1a. The slope (S_c) is the linear coefficient between intensity (I) and concentration (C). $I = S_c \times C/C_0$ ($S_c = 54{,}306$ for the calibration curve shown in Fig. 1a).

3.3 Determination of Solute Permeability of In Vitro BBB (P_{BBB})

1. Transwell filter insert with and without cultured monolayer/multilayer (in vitro BBB) is washed with PBS (*see* **Note 1**) at 37 °C.
2. The upper chamber of the Transwell filter system is filled with 0.5 mL solution with fluorescently labeled test solutes in 1% BSA at 10μM (C_0) and the bottom chamber with 1.5 mL blank 1% BSA solution (Fig. 1b) (*see* **Note 2**). The Transwell filter system is then put into the cell culture incubator (5% CO_2, 37 °C).

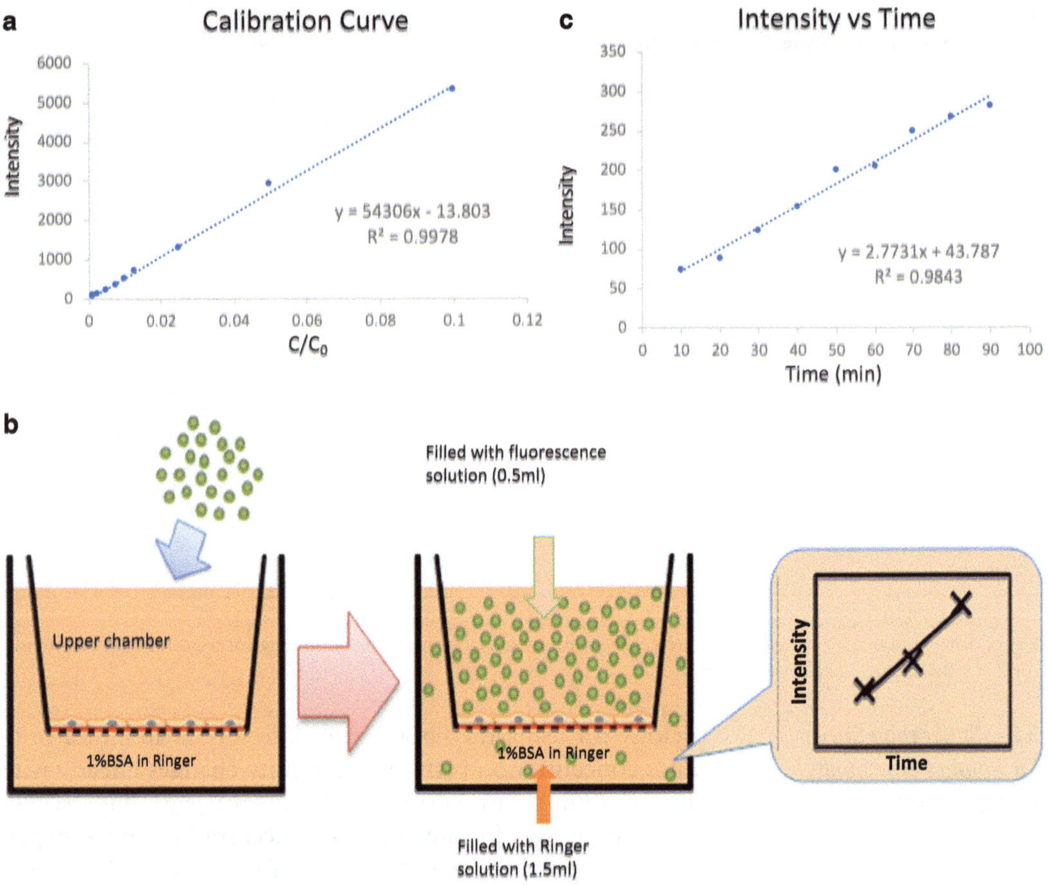

Fig. 1 Determination of solute permeability *P* of the in vitro BBB model generated on a Transwell filter. (**a**) Calibration curve of concentration vs. intensity for the fluorescence plate reader. The slope of this curve is used in determining the solute permeability *P*. (**b**) Schematic for the Transwell filter system and procedure for sample collection to determine the solute flux across the in vitro BBB. (**c**) The intensity vs. time curve from the collected samples at different times. Its slope is used to determine the solute permeability *P* of the in vitro BBB

3. The samples of 50μL are drawn every 10 min for 90 min from the bottom chamber and are replaced with the same amount of blank 1% BSA solution. The intensities of the samples collected at different times are determined by the plate reader (Fig. 1b).
4. The Intensity vs. Time curve is plotted as shown in Fig. 1c. The slope is $\Delta I_b/\Delta t$ (=2.7731 in the curve shown in Fig. 1c).
5. The solute diffusive permeability (P) of the Transwell filter with and without in vitro BBB is calculated from the relationship

$$P = \frac{\frac{\Delta C_b}{\Delta t} \times V_b}{C_u \times A}$$

where $\Delta C_b/\Delta t$ is the increase in fluorescence concentration in the bottom chamber during the time interval Δt, C_u is fluorescence concentration in the upper chamber (assumed to be constant during the experiment, thus $C_u = C_0$), V_b is the volume of the bottom chamber, and A is the surface area of the insert filter. Since we can only measure the intensity of the solution instead of concentration, from the calibration curve of Concentration vs. Intensity, we have

$$P = \frac{\frac{\Delta I_b}{\Delta t} \times V_b}{S_c \times A}$$

S_c is the slope of the calibration curve shown in Fig. 1a.

6. The Transwell filter insert itself also provides resistance on the transport of solutes. Therefore, the total resistance of this model system comprises of two parts – the resistance from the in vitro BBB and the resistance from the empty insert (*see* **Note 3**):

$$\frac{1}{P_t} = \frac{1}{P_c} + \frac{1}{P_i}$$

$$\frac{1}{P_t} = \frac{1}{P_{BBB}} + \frac{1}{P_i}.$$

Solving the above equation, we have the permeability of in vitro BBB P_{BBB},

$$P_{BBB} = \frac{P_t}{1 - \frac{P_t}{P_i}}$$

Here P_t is the measured permeability of the total system, and P_i is the permeability of the insert.

3.4 Determination of Hydraulic Conductivity of In Vitro BBB ($L_{p\ BBB}$)

The hydraulic conductivity L_p is measured using a bubble tracker system as shown in Fig. 2 [4, 18].

1. The Transwell filter with and without in vitro BBB is sealed within a transport system to form upper and lower chambers (Fig. 2) (*see* **Note 4**). The lower chamber is connected to a water reservoir by a Tygon and borosilicate glass tube. A hydrostatic pressure difference is created across the filter by adjusting the height of the reservoir with respect to the height of the fluid covering the in vitro BBB. Water flux J_v across the BBB is measured by tracking the position of an air bubble pre-inserted into the glass tube (*see* **Note 5**).
2. Initially, the reservoir is set to the same height as that of the fluid covering the in vitro BBB. At this position, the water flux across the BBB is zero as indicted by the motionless bubble (*see* **Note 6**).
3. The reservoir is then lowered by 10 cm, and a hydrostatic pressure difference Δp (10 cmH_2O) is applied across the BBB into the lower chamber of the system. To measure the

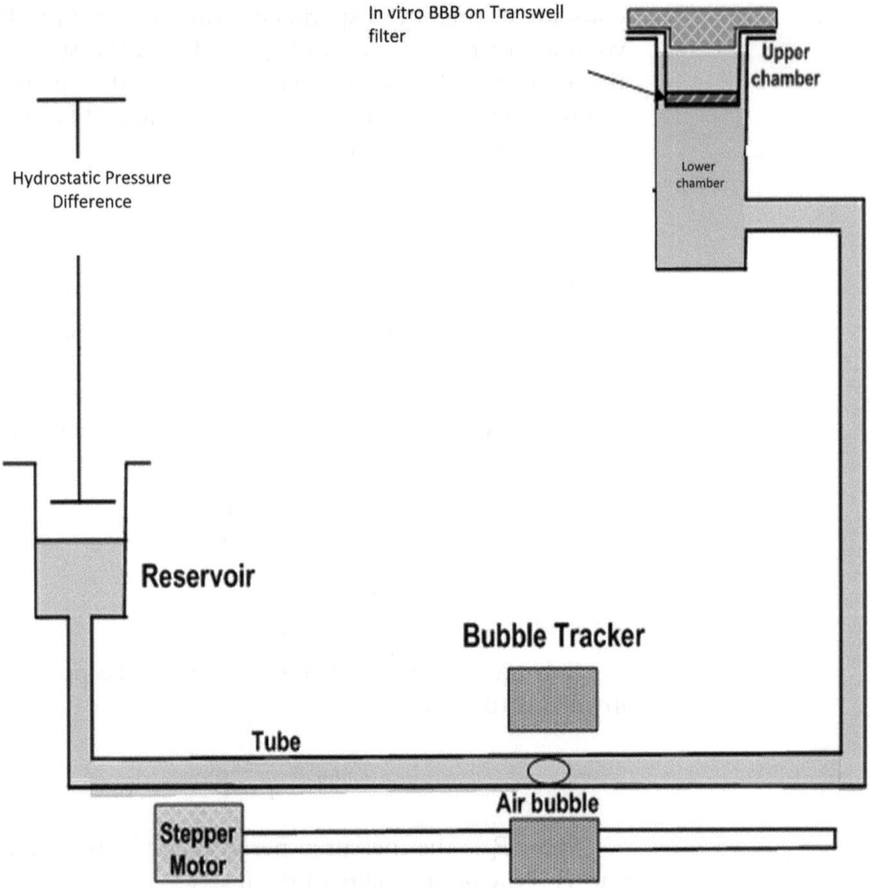

Fig. 2 Determination of hydraulic conductivity (water permeability) L_p of the in vitro BBB model generated on a Transwell filter. The schematic shows the components of each transport chamber and the components of apparatus for water flux measurement using bubble tracker previously developed in [4, 18]. (The figure is revised from Li et al. [4])

volumetric flow rate (Q_v), the air bubble displacement is tracked and converted to Q_v according to the following equation [19]:

$$Q_v = \left(\frac{\Delta d}{\Delta t}\right) \times F$$

where $\Delta d/\Delta t$ is the bubble displacement velocity and F is a tube calibration factor (fluid volume occupying a known length of tubing). The hydraulic conductivity L_p of the filter with and without in vitro BBB is calculated as

$$L_p = \frac{Q_v}{\Delta p \times A}$$

where A is the surface area of the Transwell filter and Δp is the pressure difference across the filter.

4. The Transwell filter insert itself also provides resistance on the water transport. Therefore, the total resistance of this model system comprises of two parts: the resistance from the in vitro BBB and the resistance from the empty insert (*see* **Note 3**),

$$\frac{1}{L_{pt}} = \frac{1}{L_{pBBB}} + \frac{1}{L_{pi}}$$

the hydraulic conductivity of in vitro BBB L_{pBBB},

$$L_{pBBB} = \frac{L_{pt}}{1 - \frac{L_{pt}}{L_{pi}}}$$

Here L_{pt} is the measured hydraulic conductivity of the total system and L_{pi} is that of the insert.

3.5 Determination of Trans-BBB Electric Resistance of In Vitro BBB (TER_{BBB})

We use the same solvent 1% BSA Ringer's solution as for P and L_p measurement in TER measurement (*see* **Note 7**).

1. The Transwell filter system is filled with 0.5 mL 1% BSA Ringer's solution in the upper chamber and 1.5 mL in the bottom chamber.
2. The resistance of the TER of the Transwell filter with and without in vitro BBB is measured by the resistance meter EVOM [2] (World Precision Instrument, Inc.) as demonstrated in the user's manual, quite straightforward (*see* **Note 8**):

$$TER_{BBB} = (R_t - R_{blank}) \times A$$

Here R_t is the EVOM [2] measured resistance of the Transwell filter cultured with in vitro BBB, and R_{blank} is that of the Transwell filter insert alone. The unit of R_t and R_{blank} is Ω. A is the area of the insert, in the unit of cm^2. Therefore, the unit of TER_{BBB} is Ω cm^2.

3.6 Representative Results for L_{pBBB}, P_{BBB}, and TER_{BBB}

Figure 3 demonstrates representative results for L_{pBBB}, P_{BBB} to dextran-70k, and TER_{BBB} for the in vitro BBB generated by the above protocol [4].

4 Notes

1. Can also use Ringer's solution if have enough or fresh cell culture medium.
2. Prewarm all the solutions to 37 °C before adding them into the upper and bottom chambers.
3. Need to measure permeability of at least three blank Transwell filter inserts separately under the same conditions as for the whole system. The average permeability is used for P_i.

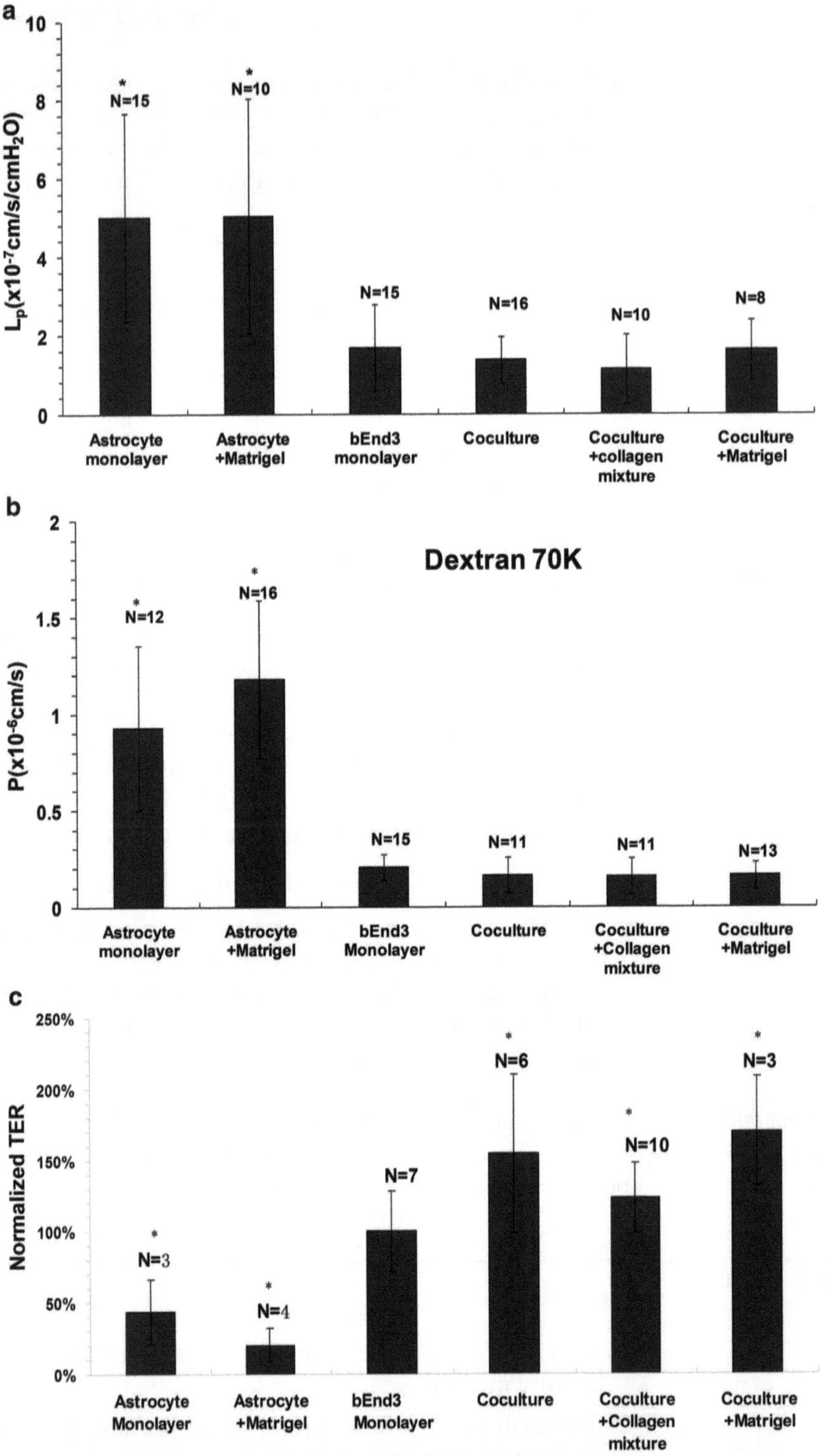

Fig. 3 Representative results for L_{pBBB}, P_{BBB}, and TER_{BBB} of the in vitro BBB model generated on a Transwell filter. (**a**) Hydraulic conductivity (L_{pBBB}), (**b**) P_{BBB} to dextran-70k, and (**c**) TER_{BBB} of the in vitro BBB models: astrocyte monoculture, astrocyte monoculture on top of a thin Matrigel coating (astrocyte + Matrigel), bEnd3 monoculture, coculture of astrocytes and bEnd3 cells on different sides of a Transwell filter (coculture), coculture with a thin coating of a mixture of collagen types I and IV on the luminal side of a Transwell filter (coculture + collagen mixture), coculture with a thin Matrigel coating on the abluminal side of a Transwell filter (coculture + Matrigel). (The figures are from Li et al. [4])

4. Before the transport experiment, the lower chamber of the system is filled with the experimental medium (1% BSA Ringer's solution or cell culture medium). The Transwell filter containing the in vitro BBB is rinsed twice with the experimental medium and then sealed in the transport chamber.
5. Both the upper and lower chambers as well as reservoir are continuously supplied with 5% CO_2—95% air to maintain the pH of the medium at 7.4. The temperature of the whole permeability measurement system was maintained at 37 °C.
6. It takes about 10 min for the bubble to be stationary.
7. Since the TER is the measurement for the electric resistance of the membrane (e.g., in vitro BBB) to ions, it is therefore dependent on the concentration of ions in the solutions across the membrane. We use 1% BSA Ringer's solution with the ion concentration at the physiological level in the blood plasma (140 mM). If using cell culture medium, its ion concentration needs to be determined first.
8. We use Epithelial Volt/Ohm Meter (EVOM [2]) (World Precision Inst. Inc. Sarasota, FL) to determine the TER of the Transwell filter with the insert area of 0.9 cm^2. The user's manual needs to be read carefully before the measurement.

Acknowledgments

The author would like to thank the funding support from the National Institutes of Health 4UH3TR002151-03.

References

1. Abbott NJ, Patabendige AK, Dolman DE, Yusof SR, Begley DJ (2010) Structure and function of the blood-brain barrier. Neurobiol Dis 37:13
2. Farkas E, Luiten PG (2001) Cerebral microvascular pathology in aging and Alzheimer's disease. Prog Neurobiol 64(6):575
3. Nicolazzo JA, Charman SA, Charman WN (2006) Methods to assess drug permeability across the blood-brain barrier. J Pharm Pharmacol 58:281
4. Li G, Simon M, Shi Z, Cancel L, Tarbell JM, Morrison B, Fu BM (2010) Permeability of endothelial and astrocyte cocultures: in vitro blood-brain barrier models for drug delivery. Ann Biomed Eng 38(8):2499
5. Fu BM (2018) Transport across the blood-brain barrier. Adv Exp Med Biol 1097:235–259
6. Deli MA, Abraham CS, Kataoka Y, Niwa M (2005) Permeability studies on in vitro blood-brain barrier models: physiology, pathology, and pharmacology. Cell Mol Neurobiol 25 (1):59
7. Brown RC, Morris AP, O'Neil RG (2007) Tight junction protein expression and barrier properties of immortalized mouse brain microvessel endothelial cells. Brain Res 1130(1):17
8. Soga N, Connolly JO, Chellaiah M, Kawamura J, Hruska KA (2001) Rac regulates vascular endothelial growth factor stimulated motility. Cell Commun Adhes 8(1):1
9. Tyagi N, Moshal KS, Sen U, Vacek TP, Kumar M, Hughes WM Jr, Kundu S, Tyagi SC (2009) H2S protects against methionine-induced oxidative stress in brain endothelial cells. Antioxid Redox Signal 11(1):25
10. Yoder EJ (2002) Modifications in astrocyte morphology and calcium signaling induced by

a brain capillary endothelial cell line. Glia 38 (2):137

11. Weksler B, Romero IA, Couraud P-O (2013) The hCMEC/D3 cell line as a model of the human blood brain barrier. Fluids Barriers CNS 10:16
12. Yuan W, Li G, Fu BM (2010) Effect of surface charge of immortalized mouse cerebral endothelial cell monolayer on transport of charged solutes. Ann Biomed Eng 38(4):1463
13. Lippmann ES, Azarin SM, Kay JE, Nessler AR, Wilson HK, Al-Ahmad A, Palecek SP, Shusta EV (2012) Derivation of blood-brain barrier endothelial cells from human pluripotent stem cells. Nat Biotechnol 30:783–791
14. Curry FE (1983) Mechanics and thermodynamics of transcapillary exchange. In: Handbook of Physiology. The cardiovascular system. The American Physiology Society, Bethesda
15. Pardridge WM (1998) CNS drug design based on principles of blood-brain barrier transport. J Neurochem 70:1781
16. Crone C, Olesen SP (1982) Electrical resistance of brain microvascular endothelium. Brain Res 241(1):49–55
17. Sugihara-Seki M, Fu BM (2005) Blood flow and permeability in microvessels. Fluid Dynamics Res 37:82–132
18. Cancel LM, Fitting A, Tarbell JM (2007) In vitro study of LDL transport under pressurized (convective) conditions. Am J Physiol Heart Circ Physiol 293:H126–H132
19. DeMaio L, Tarbell JM, Scaduto RC Jr, Gardner TW, Antonetti DA (2004) A transmural pressure gradient induces mechanical and biological adaptive responses in endothelial cells. Am J Physiol Heart Circ Physiol 286: H731–H741

Chapter 19

Subcellular Force Quantification of Endothelial Cells Using Silicone Pillar Arrays

Mohanish K. Chandurkar and Sangyoon J. Han

Abstract

Micropillar arrays are one of the tools that quantify the mechanical forces exerted by endothelial cells and any other adherent cell types. This chapter describes the fabrication and usage of the micropillars that reports spatial distribution of traction forces by endothelial cells. The fabrication begins with making a "master" pillar substrate using photolithography, followed by softlithography of the master with polydimethylsiloxane. The pillars can then be printed with an extracellular matrix protein in specific shape using stamp-off technique. Lastly, imaging of the micropillars with cells and the analysis of the traction force using Matlab-based software are described.

Key words Micropillar, Traction force microscopy, Focal adhesion, Softlithography, Extracellular matrix, Endothelial cells

1 Introduction

Endothelial cells (ECs) are under constant influence of dynamic mechanical forces from various sources. These forces include a shear stress from the blood flow, a tensile stress from the dynamic blood pressure, and a deformation of the extracellular matrix (ECM). ECs sense those forces and respond by eliciting appropriate molecular signaling cascades. The mechanotransduction signals determine the ECs' own functions as a blood barrier and inflammation preventer [1], as well as general cell functions such as survival, death, proliferation, and migration [2–4]. Integrin-based focal adhesions (FAs) provide a linkage between the cytoskeleton and the ECM, transmitting the tension in the cytoskeleton to the ECM [5]. At the same time, the FA is the main sensor for ECM stiffness sensing [2] and one of the mechano-sensors for flow shear stress [6]. While the FA's force-sensing mechanism is under active investigation, it is becoming clear that the mechano-sensation involves transmission of the force through the adhesion molecules, opening new binding

Feng Zhao and Kam W. Leong (eds.), *Vascular Tissue Engineering: Methods and Protocols*,
Methods in Molecular Biology, vol. 2375, https://doi.org/10.1007/978-1-0716-1708-3_19,

sites or phosphorylation sites within FA proteins such as talin, vinculin, or p130Cas [7]. Accordingly, it has been important to understand how the force is exerted at the FA sites under various mechanical environments. Importantly, the measurement of traction force of the ECs can help understand the relation of mechanical homeostasis and the chemical signaling governing tissue integrity within vascular graft [8]. The knowledge from the force measurement would also provide a way to assess if the ECs mimic the structure of the vascular system and successful maturation of the engineered blood vessels and vascular tissue.

Numerous techniques have been developed to measure the force exerted by a cell on the substrate. The traction forces of cells were first observed in 1980 as wrinkles in silicone thin and elastic film [9]. This tool provided the cues of the mechanical pathways and their involvement for the generation of the traction forces, focal adhesions, and cytoskeletal involvement [10, 11]. For better quantification of the silicone deformation, latex beads were introduced on top of the silicone surface of a wrinkle-free polymer [12]. Then a polyacrylamide gel substrate was used as a soft substrate with tunability of the gel stiffness [13]. While this soft gel-based assays, referred to as traction force microscopy (TFM), have been further developed for better spatial resolution [14], there has been another branch of force sensor developed using soft lithography, which consists of arrays of soft, micrometer-sized pillars [15], referred to as the micropillar arrays. One advantage of the micropillar array is that it can precisely control positions of focal adhesions as well as those of traction forces. Additionally, its ability to accommodate the ECM protein onto specific area has allowed for control over the spread area of the ECs. Using this controllability, the role of the substrate stiffness and cell spread area could be uncovered in an uncoupled manner [16].

The current chapter describes the use and fabrication of micropillar arrays and microcontact printing techniques. Micropillar arrays can be mechanically viewed as arrays of the vertical cantilevers. The deflection of the pillars can be measured through a fluorescent microscope and used to calculate traction forces. This chapter focuses on the description of the (1) microfabrication of the "master" micropillar arrays, (2) soft lithography of the PDMS micropillars, (3) micropatterning of the micropillars with stamp-off technique, (4) cell-experiment, and (5) force measurement. For each sub-topic, we list required materials and methods. This protocol is highly based on the two published protocols [17, 18]. By combining protocols for micropillar fabrication and micropatterning, we hope to provide a fuller picture of the entire pillar-based endothelial force quantification system.

2 Materials

As in every fabrication- and tissue culture-related jobs, care is needed while using the chemicals, and one should refer to MSDS. Extra precautions are also required while handling the microposts and switching between sterile and non-sterile conditions.

2.1 Microfabrication of the Micropillar Arrays

1. Silicon wafer.
2. SU-82000 series photoresist (MicroChem, Newton, MA). SU-82005 and 2010 are used.
3. Disposable wide-mouth pipettes for dispensing SU-8 on the silicon wafers.
4. SU-8 developer (MicroChem, Newton, MA).
5. Chrome mask with features for micropillars or stamp.
6. Cleanroom wipes.
7. Fluorosilane (tridecafluoro-1,1,2,2-tetrhydrooctyl)-1-trichlorosilane) (T292-KG, United Chemical Technologies).
8. 70% isopropanol alcohol (IPA).
9. 99.5% acetone.
10. Deionized water (DI).
11. Cleanroom environment (Class 1000 preferred).
12. Photoresist spincoater with adjustable acceleration and velocity.
13. UV aligner.
14. Nitrogen gas.
15. Ozone cleaner (UVO-Cleaner, Jelight, Irvine, CA).
16. Clean Pyrex dishes.
17. Hotplates.
18. Digital timers.
19. Tweezers.

2.2 Soft Lithography of PDMS Micropillars

1. SU-8 masters for the stamp and micropillar array (a result from Subheading 2.1).
2. Sylgard 184 (PDMS) base and curing agent (DOW, Midland).
3. Fluorosilane (tridecafluoro-1,1,2,2-tetrahydrooctyl)-1-trichlorosilane) (T2492-KG, United Chemical Technologies).
4. Glass pasture pipette.

5. Glass slide.
6. Aluminum foil, dish, or boat. Its size should match that of the SU-8 master.
7. Plastic cup.
8. Plastic stirrer (e.g., plastic 5 ml pipette).
9. Razor blade.
10. Digital balance.
11. Vacuum desiccator chamber.
12. N_2 air.
13. Convection oven.
14. Plasma chamber (Plasma Prep II, SPI, West Chester, PA).

2.3 Micropatterning Micropillar Arrays with Stamp-Off Technique

1. PDMS flat stamp (cross-linker: base-PDMS ratio 1:20).
2. PDMS negative stamp (cross-linker: base-PDMS ratio 1:20).
3. PDMS micropillar arrays (cross-linker: base-PDMS ratio 1:10).
4. Sterile DI.
5. 50μg/ml human fibronectin (FN) dissolved in sterile DI water (BD Biosciences).
6. Ethanol (100% and 70%).
7. 5μg/ml bovine serum albumin (BSA) conjugated with Alexa Fluor 594 (A13101, Invitrogen) or 1μg/ml fluorescent lipophilic carbocyanine (DiI) solution prepared in DI (D3886, Invitrogen).
8. 0.2% Pluronic F127 (BASF, Mount Olive, NJ).
9. Phosphate-buffered solution (PBS) pH 7.4.
10. Aluminum foil.
11. 100 mm petri dish.
12. Sterile biosafety cabinet.
13. UV-ozone (UVO-Cleaner, Jelight, Irvine, CA).
14. Nitrogen gas.
15. Sterile tweezers.

2.4 Cell Seeding

1. Human umbilical vein endothelial cell lines (HUVECs).
2. 0.05% trypsin-ethylenediaminetetraacetic acid (EDTA).
3. Growth medium.
4. Tissue culture dish.
5. Phosphate-buffered solution (PBS) pH 7.4.
6. Sterile tweezers.

2.5 Imaging

1. Inverted confocal or epifluorescence (wide-field) microscope.
2. 60× (or 100×) objective.

2.6 Traction Force Analysis

1. Workstation (a regular PC or a laptop is fine).
2. Matlab ^{R} (2015b or higher).
3. Matlab-based micropillar analysis package software.

3 Methods

3.1 Microfabrication of the Micropillar Arrays

The steps describe the process of fabrication of SU-8 master for the micropillar or stamp. The features of the stamps are made using SU master by two layers of SU-8 for micropillars or stamp. The first layer works as an adhesion layer to silicon wafer, whereas the second layer makes the features of micropillars or stamp (Fig. 1a).

1. Clean the silicon wafer surface using 99.5% acetone for 10 s, rinse off the acetone, followed by a rinse with 70% IPA for 10 s, and immerse the wafer in DI water three times to remove residual particles that may be present on the wafer surface.
2. Preheat the hotplate to 200 °C. Place the wafer on a preheated hotplate for wafer dehydration for 10 min. This step ensures SU-8 adhesion to the wafer surface. Allow the wafer to air cool at room temperature before the next step.
3. Spincoat a flat layer of SU-82005 on the wafer base:
   (a) Fix the wafer on the spincoater, and follow the manufacturer specifications for the chosen viscosity of SU-82000 series to set acceleration, rate, and time for spincoating. Obtaining a base layer of <5 μm will require optimization of spincoating process with initial trials.
   (b) Pour the photoresist on the wafer to cover 1 ml/in. area with even distribution throughout. Make sure the photoresist is dispensed in a smooth motion onto the wafer. Allow the photoresist to sit for a while before spincoating for proper coverage.
4. After spinning, remove excess of the edge photoresist using automatic edge removal cycle available on recent spincoaters or by using a cleanroom wipe lightly wetted with SU-8 developer. This step ensures good-quality photomask alignment.
5. Heat the hotplate to 95 °C, and place the wafer on the hotplate according to manufacturer's recommended pre-exposure bake time. It helps to dry and set SU-8 in place preventing unwanted free flow. An approximate cycle time is 3–5 min.
6. Set up the wafer on the UV aligner. Treat the wafer under UV light according to manufacturer's recommended power and

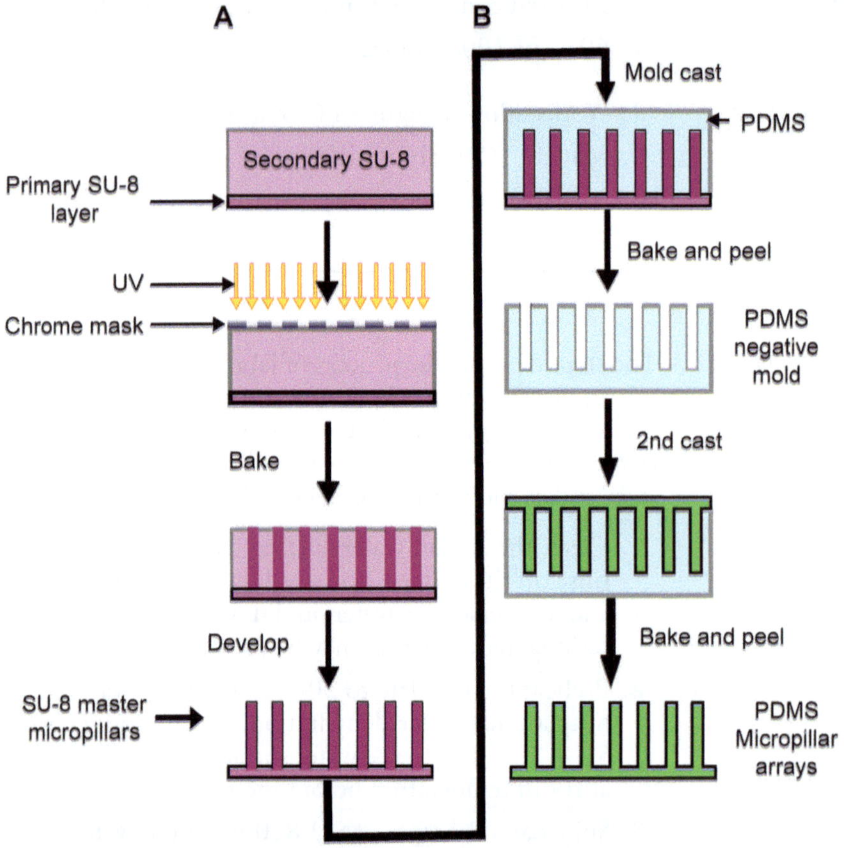

Fig. 1 Microfabrication of the micropillar arrays. (**a**) Fabrication of the "master" micropillar arrays via photolithography using SU-8 photoresist. (**b**) Soft lithography of the PDMS micropillars from the master pillar arrays. The double-casting method, i.e., making a negative PDMS mold, helps maintenance of the master for a longer time. The negative mold can be used 5–8 times depending on the condition

time. For the base layer, photomask is not required as no patterns are expected. The base layer enables strong adhesion with the second layer. Typically, exposure time varies between 3 and 10 s.

7. Place the base layer wafer on a hotplate preheated to 95 °C for manufacturer's recommended post-exposure bake time. Expected cycle time varies from 3 to 5 min. To avoid thermal stress, avoid quick cooling of the wafer by placing it on nonwoven wipe. Cracks in the SU-8 film can be removed by annealing the wafer cylindrically on a hotplate at 95 °C and cooling it several times.
8. Spincoat a second layer of SU-8 onto the base wafer to create the micropillar of the stamp pattern:
   (a) Fix the wafer on a spincoater, and set the acceleration, rate, and time for the photoresist in the spincoater. The

micropillar layer thickness ranges from 5 to 15 μm, whereas stamp pattern ranges between 5 and 50 μm.

(b) Repeat **step 3b** to dispense photoresist. For thicker photoresist layer, adjust the volume and time for uniform spreading.

9. Repeat **step 4** to remove the excessive edge photoresist.
10. Repeat **step 5** for pre-exposure bake.
11. Transfer the wafer on the UV aligner. Place the chrome mask on the wafer such that full contact is made between the mask and SU-8 photoresist without any gaps. Incomplete removal of the edge photoresist can cause gaps between the photoresist and the mask, which leads to poor pattern transfer due to diffraction of the UV light.
12. Expose the wafer to UV for the manufacturer's recommended power and time. The micropillar layer requires between 5 and 15 s, while a thicker stamp layer will require between 5 and 60 s.
13. Repeat **step** 7. Annealing helps remove thermal stress.
14. To develop features of micropillars, submerge the wafer in SU-8 developer. Whisk for recommended time. In this case it is 5–15 min depending on the thickness, feature density, and size of the SU-8. High density and thickness of the micropillar require more time for the developer to penetrate inside the wafer.
15. Clean the wafer with fresh SU-8 developer before drying to remove the residue of the SU-8.
16. Rinse with IPA for 10 s.
17. Use filtered nitrogen gas to dry the wafer.
18. If cracks are visible on the structure, use annealing steps till cracks are removed. To remove the cracks, place the wafer on the 150 °C hotplate for 5 min, and cool on a nonwoven wipe. Annealing process can affect the structural integrity of the wafer due to SU-8 photoresist which deforms due to high temperature.
19. After the annealing process and removal of cracks, the final hard bake step cross-links the SU-8 features on the master wafer. This hard-baked master can be repeatedly used for polydimethylsiloxane (PDMS) casting without damaging the SU-8 structure.
20. For that, initially place the wafer on hotplate at 150 °C for 30 min then on 200 °C for 30 min.
21. Cool on nonwoven wipe.

22. Glue the wafer on the glass side for ease of handling, and avoid breakage of master.
23. Coat the master wafer with fluorosilane (tridecafluoro-1,1,2,2-tetrhydrooctyl)-1-trichlorosilane) (T292-KG, United Chemical Technologies); this will avoid bonding of PDMS with master wafer surface.
    (a) Treat master in oxygen or air plasma for 60–120 s.
    (b) Place SU-8 master with 50 μl of 100% fluorosilane and in a silane desiccator under vacuum. This helps to vaporize silane and bind with SU-8 surface. 1 h is the recommended desiccation time for throughout coverage.

3.2 Soft Lithography of PDMS Micropillars

The steps include the process of PDMS fabrication of micropillar arrays from the SU-8 master. PDMS stamps are fabricated through a single-casting process using the SU-8 master, while the micropillar array fabrication requires a double-casting process from the master (Fig. 1b).

3.2.1 Methods for Stamp Fabrication

1. Prepare 20:1 mixture ratio of base with curing agent by weight in a plastic cup using a digital scale.
2. Stir PDMS mixture for 5–10 min with a plastic stirrer for thorough mixing.
3. Use desiccator to degas PDMS mixture to remove all air bubbles caused by stirring.
4. Place the SU-8 master in an aluminum dish or boat.
5. Add degassed PDMS over the master to the thickness of 1 cm. Poured PDMS can cause elevation of the SU-8 master; use the disposable pipette to position master all the way to the bottom of the dish.
6. Degas the PDMS again until all bubbles disappear. Also, a slight puff of N_2 gas pops air bubbles that are trapped at the surface.
7. Place the aluminum dish in a convection oven at 110 °C for 15–20 min or until the PDMS reaches firmness.
8. Cut away the dish using razor blade.
9. Cut excess of the PDMS around and below the SU-8 master.
10. Peel away the PDMS from the dish carefully.
11. Cut PDMS to form 1 cm × 1 cm × 1 cm (width × height × thickness) stamp. Mark with a notch on one of the stamp corner on the side that is opposite to stamp feature.

3.2.2 Method for Negative Molds for Micropillar Arrays

1. Prepare 10:1 mixture of base with curing agent by weight in a plastic cup using a digital scale.
2. Repeat steps in Subheading 3.1 to obtain negative stamp copies of SU-8 master with micropillar arrays.
3. The negative mold is used to make PDMS micropillar arrays (*see* **Note 1**).
4. Plasma treat the negative molds at 100 W for 90 s inside the plasma chamber to activate the surface of the PDMS. Power and exposure times vary with different plasma chambers.
5. Use the glass-based gas desiccator to place negative molds in feature up orientation. Use tweezers for handling. Cover the glass slide with 25–50 ml of fluorosilane, and place it in the center of the desiccator with the help of glass pipette. Also place the glass pipette inside the desiccator with small amount of fluorosilane in its tip inside the desiccator along with glass slide.
6. Seal the gas desiccator, and apply the vacuum inside.
7. Expose the fluorosilane vapors over the negative mold for 1 h.

3.2.3 Method for Micropillar Arrays

1. Prepare the 10:1 mixture of base with curing agent by weight in a plastic cup using a digital scale.
2. Stir PDMS mixture for 5–10 min with a plastic stirrer for thorough mixing.
3. Degas PDMS mixture inside the desiccator till all air bubbles are removed caused by stirring.
4. Plasma treat the glass slide in the presence of oxygen inside the plasma chamber at 100 W for 90 s.
5. Add the degassed PDMS to the negative mold with the help of pipette till all the features are covered with PDMS. Remove all the air bubbles by degassing or with gentle N_2 gas puff.
6. Place the plasma treated glass side on the PDMS-filled negative mold carefully such that plasma treated side of the glass slide faces the negative mold. Start placing the slide from one edge slowly moving towards the other side such that no air bubble is trapped between the negative mold and the glass slide.
7. Heat the negative mold into a convention oven at 110 °C for 6 h for the PDMS to fully cure.
8. Once cured, remove and cool the negative mold at room temperature.
9. Peel off the glass slide away from the negative mold slowly to release the micropillar arrays on a glass slide. Peeling along the

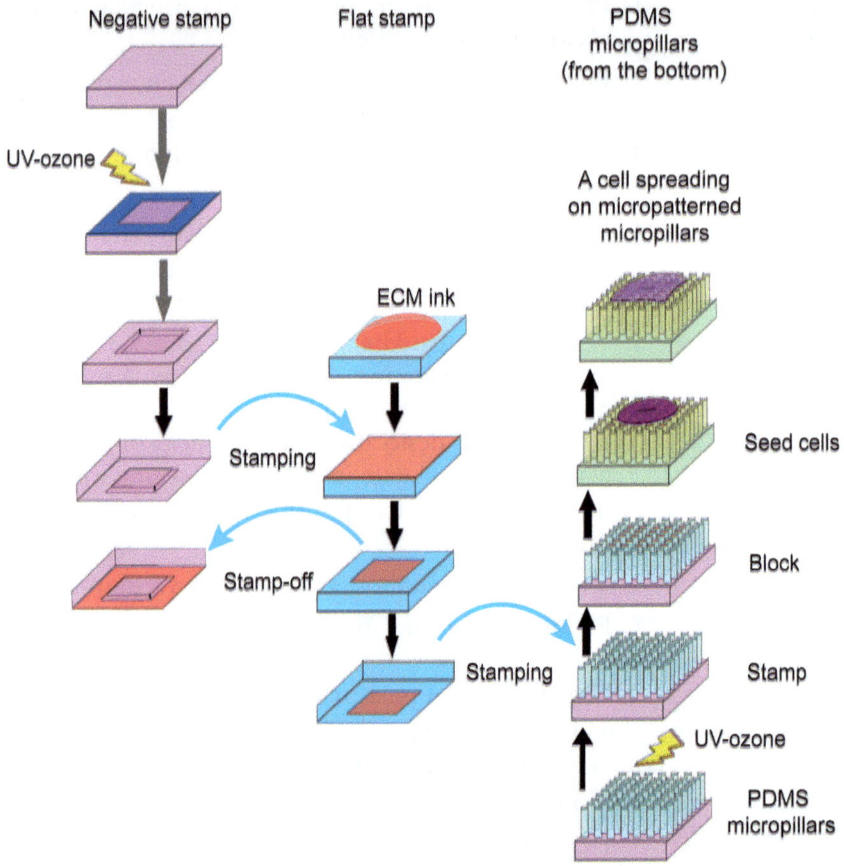

Fig. 2 Microcontact "stamp-off" printing procedure. See text for details

diagonal direction is recommended of the micropillar array to prevent the micropillars from sticking to each other.

10. Inspect the micropillar arrays for collapsed micropillars under light microscope to determine the usability.
11. Remove excess PDMS from the edges and side that could possibly interfere with the stamping.

3.3 Micropatterning Micropillar Arrays with Stamp-Off Technique

The micropillar tips can be coated with a pattern of the ECM using stamp-off method. The two types of stamps are required, a flat stamp coated with the ECM protein (e.g., fibronectin or FN) and a negative stamp that has negative pattern (inverse of a pattern of interest, e.g., square shape) to remove the excess of the protein (Fig. 2). It is required to complete all the steps under sterile conditions inside a biosafety cabinet (*see* **Note 5**).

1. Place the flat stamp in a petri dish facing the working side up.
2. Coat the stamp with FN droplets on the corners first, then on the edges, and lastly on the inside region for throughout

covering. After the adsorption of the FN on the stamp surface, the surface becomes hydrophilic. This process requires less volume of the FN for overall coating.

3. Incubate the stamps with FN solution for 1 h.
4. Rinse the stamps with DI water. Avoid direct water on the stamps; instead, pour water at the edges, and let the water level rise so that stamps submerge in water.
5. Transfer the stamps to other petri dish filled with DI water. Repeat one more time.
6. After the wash, dry the stamps with N_2 gas.
7. Expose the negative stamp to the UV-ozone for 7 min to increase its hydrophilicity. This step can be performed outside the biosafety cabinet.
8. With the help of tweezers, place the flat stamp side coated with FN on top of the patterned side of the negative stamp. Make sure both the stamps are in contact by gentle tap using tweezers. Hold for 5 s, and remove the stamps. This will allow the flat stamp for further patterning on the micropillar.
9. Similar to the flat stamp, expose the micropillar arrays to the UV-ozone for 7 min to activate the surface and make it hydrophilic. This step can be performed outside the biosafety cabinet.
10. Now place the flat stamp on the micropillar. Tap gently with the tweezers to allow both the surface to contact each other and not deform the micropillars. Practicing under inverted light microscope can be helpful to understand the tapping force required to form good contact without deforming the surface of micropillars.
11. Submerge the micropillars in 200 proof ethanol for 10 s in a petri dish to allow wetting of the surface.
12. Treat the micropillars with 140 proof ethanol in another petri dish for 10 s. This step ensures the sterility of the micropillar while handling. Further steps should be performed under the biosafety cabinet using the aseptic techniques.
13. Transfer the micropillars to sterile DI water, and wash for 10 s to ensure removal of ethanol. Take extra precaution while transferring the micropillars between the dishes as hydrophobic micropillars under surface tension of water may damage the stamps. This can be avoided by extra careful handling and keeping the stamp arrays from dewetting during the transfers.
14. Repeat **step 13**, two more times.
15. Transfer the micropillar arrays to a petri dish with the fluorescent dye.

(a) For BSA-based method, put droplets of 5 μg/ml bovine serum albumin (BSA) conjugated with Alexa Fluor 594 (A13101, Invitrogen) onto the petri dish, and place the micropillars on each droplet for coating.

(b) For DiI-based method, submerge the micropillar arrays upside down in a petri dish with 1 μg/ml fluorescent lipophilic carbocyanine (DiI) solution prepared in DI (D3886, Invitrogen). DiI produces stronger fluorescent signal than BSA.

16. Incubate the petri dish containing dye and cover with aluminum foil for 1 h.
17. Use sterile DI water to wash the micropillars for 10 s for BSA. For DiI, rinse the micropillars with sterile DI water to remove excess of DiI solution.
18. For Pluronic treatment, place the micropillars onto the Pluronic solution in upside down position. Incubate for 30 min. This allows the unstamped micropillars to block the attachment of the cells on the undesired areas including the base and the sides.
19. Repeat **step 13**. Pluronic treatment causes the PDMS surface to become hydrophilic, so less precaution is needed for micropillar transfer between dishes as dewetting issue is greatly reduced.
20. Finally, transfer the micropillars into dish with PBS. These PBS dishes can be incubated for a period of 1 week. Better results are obtained after immediate use for cell culture.

3.4 Cell Seeding

1. Cells should be near confluent or fully confluent in the tissue culture dish or flask. This helps to ensure enough cells are available to land on the micropillars.
2. Micropillar substrates should be placed onto a new tissue culture dish containing prewarmed media and allowed to incubate for 1 h prior to seeding.
3. In the cell-seeded tissue culture dish or flask, aspirate away the growth media.
4. Rinse the cells twice with 1× PBS of pH 7.4 to remove residual media in the dish.
5. Use trypsin/EDTA solution to detach the cell from the culture dish. Add 1 ml of 0.05% trypsin/EDTA to culture dish or flask, and rock the dish or flask back and forth so that trypsin covers the entire surface area.
6. Immediately aspirate the trypsin, and add 1 ml of fresh 0.05% trypsin/EDTA.

7. Place the cells with the trypsin/EDTA solution in CO_2 incubator for 5 min, and check if they have detached from the surface and started looking rounded under microscope.
8. Immediately add 2–5 ml of cell media to inactivate trypsin. Transfer them to 15 ml centrifuge tube.
9. Centrifuge the cell suspension for 5 min at 2000 rpm (490 × *g*) to form pellet.
10. Resuspend the cells in appropriate amount of prewarmed media, and transfer 10^4 to 10^6 cells/ml, depending on single-cell vs. multi-cell force measurement, to the dish containing the micropillar substrate using pipette. For example, for single-cell traction force experiment on 1 cm × 1 cm-sized micropillar template, use 10^4 cells/ml (using automated cell counter or using manual counting via hemocytometer). Pay special attention to aspirate back and forth in a pipet at least ten times to break up potential cell clumps within the cell solution.
11. Place this dish into the incubator, and allow cells to settle onto the micropillars.
12. Periodically observe the density of cells that have landed and settled on the micropillars. If a significant portion of them have settled, then a rinse with fresh prewarmed media can be done to remove unattached cells. This step may or may not be necessary depending on the type of cell being seeded.
13. Allow at least 4 h for the cells to spread and stabilize on the micropillar tips before traction force measurements are taken (*see* **Note 6**).

3.5 Imaging

1. Place the micropillar (either fixed or live) on the stage top incubator of the microscope (*see* **Note 2**).
2. Find the single cell on the micropillar under Alexa Fluor 594 channel.
3. Focus on the micropillar pillars' contact to the cell, and image the location.
4. Adjust the *z*-direction control, focus on the bottom of the micropillar in contact with the stamp, and capture the image for the same.
5. Capture images for other fluorescent channels if any. For example, actin image (via phalloidin staining) is highly recommended for cell area recognition. Additionally, immunofluorescently stained vinculin or fibronectin (FN) channel can be imaged at the *z*-location of tips of the pillars.

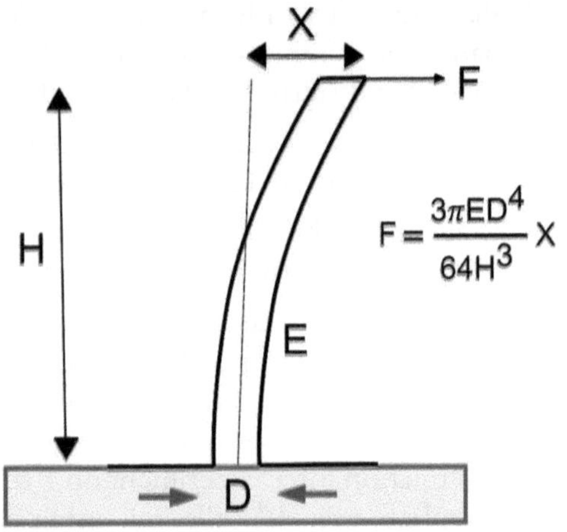

Fig. 3 Calculation of the micropillar spring constant. Single cylindrical micropillar with traction force *F* linearly proportional to the tip deflection *X*, the elasticity modulus *E* of the silicone (for PDMS $E = 2$–2.5 MPa), and to the fourth power of the diameter *D* with an inverse relationship with the third power of the height, *H*

6. After imaging, store images in an organized manner into a computer, and clean the microscope accordingly.

3.6 Traction Force Analysis

1. Launch Matlab, and make sure that a directory of containing Matlab codes is in Matlab paths.
2. Name the image files for the program to be able to identify them, by attaching "top," "bottom," and "actin" at the end of corresponding file names.
3. Set up initial parameters such as pillar diameter, post spacing, and existence of other image channels (e.g., vinculin and/or fibronectin channels). Enter also prefix identifier, i.e., the name before top or bottom, etc.
4. The program first segments "actin" image using Otsu thresholding. If the segmentation looks okay, click "yes," or click "no" to proceed with manual adjustment of cell segmentation.
5. The software shows a composite image of top pillars with actin overlay. Rotate the image to an upright orientation by clicking an upper-left pillar and an upper-right pillar. Type how many posts there are in between the two clicked posts, including the clicked posts.
6. Crop the image by drawing a rectangle around a cell shown (by actin). The cropped region will be dissected into sub-blocks containing one pillar in the center per sub-block. After this, the program will go through identification of centers of each pair of pillar sub-block images, i.e., top and bottom, and obtain a deflection vector field. The obtained deflection vectors will be

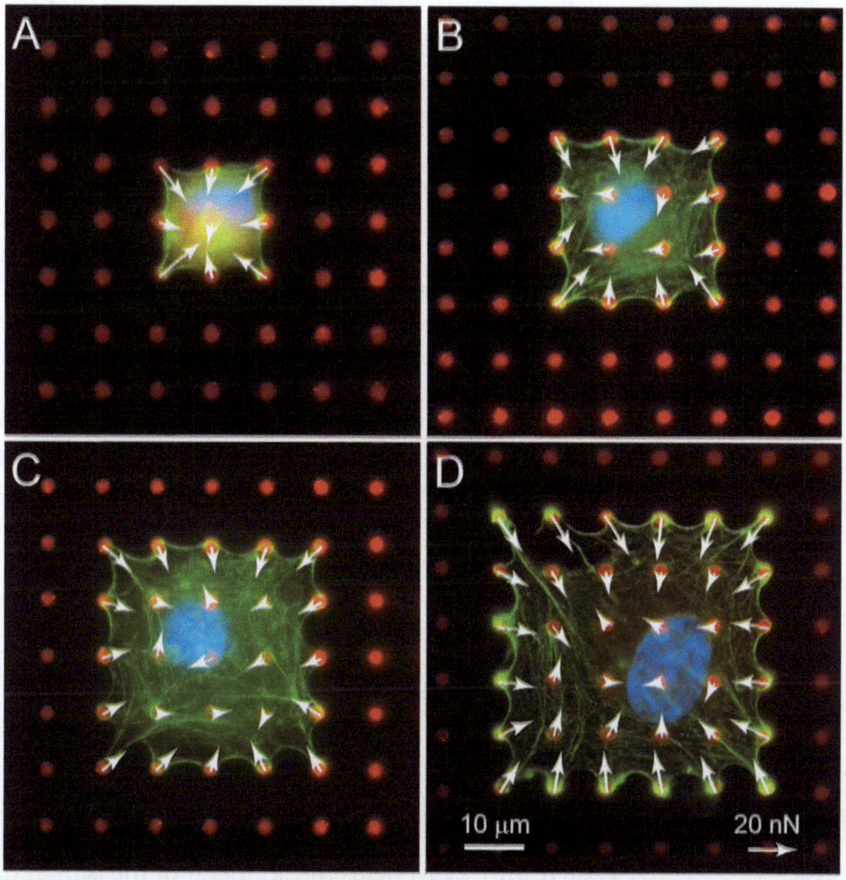

Fig. 4 Images of single human pulmonary artery endothelial cells on the micropillars taken under epifluorescence microscope. *Green* actin, *red* micropillars, *blue* nuclei. Arrows represent the direction and magnitude of the traction forces. Note that cells were spread in a shape of squares with different areas ranging from 18 μm × 18 μm (**a**), 27 μm × 27 μm (**b**), 36 μm × 36 μm, to 45 μm × 45 μm. (Adopted from Han et al. 2012)

converted to a force field by being multiplied with a spring constant of the cantilever shown in Fig. 3 (*see* **Note 3**).

7. A composite image with force vector overlay will be shown in the next figure window. Save the image, necessary output parameters, and raw comma-separated data for individual force vectors. An example figure is shown in Fig. 4 (*see* **Note 4**).

4 Notes on Applications of Micropillars

1. For repeated use of negative mold, mold needs to be treated with plasma and fluorosilane after two castings. Inspection of the mold is necessary after every casting under light microscope to check its intactness or disorientation.

2. To watch cells under inverted microscope, glass coverslip should be used instead of glass slide. The total thickness from the top of the stamp to that of the objective lens should be less than the working distance of the objective.
3. The micropillar substrate works as a force sensor; each deflection of the micropillar represents the local traction force. The relationship of the traction force with the deflection of the micropillar is obtained by the spring constant. Considering the uniform material properties and beam bending theory for the micropillar as a cantilever beam, the one end is fixed at the bottom, while other free end is loaded with bending force under no friction (Fig. 3), the spring constant is obtained by the dimensions of the pillars and Young's modulus of the PDMS material. Calibrations are required to obtain a more correct spring constant of the pillars, e.g., via using the deflection of the pulled glass microneedles which are already calibrated [17].
4. The endothelial cells exert the forces in the form of traction over the individual micropillars in order of nanonewtons. Observations have shown that endothelial cell attached over the individual posts showed larger force at the peripheral edges while the force was small at the central areas—the smallest force limit considered to be on the micropillars without cells.
5. One of the modifications of the technique involved restricting cell adhesion to a confined area. A small area of the micropillar coated with ECM helps the adhesion of the cell on the micropillars in the desired geometry (square, rectangle, circle, etc.). To obtain that, a flat stamp in extruded microfeature is loaded with ECM and transferred on the micropillar tops for cell patterning. This technique helps to show the correlation of traction force with micropillars. On a large scale, the same micropillars can also be used to study the collective effect of the cell monolayer on the traction force generation.
6. It is very important to note that micropillars have different topology and structural surrounding than that of the endothelial cells' own microenvironment. The micropillar topography may add additional signals to the endothelial cell behavior, and they need to be investigated further. More investigation based on different systems, *i.e.*, on micropillars *vs.* on a flat gel-based substrate, can be helpful for elucidating potential differences of endothelial cells' cytoskeleton, adhesion, signaling and traction force. Current system can be an important tool to shed light on the mechanics of the endothelial cells in vitro, which can be easily extendable to monolayer mechanics studies.

Acknowledgments

This work was supported by NIH 1R15GM135806-01.

References

1. Pober JS, Sessa WC (2007) Evolving functions of endothelial cells in inflammation. Nat Rev Immunol 7:803–815
2. Discher DE, Janmey P, Wang Y-L (2005) Tissue cells feel and respond to the stiffness of their substrate. Science 310:1139–1143
3. Vogel V, Sheetz M (2006) Local force and geometry sensing regulate cell functions. Nat Rev Mol Cell Biol 7:265–275
4. Schwartz MA (2010) Integrins and extracellular matrix in mechanotransduction. Cold Spring Harb Perspect Biol 2:a005066
5. Geiger B, Spatz JP, Bershadsky AD (2009) Environmental sensing through focal adhesions. Nat Rev Mol Cell Biol 10:21–33
6. Tzima E et al (2005) A mechanosensory complex that mediates the endothelial cell response to fluid shear stress. Nature 437:426–431
7. Hoffman BD, Grashoff C, Schwartz MA (2011) Dynamic molecular processes mediate cellular mechanotransduction. Nature 475:316–323
8. Zhang Y et al (2014) Collective cell traction force analysis on aligned smooth muscle cell sheet between three-dimensional microwalls. Interface Focus 4:20130056–20130056
9. Harris AK, Wild P, Stopak D (1980) Silicone rubber substrata: a new wrinkle in the study of cell locomotion. Science 208:177–179
10. Chrzanowska-Wodnicka M, Burridge K (1996) Rho-stimulated contractility drives the formation of stress fibers and focal adhesions. J Cell Biol 133:1403–1415
11. Helfman DM et al (1999) Caldesmon inhibits nonmuscle cell contractility and interferes with the formation of focal adhesions. Mol Biol Cell 10:3097–3112
12. Lee J, Leonard M, Oliver T, Ishihara A, Jacobson K (1994) Traction forces generated by locomoting keratocytes. J Cell Biol 127:1957–1964
13. Dembo M, Wang Y-L (1999) Stresses at the cell-to-substrate interface during locomotion of fibroblasts. Biophys J 76:2307–2316
14. Bachir AI et al (2015) Correlative traction force microscopy and fluorescence fluctuation analysis of molecular aggregation and complex formation in cell adhesions in distinct microenvironments. Biophys J 108:493a–494a
15. Tan JL, Tien J, Pirone DM, Gray DS, Bhadriraju K, Chen CS (2003) Cells lying on a bed of microneedles: an approach to isolate mechanical force. Proc Natl Acad Sci U S A 100:1484–1489
16. Han Sangyoon J, Bielawski Kevin S, Ting Lucas H, Rodriguez Marita L, Sniadecki NJ (2012) Decoupling substrate stiffness, spread area, and micropost density: a close spatial relationship between traction forces and focal adhesions. Biophys J 103:640–648
17. Sniadecki NJ, Chen CS (2007) Microfabricated silicone elastomeric post arrays for measuring traction forces of adherent cells. In: Methods in cell biology. Academic Press, Cambridge, Massachusetts
18. Sniadecki NJ, Han SJ, Ting LH, Feghhi S (2013) Micropatterning on Micropost Arrays, Methods in Cell Biology. 121:61–73

Chapter 20

Computational Assessment of Hemodynamics Vortices Within the Cerebral Vasculature Using Informational Entropy

Kevin Sunderland, Feng Zhao, and Jingfeng Jiang

Abstract

Propper assessment of hemodynamic swirling flow patterns, vortices, may help understand the influence of disturbed flow on arterial wall pathophysiology and remodeling. Studies have shown that vortices trigger pathologic cellular changes within the vasculature such as increased inflammation and cellular apoptosis, leading to weakening of the vessel wall indicative of aneurysm development and rupture. Yet many studies qualitatively assess the presence of vortices within the vasculature or assess only their centermost region (critical point analysis) which overlooks the broader characteristics of flow, leading to a narrow view of vortices. This chapter provides a protocol for utilizing commercially available computational fluid dynamic software (ANSYS-FLUENT) to simulate realistic hemodynamic flow patterns, fluid velocity, and wall shear stress in the complex geometry of the cerebral vasculature, as well as an innovative method for assessing flow vortices. This innovative analytic methodology can identify areas of flow vortices and quantify how the broader bulk-flow (opposed to critical point) characteristics change in space and time over the cardiac cycle. Analysis of such flow structures can be used to identify specific characteristics such as vortex stability and the portion of an aneurysmal sac that is dominated by swirling flow, which may be indicative of vascular pathologies.

Key words Computational fluid dynamics, Aneurysms, Swirling flow vortices, Informational entropy

1 Introduction

Computational fluid dynamics (CFD) applies principles of computational physics and numerical analysis to simulate flow within a system. Advances in computer hardware and commercially available CFD solvers have enabled clinical and basic science researchers to understand how hemodynamic flow plays a role in vascular pathologies. Hemodynamic flow conditions regulate cellular vascular

---

Supplementary Information The online version of this chapter (https://doi.org/10.1007/978-1-0716-1708-3_20) contains supplementary material, which is available to authorized users.

Feng Zhao and Kam W. Leong (eds.), *Vascular Tissue Engineering: Methods and Protocols*,
Methods in Molecular Biology, vol. 2375, https://doi.org/10.1007/978-1-0716-1708-3_20,

function and vascular mechanical properties, with uni-directional laminar flow and wall shear stressors (WSS) around 15 dynes/cm^2 seen as modulating vascular endothelial cell morphology and function which help regulate vascular tone and inflammatory processes [1]. While the assessment of WSS has led to significant increases in understanding vascular pathologies, the "near-wall" values overlook the broader flow patterns within the vascular intima. Areas of vortices are shown to trigger pathological problems within the vasculature due to their non-laminar recirculating flow directionality, as well as significant lowering of localized WSS intensity (<2 dynes/cm^2). These flow conditions are antithetical to vascular health, promoting increased vascular inflammation [2, 3], cellular apoptosis [4, 5], and weakening of the vascular wall. Earlier studies used critical point analysis to determine areas of vortices, focusing on the centroid region of said flow patterns [6, 7]. Such methods give a limited assessment of vortices as the centermost region overlooks the majority of flow that makes up a vortex pattern which may hold information concerning vortices relation to vascular changes. Our earlier work has applied methods of Shannon's (informational) entropy alongside critical point analysis to assess the bulk-flow spatiotemporal characteristics of vortices to both improve statistical models to predict areas of likely aneurysm development in cerebral vasculature [8] and determine how the surgical repair of an aneurysm induces detrimental hemodynamic conditions within any remaining untreated aneurysms [9]. Shannon's entropy expresses the information of a system, specifically how deterministic an event is within said system. From the perspective of flow analysis, Shannon's entropy can determine if the flow velocity vectors within a given area of the vasculature are equally likely to point in any direction as would be the case in swirling flow, or if they all gravitate toward one direction as in laminar flow. Shannon's entropy allows the identification of the larger area of flow making up a vortex as opposed to only its centermost region.

This method for vortex identification allows for a more robust analysis of their characteristics within the vascular system. As several investigators have assessed vortex patterns to better understand changes to vascular cells [10], predict areas likely to develop aneurysms, and predict aneurysmal rupture risk [11], a reliable way to both identify the broader area of flow making up vortices and assess their spatiotemporal characteristics is critically important. Understanding how vortex characteristics such as vortex stability trigger different degrees of cellular changes indicative of aneurysm development and rupture could 1 day improve the clinical management of this vascular pathology. In this chapter, we provide a protocol on the computational procedures and software used from previous publications to analyze flow vortices.

2 Materials

2.1 Computational Workstation

1. Ubuntu 16.04 64-bit operating system.
2. Intel Xenon® E52-2643 v3 CPU.
3. 32 GB Ram.
4. NVIDIA Quadro K4000 graphics processing unit (*see* **Note 1**).

2.2 Computational Software and Programing Languages

1. Mimics 22.0: Mimics Innovation Suite (Materialise Inc., Leuven, Belgium).
2. 3-matic 14.0: Mimics Innovation Suite (Materialise Inc., Leuven, Belgium).
3. FLUENT version 17.0: ANSYS, Inc., Canonsburg, PA.
4. ENSIGHT 10.1.6(d) 3D post-processing software: ANSYS, Inc., Canonsburg, PA.
5. Paraview data analysis and visualization application: https://www.paraview.org/
6. Python 2.7.6 and C++ (gcc 4.8.4 compiler).
7. Visualization and Vascular Modeling Toolkit (VTK/VMTK) open-source programming libraries: http://www.vmtk.org/.

3 Methods

3.1 Creation of 3D Patient-Specific Vasculature Computational Structure from 3D Digital Subtraction Angiography Images in DICOM Format (https:/www.dicomstandard.org/)

1. Import DICOM files into Materialise Mimics software using the "New Project" function. Select the folder containing DICOM files, and import the folder.
2. Isolate the area of vasculature using the New Mask tool. Set the pixel intensity Threshold to 2500GV (min)—Max GV (max) (*see* **Note 2**). Alternatively, the Region Growing function can be performed: right clicking on an area of the vasculature in the DICOM images and setting 26-point connectivity. This step produces a set of 2D masks each delineating blood flowing regions.
3. Create a 3D surface structure (Calculate Part) from a set of isolated 2D Masks. Set structure calculation to "High Quality," minimizing matrix size, maximizing triangle reduction: edge tolerance 0.05 mm, edge angle 10.0°, iterations 10. Save structure as an STL+ file: Binary STL file, scale factor 1.0.
4. "Remesh" the object, which will import the STL object into the Materialise 3-matic software.
5. Use the "Trim" function under the "Finish" tab to remove extraneous portions of the vasculature distal to the area of the interest (i.e., distal to a stenosis or aneurysms), keeping at

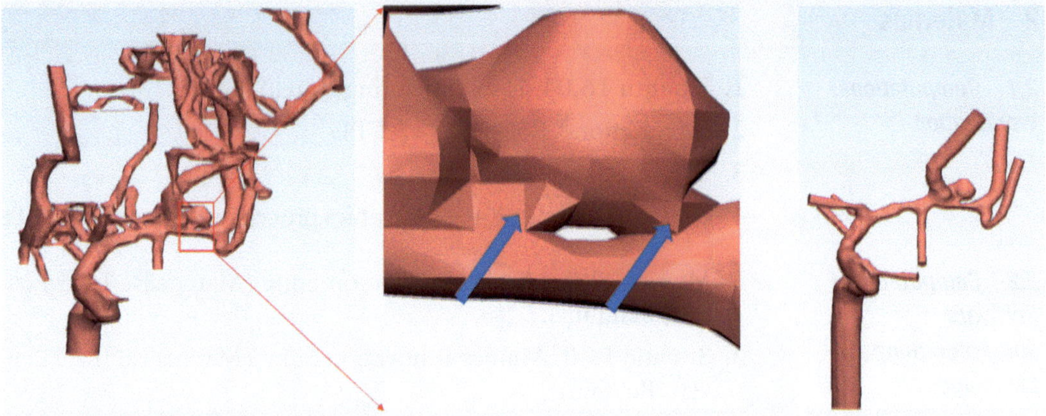

Fig. 1 (**Left**) Extracted vessel structure from DICOM imaging files: area of "merged" vessel highlighted by the red box. (**Center**) Close-up of vessel areas (blue arrows) that have "merged" together, requiring separation. (**Right**) Example of edited vessel structure used for FLUENT simulation

minimum a distal portion of the parent vessel at length equal to the diameter of the parent vessel, as a fluid outlet. Leave the available vessel proximal to the area of interest intact when applicable, trimming away areas of significant errors in the vessel structure (due to imaging artifacts).

6. At both the trimmed inlet and outlet portions of the vessel, add straight (in relation to the direction of the trimmed vessel) extensions using the "Extend Surface" function in 3-matic with a length six times the circumference of the vessel to which it is attached (*see* **Note 3**).
7. Vessels in proximity may have "merged" together during model creation (Fig. 1b). Fix these errors by deleting the shared surface triangles between the two vessel areas. Fill the resultant holes using the "Fill Hole Normal" functionality (*see* **Note 4**).
8. Assess and fix errors in the edited structure using the 3-matic "Fix Wizard," under the "Fix" tab. Select the model as the "Part," choose "Full Analysis," and then "Update" to scan for errors in the model. "Follow Advice" and "Apply" the Fix Wizard's different advice to correct errors, with a model being corrected once the following values are achieved: inverted normals, bad edges/planar holes, shells, overlapping surface triangles, and intersecting triangles at 0 0 1 0 and 0, respectively (*see* **Note 5**).
9. Smooth any remaining surface irregularities using the "Surface smoothing" functionality, setting smooth factor to 0.700 and allowing the software to perform post-processing.
10. Check the resultant quality of the generated surface mesh using the "Create Inspection Scene" tool in the "Remesh" tab. Too

many surface elements with large aspect ratios (height/base) could impact simulation accuracy. Surface remeshing using a smaller max edge length (0.05) or adjusted "shape measure" (0.5 height/base (A)) can reduce the number of elements with large aspect ratios.

11. Trim a small section of the ends of the inlet and outlet vessels tangential to vessel directionality. Save the edited vessel structure (Fig. 1c) as a stereolithography file (.stl).
12. Start ANSYS Workbench and the "Fluent (with Fluent Meshing)" toolbox. Right click in the "Mesh" portion of the toolbox, and choose "Edit." Import the created STL file through the "Import→CAD" function, import as a single file, set the unit length to mm (as DICOM files' unit length is typically in millimeters), and set feature angle to 40.
13. In the "Outline View" tab, right click on "Geometry Objects" and Draw All. Right click on the drawn surface, and use the "Separate" function to differentiate the cut planes for inlet and outlet portions of the vessel. Right click on the inlet and outlet (s) individually, and use the "Rename" function to name each portion, and to "Set Boundary Type," set inlet as velocity-inlet and outlet(s) as pressure-outlet. In the Outline View, pull down the "Geometry objects" tab and right click on "zone-solid" and choose "merge walls."
14. Right click all sections (wall, inlet, outlet(s)), and "Remesh," setting a Sizing (constant-size), Feature Angle (40), and Constant Size, setting the value to half of the size of default calculation (*see* **Note 6**). Right click on "zone-solid," and choose "convert to mesh object."
15. Pull down "Mesh Objects: zone-solid" tab, and right click on the "Volumetric Region," and compute the volumetric region of the vascular model. Then "Setup Scoped Prisms" and set the following parameters (*see* **Note 7**): Offset Method aspect-ratio, First Aspect Ratio 10, Number of Layers 3, Object zone-solid, Scope To solid-regions, and Grow On only-walls. Click "Create" to add these parameters to the "Scoped Prisms List."
16. Right click "Volumetric Region," and choose "Auto Mesh" to perform the volumetric meshing of the vasculature model. Set the following parameters: Object zone-solid, Keep Solid Cell Zones, Grow Prisms scoped and 'set' to choose the parameters set from the previous step, Quad Tet Transition Pyramids, Volume Fill Tet, Volume Fill Options Cell Sizing Size Field, max Cell Length → Compute to set the max cell size based on the mesh object (*see* **Note 8**), Merge Cell ones within Regions. Apply, then Mesh.

17. Export the mesh file (.msh) then "Switch to Solution" to import the volumetric mesh into the ANSYS-FLUENT solver.

3.2 Computational Fluid Dynamic Simulation and Flow Vortex Identification

1. In the "Setup" tab, set the view length unit for the simulation as mm, and set the "mesh was created in" value as mm. Choose the "Pressure-based" type for the solver, and set the time as "Transient." Choose the "Viscous—Laminar" flow model and fluid as Newtonian for vascular flow in larger vessels (*see* **Note 9**). Choose the appropriate material properties for both the vessel wall and the fluid (blood) component. In our previous work, the vessel wall material properties were left as the default solid properties (aluminum), and the fluid mass density was set as 1040 kg/m^3 and viscosity as 0.004 kg/m-s. Under "Cell Zone Conditions," change the internal zone from type solid to type fluid, and use the pull-down menu in the resultant pop-up window for "Material Name" to choose the fluid material.
2. Set the "Boundary Conditions" for the wall as stationary (rigid) with no-slip shear conditions (*see* **Note 10**).
3. Import a user-defined function (UDF) in FLUENT to be used as the inflow velocity values mimicking flow within the internal carotid artery (*see* **Note 11**). In the "User-Defined" tab, open "functions—compiled," "Add" a created UDF (.c) source files, "build" then "load."
4. Set the "Boundary Conditions" for inlet velocity magnitude (m/s) using the UDF built-in **step 3**. Choose the inlet "Zone," and pull down the "Type" tab, and choose the velocity-inlet option. In the resultant pop-up window, choose the pull-down menu next to "Velocity Magnitude (m/s)," and select the loaded UDF, and set the "Supersonic/Initial Gauge Pressure (pascal)" as 0. For the outlet zones, choose the "type" pressure-outlet, and leave as default (gauge pressure (pascal)) as 0 unless an outlet back pressure is desired. Leave the fluid zone and wall "Zone" as default.
5. "Dynamic Mesh" can be left unchecked. The "Reference Values" need to be computed from the inlet zone, with the reference zone as the fluid.
6. Set the "Solution Methods" as follows: "Pressure-Velocity Coupling Scheme" as SIMPLEC [12], "Spatial Discretization Gradient" as least-squares cell-based, "Pressure" as second-order, and "Momentum" as second-order upwind. Leave the "Solution Controls (under relaxation factors)" (*see* **Note 12**), "Monitors," "Report Definitions," and "Report Plots" as default. Set the "Solution Initialization → Compute from" as the velocity inlet, and then choose to "Initialize" these settings.
7. "Calculation activities" will need to be set according to the nature of the study, but as an example per our previous work:

"autosave every (time step)" to 50, "Edit" the autosave values to "Retain only the most recent files" with a maximum number of data files as 20. Set the "Run Calculation" at a timestep of "Time Step Size (s)" 0.001, "Number of Time Steps" as 4000 simulating 4 cardiac cycles (at 1 s/cycle) (*see* **Note 13**). Increase the "Max Iteration/Time Step" to 1000 then "Calculate." *A video showing the process for Subheading 3.1,* ***step 12*** *through Subheading 3.2 (to this point) can be seen in the Supplementary Material.*

8. Upon simulation completion, "Export" the data as the EnSight Case Gold filetype, with the data at the "Node" location, and in binary format. The "Quantities" of exported data vary per study. For vortex analysis, the following quantities are needed: Velocity Magnitude, *X* Velocity, *Y* Velocity, and *Z* Velocity. This should be done for each of the 20 saved data points from the final simulated cardiac cycle.
9. Open the data (.encase) in EnSight. Highlighting all areas of the vessel geometry (in the "Parts" window) and choosing velocity under "Vectors" and the *X Y* and *Z* velocity "Scalars." Export the data using: "Files → Export → Geometric Entities → Output formats: Case (Gold)". "Save as binary file(s)" from "Being time step (0)" and "End time step" as a total number of timesteps, "Step by (1)."

   Currently, no commercially available software exists to identify areas of swirling flow within the vasculature. In-house computational scripts will be needed to use our method of identifying vortices. A description of the fundamental steps to create a code capable of identifying vortices will be provided.
10. Open the .case file in Paraview, use the "Transform" filter, and scale the data by 1000, 1000, 1000. Save the scaled data in the "VTK Multi-Block Files(*.vtm) format, writing all timesteps as file-series. This also creates unstructured grid (.vtu) files of the data, with a folder per timestep. Extract the surface mesh using the "Extract Surface" filter, saving the surface as a .stl file.
11. Move the .vtu files from each timestep folder, and place them into one folder (*see* **Note 14**).
12. Isolate the area of interest in the surface mesh using the "Clip" function in Paraview, then "Extract Surface" to extract the resultant surface mesh, and save the area as a .stl file (*see* **Note 15**). This isolated geometry can be considered a surface mask.
13. Resample the data onto a 3D rectilinear computing grid with voxels of 0.2 mm edge length (.vti file) using the VTK/VMTK vtkProbeFilter function; then using the surface mask, isolate the data for the area of interest using the SelectEnclosedPoints function in VTK/VMTK.

14. Convert the 3D angular space of the flow field into a spherical area with 360 bins of equal area, resulting in "cones" connecting at the unit sphere's center and creating surface patches [13]. For each of the flow vectors surrounding a voxel, determine the bin to which the vector aligns: normalize both the velocity vectors and bin directions, and determine which bin is the closest in value to the normalized vector. Calculate the probability ($p(x_i)$ of the velocity field (X) direction ($x \in (x_1, x_2, \ldots x_n)$) to concentrate on any bin.

$$H(X) = -\sum_{x \in X} p(x_i) \, log_2 p(x_i)$$

Dividing the probability by the maximum possible entropy ($\log_2(N)$), normalizes the entropy between 0 and 1: with 1 as all vectors aligned in the same direction (perfectly laminar flow) and 0 as the likelihood that any vector could be in any bin (swirling or random flow; *see* Fig. 2). For our work with studying flow in aneurysms, a fixed volume of interest ($N_x = N_y = N_z = 11$) was selected around each voxel to analyze localized flow areas.

15. Calculate the critical point of flow field(s) (*see* **Note 16**). Use the λ_2 method [14] to decompose the velocity vector tensor into its strain-rate (symmetric(S)) and spin (asymmetric(Ω)) tensors.

$$S = \frac{\nabla\vec{v} + \nabla\vec{v}^T}{2}, \Omega = \frac{\nabla\vec{v} - \nabla\vec{v}^T}{2}$$

When $S^2 + \Omega^2$ has 2 negative eigenvalues ($\lambda_1 > \lambda_2 > \lambda_3$), calculate the dot product of the normalized velocity vector $\left\|\vec{v}\right\|$ and the λ_2 value.

16. Apply a 5 × 5 × 5 median filter to the critical point data.
17. Multiply the normalized entropy to the (filtered) critical point data.
18. Apply the VMTK marching cube function to the combined normalized entropy-critical point outcome, setting the value threshold to 0.3. This will map areas with the (combined) values ≥ 0.3 to 3D *iso-surfaces*, connecting adjacent areas into a combined surface (Fig. 3). Save *iso-surface(s)* as .stl files.
19. Remove any *iso-surfaces* with a volume of <0.5 mm^3.
20. Repeat **steps 4–10** for all datafiles per simulation.
21. Calculate the spatiotemporal characteristics of vortex *iso-surfaces*:
    (a) Number of *iso-surfaces* per timestep.
    (b) Degree of volume overlap of *iso-surfaces* between timesteps (a surrogate of vortex stability) (*see* **Note 17**).

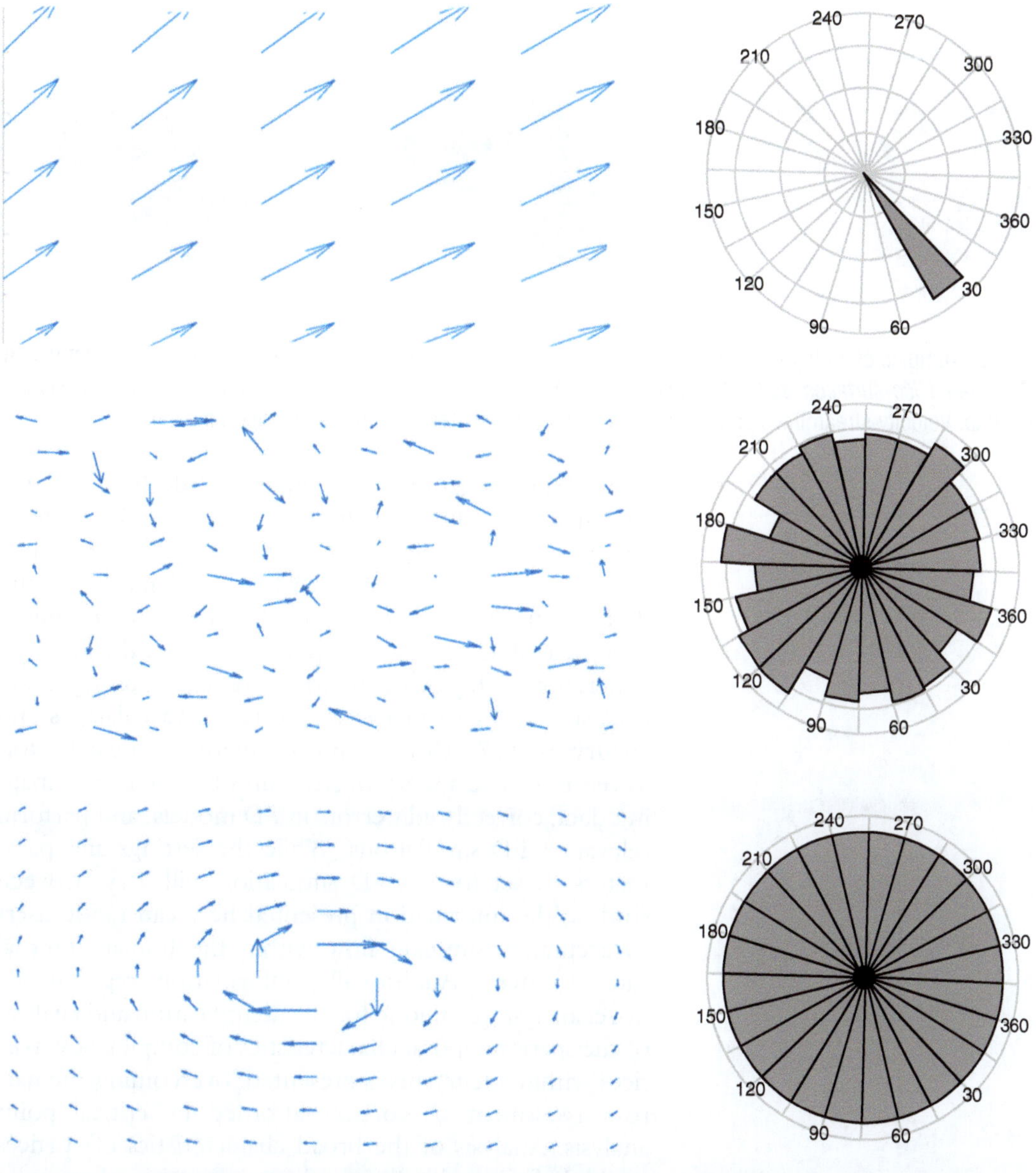

Fig. 2 Three examples showing the relationship between the 2D angular histogram and entropy: (**top**) laminar flow with and entropy close to 1, (**middle**) Brownian (random) flow, (**bottom**) perfectly rotational flow with entropies close to 0. The left and right plots are the 2D vector flow field and their histogram of angular vector direction respectively. Both random and rotational flow have similar histogram frequency distribution

(c) Volume of *iso-surface* in relation to the volume of their aneurysm sac. The volume of the sealed aneurysm mask and vortex *iso-surfaces* can be calculated in Paraview using the "Integrate Variables" function.

Proper assessment of hemodynamic conditions within aneurysms helps improve the understanding of the conditions indicative of aneurysm growth and rupture. In the

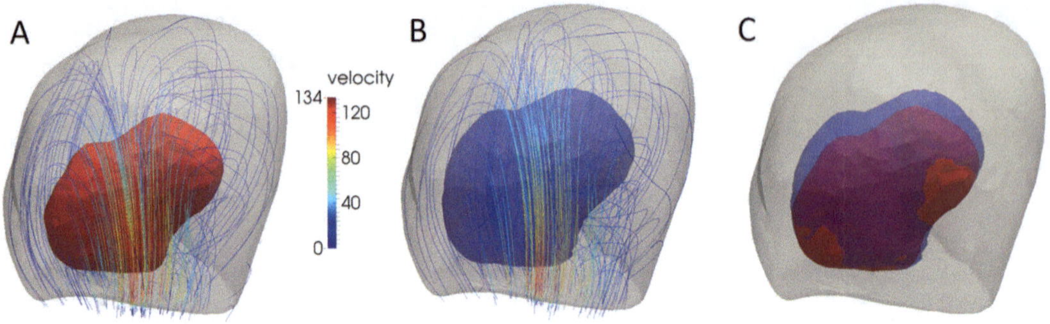

Fig. 3 Examples of vortex *iso-surfaces* and their spatiotemporal variation between subsequent timesteps: (**a**) Timestep *i iso-surfaces* (red), (**b**) timestep *i* + 1 *iso-surfaces* (blue), (**c**) both timesteps with *iso-surface* overlap. Velocity streamlines show overall flow pattern and velocity (mm/s) at each timestep

past, CFD simulations were applied to idealized vasculature structures, limiting the evaluation of individualized patient hemodynamic characteristics [15, 16]. Continued advancements in computer hardware, 3D modeling and design software, and commercially available CFD solvers have helped accelerate research in this area through the generation of high-quality flow data in complex geometries such as "patient-specific" cerebral vasculatures and aneurysms [17, 18]. Yet proper methods must be followed to isolate aneurysm structures from medical imaging data, correctly edit errors in 3D models, and perform relevant CFD simulations. While the settings and parameters to set for a CFD simulation will vary between studies, the information presented here can guide users to accurately simulate flow within the human internal carotid artery. Additionally, information regarding a novel analysis technique for the identification and analysis of the spatiotemporal characteristics of complex flow vortices within aneurysms is presented, overcoming the narrow assessment of vortices afforded by critical point analysis. Analysis of the broad characteristics of vortices could help identify flow indices indicative of vascular pathologies such as aneurysm initiation [8], growth, and rupture [19, 20].

4 Notes

1. For our work we used four cores per simulation, allowing multiple simulations to be run at once on the same workstation while still reducing simulation time. Testing of simulation speed vs the number of GPU processors used should be

performed to maximize the performance (speed vs. the number of concurrent simulations) of a workstation.

2. In this protocol, DICOM images are in Hounsfield scale. The minimum gray value (GV) of 2500 GV was found to be ideal for the identification of the cerebral vasculature, but this may need to be adjusted depending on the area of vasculature analyzed or quality of imaging data. Values below 2500 GV begin to identify areas of bone, and if values are too low then the background of the image will be identified.
3. Initial flow velocity generated in simulations typically follows a "plug-flow" pattern, uniform across the cross-sectional area of the vasculature. A vessel extension gives the simulation distance to correct the plug flow to a parabolic flow shape indicative of actual flow in the vasculature, before reaching area(s) of interest.
4. If "merged" vessels occur at the areas of secondary outlet vessel and parent vessel connection, separating the two will result in a single hole in which the "Filling Hole Normal" function will be unable to properly correct as it will re-merge the vessels. To properly correct the resultant hole, the hole must be "split" into two distinct holes using the "Add triangle" function. This adds a surface triangle across the diameter of the hole (tangential to the parent vessel direction). After choosing the "Add Triangle" function, click on two connected triangle vertices along one side of the hole and another vertex on the opposite side of the hole. If a yellow border appears at the boundary of the added triangle and the first two vertices, delete the triangle, and try again as the surfaces are not connected. These two holes can now be filled using the "Fill Hole Normal" functionality.
5. The 3-matic "Fix Wizard" has the potential to recursively fix one problem then recreate a previously fixed problem: deleting overlapping triangle, then creating overlapping triangles when filling the resultant hole. If this happens, use the "Fix Wizard" to only delete overlapping triangles, intersecting triangles, and inverted normals. Then, close the "Fix Wizard," and repair the resultant holes using the "Fix holes normal" and "Add triangles" 3-matic functions as needed.
6. The needed size of surface mesh elements can vary depending on the requirements of a given study and vascular area of interest. A mesh sensitivity analysis can determine the fewest number of elements needed to achieve similar simulation results.
7. The Scoped Prism layer is a specialized element layer needed for the accurate calculation of WSS values. A greater number of elements may be needed in large vessels such as the aorta.
8. Similar to surface mesh element size, volumetric mesh sensitivity tests should be performed to determine the element size/number of elements needed to achieve accurate simulation

outcomes. Our study used 1.5–2 million total elements per simulation.

9. The shear thinning (non-Newtonian) properties of blood are shown to have a limited impact on simulated flow patterns [21], though it does increase wall shear stress values.
10. Vessel rigidity causes increases in simulated WSS, though changes to overall flow patterns are not significantly impacted in much of the vasculature [22]. Care must be taken to ensure if vessel compliancy and mechanical properties should be more accurately represented in vessels of high compliance (i.e., aorta) or around cardiac valves. The no-slip boundary condition is the typical setting for viscous fluid simulations. This property may need to be adjusted depending on what is being studied/ simulated.
11. Our UDF converted a volumetric flow rate waveform (Gwilliam et al., averaged 280 ml/min in the internal carotid artery) to flow velocity for 30 points over a 1-s flow waveform. The UDF uses the inlet cross-sectional area of a vessel to ensure the same volumetric flow rate regardless of inlet size for our simulations: $\textit{flow velocity} = \frac{\textit{flow rate}}{\textit{inlet cross setional area}}$. Interpolating the flow velocity between the 30 points must be performed in the UDF to account for the timestep size in the simulation. Detailed information on the writing of a UDF can be found in the ANSYS-FLUENT user manual.
12. Our values were left as default: Pressure 0.3, Density 1, Body Forces 1, Momentum 0.7. For each calculation iteration, the values obtained should attempt to converge (aided by the "relaxation factors"), yet equations may become unstable in their convergence, impacting simulation accuracy. If a simulation has trouble converging (typically Momentum), then decrease the "Momentum" under-relaxation factor to 0.6, increase the "Pressure" under-relaxation factor to 0.4, and re-test the simulation.
13. The choice of timestep size for simulation will need to be increased or decreased depending on what is being simulated and the complexity of the simulated geometry. The smaller step size can improve the accurate simulation of small sudden spatiotemporal changes in fluid velocity and directionality. Yet too small of a step size causes a dramatic increase in computational time with little to no change in simulation accuracy.
14. The folders created by Paraview for .vtu files contain files for each surface in the simulation: inlet, outlet(s), wall, and one for all surfaces total (including volumetric flow information). Typically, the files with the naming structure "file name_timestep #_0_0.vtu" are the "all surfaces" files: file_0_0_0.vtu,

file_1_0_0.vtu, etc. These are the files that should be analyzed for the assessment of hemodynamic vortex patterns.

15. Vortex flow in aneurysms with a width of <4 mm^3 had a portion of their resultant vortex extend outside the aneurysmal sac, especially for bifurcation aneurysms. Visual analysis of flow streamlines (in Paraview) can help determine if vortex patterns exist outside the area of a chosen surface mask.
16. The adoption of critical point analysis is needed to help the normalized entropy method to distinguish between swirling flow and random (Brownian) flow. From a mathematical/vector directionality perspective, the probability of vectors aligning with any bin is the same in swirling and Brownian flow, yet Brownian flow is unlikely to have a centralized critical point.
17. The degree of volume overlap between *iso-surfaces* of subsequent timesteps of the cardiac cycle can be determined using the "vtkSelectEnclosedPoints" function, identifying the number of points of one *iso-surface* existing at or within the same spatial location as the next timestep's *iso-surface.*

Acknowledgments

The study was partially funded by a research contract through Siemens Medical Solution (USA) Inc. Kevin Sunderland was supported by a pre-doctoral fellowship from the American Heart Association [18PRE33990321] and grants from the National Institutes of Health [R01HL146652-01A1, R15HL145654-01 and 1R01EB029570-01A1].

References

1. Zhou G, Zhu Y, Yin Y, Su M, Li M (2017) Association of wall shear stress with intracranial aneurysm rupture: systematic review and meta-analysis. Sci Rep 7:5331. https://doi.org/10.1038/s41598-017-05886-w
2. Baeriswyl DC, Prionisti I, Peach T, Tsolkas G, Chooi KY, Vardakis J, Morel S, Diagbouga MR, Bijlenga P, Cuhlmann S, Evans P, Kwak BR, Ventikos Y, Krams R (2019) Disturbed flow induces a sustained, stochastic NF-κB activation which may support intracranial aneurysm growth in vivo. Sci Rep 9(1):4738. https://doi.org/10.1038/s41598-019-40959-y
3. Bailey KA, Moreno E, Haj FG, Simon SI, Passerini AG (2019) Mechanoregulation of p38 activity enhances endoplasmic reticulum stress–mediated inflammation by arterial endothelium. FASEB J 33(11):12888–12899
4. Heo K-S, Le N-T, Cushman HJ, Giancursio CJ, Chang E, Woo C-H, Sullivan MA, Taunton J, Yeh ETH, Fujiwara K, Abe J-I (2015) Disturbed flow-activated p90RSK kinase accelerates atherosclerosis by inhibiting SENP2 function. J Clin Invest 125(3):1299–1310. https://doi.org/10.1172/JCI76453
5. Pan L, Hong Z, Yu L, Gao Y, Zhang R, Feng H, Su L, Wang G (2017) Shear stress induces human aortic endothelial cell apoptosis via interleukin-1 receptor-associated kinase 2-induced endoplasmic reticulum stress. Mol Med Rep 16(5):7205–7212
6. Byrne G, Mut F, Cebral J (2014) Quantifying the large-scale hemodynamics of intracranial aneurysms. Am J Neuroradiol 35(2):333–338. https://doi.org/10.3174/ajnr.A3678

7. Oeltze-Jafra S, Cebral J, Janiga G, Preim B (2015) Cluster analysis of Vortical flow in simulations of cerebral aneurysm hemodynamics. IEEE Trans Vis Comput Graph 21:757–766. https://doi.org/10.1109/TVCG.2015.2467203
8. Sunderland K, Jiang J (2019) Multivariate analysis of hemodynamic parameters on intracranial aneurysm initiation of the internal carotid artery. Med Eng Phys 74:129–136
9. Sunderland K, Huang Q, Strother C, Jiang J (2019) Two closely spaced aneurysms of the Supraclinoid internal carotid artery: how does one influence the other? J Biomech Eng 141 (11)
10. Chiu J-J, Chien S (2011) Effects of disturbed flow on vascular endothelium: pathophysiological basis and clinical perspectives. Physiol Rev 91(1):327–387. https://doi.org/10.1152/physrev.00047.2009
11. von Spiczak J, Crelier G, Giese D, Kozerke S, Maintz D, Bunck AC (2015) Quantitative analysis of Vortical blood flow in the thoracic aorta using 4D phase contrast MRI. PLoS One 10(9):e0139025. https://doi.org/10.1371/journal.pone.0139025
12. Versteeg HK, Malalasekera W (2007) An introduction to computational fluid dynamics: the finite volume method. Pearson education, New York
13. Leopardi P (2006) A partition of the unit sphere into regions of equal area and small diameter. Electron Trans Numer Anal 25 (12):309–327
14. Jiang M, Machiraju R, Thompson D (2005) Detection and visualization of. The visualization handbook, vol 295. Elsevier, Amsterdam
15. Hoi Y, Woodward SH, Kim M, Taulbee DB, Meng H (2006) Validation of CFD simulations of cerebral aneurysms with implication of geometric variations. J Biomech Eng 128 (6):844–851. https://doi.org/10.1115/1.2354209
16. Farnoush A, Avolio A, Qian Y (2012) Effect of bifurcation angle configuration and ratio of daughter diameters on hemodynamics of bifurcation aneurysms. AJNR Am J Neuroradiol 34:391–396. https://doi.org/10.3174/ajnr.A3222
17. Jou L-D, Quick CM, Young WL, Lawton MT, Higashida R, Martin A, Saloner D (2003) Computational approach to quantifying hemodynamic forces in Giant cerebral aneurysms. Am J Neuroradiol 24(9):1804–1810
18. Steinman DA, Milner JS, Norley CJ, Lownie SP, Holdsworth DW (2003) Image-based computational simulation of flow dynamics in a Giant intracranial aneurysm. Am J Neuroradiol 24(4):559–566
19. Sunderland K, Wang M, Pandey A, Gemmete J, Huang Q, Jiang J (2019) Abstract 16063: quantitative analysis of flow vortices for predicting middle cerebral aneurysm rupture. Circulation 140(Suppl_1):A16063. https://doi.org/10.1161/circ.140.suppl_1.16063
20. Varble N, Trylesinski G, Xiang J, Snyder K, Meng H (2017) Identification of vortex structures in a cohort of 204 intracranial aneurysms. J R Soc Interface 14(130):20170021. https://doi.org/10.1098/rsif.2017.0021
21. Kumar D, Vinoth R, Raviraj Adhikari VS (2017) Non-Newtonian and Newtonian blood flow in human aorta: a transient analysis. Biomed Res 28(7):3194–3203
22. Siogkas P, Sakellarios A, Exarchos T, Stefanou K, Fotiadis D, Naka K, Michalis L, Filipovic N, Parodi O (2011) Blood flow in arterial segments: rigid vs. deformable walls simulations. J Serbian Soc Comput Mech 5 (1):69–77

Index

Feng Zhao and Kam W. Leong (eds.), *Vascular Tissue Engineering: Methods and Protocols*, Methods in Molecular Biology, vol. 2375, https://doi.org/10.1007/978-1-0716-1708-3,

I

L

M

P

R

S

T

V

MIX
Papier aus verantwortungsvollen Quellen
Paper from responsible sources
FSC® C105338

If you have any concerns about our products,
you can contact us on
ProductSafety@springernature.com

In case Publisher is established outside the EU,
the EU authorized representative is:
Springer Nature Customer Service Center GmbH
Europaplatz 3, 69115 Heidelberg, Germany

Printed by Libri Plureos GmbH
in Hamburg, Germany